Differential Equations

Theory, Numerics and Applications

Differential Equations
Theory, Numerics and Applications

Proceedings of the ICDE '96
held in Bandung, Indonesia

Edited by

E. van Groesen
Faculty of Applied Mathematics,
University of Twente,
Enschede, The Netherlands

and

E. Soewono
Department of Mathematics,
Institut Teknologi,
Bandung, Indonesia

SPRINGER-SCIENCE+BUSINESS MEDIA, B.V.

A C.I.P. Catalogue record for this book is available from the Library of Congress.

ISBN 978-94-010-6168-1 ISBN 978-94-011-5157-3 (eBook)
DOI 10.1007/978-94-011-5157-3

Printed on acid-free paper

TABLE OF CONTENTS

PREFACE

The **International Conference on Differential Equations**, *theory, numerics and applications*(**ICDE'96-Bandung**) was held successfully at the West Aula of Institut Teknologi Bandung on September 29 - October 2, 1996, hosted by the Center of Mathematics and the Department of Mathematics ITB. This was the first international conference on differential equations in the region and attended by participants from 12 countries: Australia, Cambodia, Hong Kong, France, Indonesia, Malaysia, Netherlands, Philippine, Thailand, Singapore, USA and Vietnam.

We would like to express our gratitude to the following organizations and institution: Directorate General of Higher Education of Indonesia (through Center Grant Project), Institut Teknologi Bandung, UNESCO (through Participating Programme and ROSTSEA Programme), European Economic Community(through the Joint Research Project between the Faculty of Applied Mathematics, Universiteit Twente and the Department of Mathematics, Institut Teknologi Bandung), South East Asian Mathematical Society (SEAMS), French Embassy in Jakarta and local sponsors for their generous support which have made this conference possible.

Editors

Part I: Invited Lectures

PARAMETRIC EXCITATION IN MECHANICAL SYSTEMS

A.H.P. VAN DER BURGH
Delft University of Technology
Faculty of Technical Mathematics and Informatics
P.O. Box 5031, 2600 GA Delft, The Netherlands

Abstract. Parametric excitation in a mechnical system may occur if a parameter of the system becomes time-dependent. The mathematical model for this type of excitation is characterized by terms in the differential equations which have time-dependent coefficients. A standard example of an equation which displays parametric excitation is the Mathieu equation. In this paper two systems will be considered in more detail: a pendulum and a stretched string both with varying length. The attention will be focused to the (unstability) of the equilibrium position (trivial solution). Also finite amplitude motions will be considered, for the description of which however additional nonlinear terms are taken into account.

1. Introduction

Although the concept of parametric excitation seems not to be very well-known among scientists in the exact sciences, almost everybody has experience with parametric excitation. The mechanism that causes unstability of the equilibrium position of a swing, the moving body of a child on the swing, is a practical example of parametric excitation to which can be referred as a phenomenon which almost everybody has experienced in his life. From mechnical point of view a person moving vertically on a swing may be viewed as a pendulum with a variable length. As the length is a parameter of a pendulum one may identify a pendulum with a varying length as an example of a system with parametric excitation. In the real-life situation it is well-known that an appropriate vertical motion of the body of a person on the swing will cause an unstability of the equilibrium position and will hence induce motion of the swing. In section 3. it will be shown that the Mathieu equation is an adequate model for the study of the stability of the equilibrium position of this system. A second example of a system that displays parametic excitation is related to Melde's [4.] experiment and is discussed in section 4. This example is interesting because it concerns a continuous system which is still not understood very well nowadays. Melde performed experiments with a string of silk, fixed on one side at the end of the prong of a tuning fork and on the other side to a peg, allowing to vary the tension in the string by turning the peg. The position of the peg and the tuning fork is such that the oscillating tuning fork is only able to induce pure horizontal deflections in the string. Melde observed that for a critical value of the tension in the string the longitudinal horizontal excitation by the tuning fork induced a vertical, that is a transverse motion of the string. A motivation to study this problem is that adequate model equations for the study of Melde's phenomenon seem not available in the literature. Moreover it is believed that the understanding of Melde's experiment is of relevance for the understanding of wave propagation in the hull of a ship and in walls and floors of buildings. For convenience of the reader some qualitative and quantitative methods and results needed for the study of parametric excitation are presented in section 2.

2. Some theory

The most simple equation which displays parametric excitation is the Mathieu equation:

$$\ddot{u} + (\delta + \varepsilon \cos 2t)u = 0, \tag{1}$$

where $u = u(t)$ and δ and ε are real parameters.
A Mathieu equation with a $\cos wt$ forcing term is by no means more general than (1). By an appropriate time and parameter transformation one easily shows that limiting the attention to $w = 2$ is no restriction of the generality. One easily verifies that $u \equiv 0$ is a solution of (1). Depending on the values of δ and ε this solution may or may not be stable. This can be made clear by application of a theorem by Floquet.
If we set $u = x_1$ and $\dot{u} = x_2$ then (1) can be written as:

$$\dot{\mathbf{x}} = A_2(t)\mathbf{x}, \tag{2}$$

where $\mathbf{x} = (x_1, x_2)^T$ and $A_2(t) = \begin{pmatrix} 0 & 1 \\ -\delta - \varepsilon \cos 2t & 0 \end{pmatrix}$.
Instead of (2) we consider now the system

$$\dot{\mathbf{x}} = A(t)\mathbf{x}, \tag{3}$$

where $A(t)$ is now a continuous T-periodic $n \times n$-matrix.

Theorem (Floquet):

Each fundamental matrix $\Phi(t)$ of (3) can be written as:

$$\Phi(t) = P(t)e^{Bt},$$

where $P(t)$ is T-periodic and B a constant $n \times n$-matrix.

Proof: If $\Phi(t)$ is a fundamental matrix of (3) then $\Phi(t+T)$ is also a fundamental matrix.
$\Phi(t)$ and $\Phi(t+T)$ are linearly dependent as they contain both n linearly independent solutions of (3).
This means that there exists a nonsingular constant matrix C such that

$$\Phi(t+T) = \Phi(t).C.$$

Let B a constant $n \times n$-matrix be defined by

$$C = e^{BT}.$$

Now we show that $P(t) = \Phi(t)e^{-Bt}$ is T-periodic.

$$\begin{aligned} P(t+T) &= \Phi(t+T)^{-B(t+T)} = \Phi(t).Ce^{-B(t+T)} = \\ &= \Phi(t)e^{-Bt} = P(t) \end{aligned}$$

Some remarks:

- The matrix C and its eigenvalues are usually called the monodromy-matrix respectively the characteristic (Floquet) multipliers of equation (3).
  From Floquet's theorem it follows that the general solution of (3) can be written as:
  $$\mathbf{x}(t) = \Phi(t)\mathbf{c} = P(t)e^{Bt}\mathbf{c},$$
  where $\mathbf{c} \in \mathbb{R}^n$.
  It is not difficult to show that $\mathbf{x}(t+T) = P(t)e^{Bt}C\mathbf{c}$.
  Let now $\mathbf{c}$ be an eigenvector of C: $\mathbf{c} = \mathbf{c}_j$ with $C\mathbf{c} = \lambda_j\mathbf{c}_j$. Then it follows that:
  $$\mathbf{x}(t+T) = \lambda_j\mathbf{x}(t).$$
  From this property it can be understood that $\lambda_j \ldots j = 1, \ldots, n$, the eigenvalues of C are called the Floquet multipliers of equation (3).
- Suppose that $P(t)$ would be known. Then it follows that $\mathbf{x} = P(t)\mathbf{y}$ transforms equation (3) in:
  $$\dot{\mathbf{y}} = B\mathbf{y}$$
  which is an equation with constant coefficients.

- It may be clear that the eigenvalues of B determine the existence of periodic solutions of (3) respectively the stability of the trivial solution.
- In the special case that $A(t) = A_2(t)$ the eigenvalues of B depend on δ and ε. Regretfully there are no general methods available to compute P and B. Each equation requires a special study as shall be shown in what follows.

3. The averaging method

A basic result from the theory of averaging can be formulated as follows. Consider an initial problem for the system:

$$\begin{aligned}\dot{\mathbf{x}} &= \varepsilon \mathbf{f}(\mathbf{t}, \mathbf{x}) \\ \mathbf{x}(0) &= \mathbf{x}_0,\end{aligned} \tag{4}$$

where $\mathbf{f} : \mathbb{R}^{n+1} \to \mathbb{R}^n$, $x \in \mathbb{R}^n$, $\mathbf{f}$ is T-periodic in t and $|\varepsilon|$ a small parameter.
The average value $\mathbf{f}_0(\xi)$ of $\mathbf{f}(t, \xi)$ is defined as follows:

$$\mathbf{f}_0(\xi) = \frac{1}{T}\int_0^T \mathbf{f}(t, \xi)dt,$$

where in carrying out the integration: ξ is kept constant.
Consider additionally the initial value problem for the averaged system:

$$\begin{aligned}\dot{\xi} &= \varepsilon \mathbf{f}_0(\xi) \\ \xi(0) &= \mathbf{x}_0\end{aligned} \tag{5}$$

The following theorem states in what sense $\xi(t)$ is an approximation of $\mathbf{x}(t)$.

Theorem:
Let $\mathbf{f}$ and $\frac{\partial \mathbf{f}}{\partial \mathbf{x}}$ be continuous and bounded in $\mathbb{R}^n \times D$ where $D \in \mathbb{R}^n$. Let T be a constant independent of ε. Then we have:

$$\mathbf{x}(t) - \xi(t) = O(\varepsilon) \text{ on a time-scale of } O\left(\frac{1}{\varepsilon}\right)$$

For the proof of this theorem we refer to [1].
In order to obtain quantitative results, the question raises if we can apply the averaging method to the Mathieu equation.
The first step is to put the Mathieu equation in the standard form (4). This can be done by considering a rather general non-linearly perturbed equation of the form:

$$\ddot{u} + \omega^2 u = \varepsilon f(u, \dot{u}, t) \tag{6}$$

The Mathieu equation is a special case of (6) by setting $\delta = \omega^2 - \varepsilon\beta$ and $f = -u\cos 2t + \beta u$.
The reason for setting $\delta = \omega^2 - \varepsilon\beta$ is that we shall study the stability properties of the solutions of the Mathieu equation in the $\delta - \varepsilon$ plane in the neighbourhood of $\omega \in \mathbb{R}$ arbitrary but fixed.
Now we will write (6) as a system:

$$u = u_1 \quad , \quad \dot{u}_1 = u_2 \quad \text{which produces}$$

$$\begin{aligned} \dot{u}_1 &= u_2 \\ \dot{u}_2 &= -\omega^2 u_1 + \varepsilon f(u_1, u_2, t) \end{aligned} \tag{7}$$

For $\varepsilon = 0$ the solutions are:

$$\begin{aligned} u_1 &= x_1 \cos\omega t + x_2 \sin\omega t \\ u_2 &= -x_1\omega\cos\omega t + x_2\omega\cos\omega t, \end{aligned} \tag{8}$$

where x_1 and x_2 are real constants.
For $\varepsilon \neq 0$ we use (8) as a transformation where now $x_1 = x_1(t)$ and $x_2 = x_2(t)$.
Substitution of (8) in (7) gives:

$$\begin{aligned} \dot{x}_1\cos\omega t \quad &+ \quad \dot{x}_2 \sin\omega t = 0 \\ -\dot{x}_1\omega\sin\omega t \quad &+ \quad \dot{x}_2\cos\omega t = \varepsilon f. \end{aligned}$$

Solving this system with respect to $\dot{x}_1$ and $\dot{x}_2$ yields:

$$\begin{aligned} \dot{x}_1 &= \tfrac{\varepsilon}{\omega} f \sin\omega t \\ \dot{x}_2 &= \tfrac{\varepsilon}{\omega} f \cos\omega t \end{aligned} \tag{9}$$

Where $f = f(x_1\cos\omega t + x_2\sin\omega t, \; -x_1\omega\sin\omega t + x_2\omega\cos\omega t, \; t)$ (9) is of the standard form (4) where f can be made T-periodic with respect to t by choosing an appropriate value of ω.
For the case of the Mathieu equation (9) reads:

$$\begin{aligned} \dot{x}_1 &= -\tfrac{\varepsilon}{\omega}(\beta - \cos 2t) \quad (x_1\cos\omega t + x_2\sin\omega t)\sin\omega t \\ \dot{x}_2 &= \tfrac{\varepsilon}{\omega}(\beta - \cos 2t) \quad (x_1\cos\omega t + x_2\sin\omega t)\cos\omega t \end{aligned} \tag{10}$$

We investigate the case $\omega = 1$, the so-called principle resonance, and apply the averaging method:

$$\begin{aligned} \dot{\xi}_1 &= -\varepsilon\tfrac{1}{2}\xi_2\left(\beta + \tfrac{1}{2}\right) \\ \dot{\xi}_2 &= \varepsilon\tfrac{1}{2}\xi_1\left(\beta - \tfrac{1}{2}\right) \end{aligned} \tag{11}$$

where $\xi_1(t)$ and $\xi_2(t)$ are accurate $O(\varepsilon)$ approximations for $x_1(t)$ and $x_2(t)$ if they satisfy the same initial conditions.
We remark that (11) can be written as:

$$\dot{\xi} = \mathbf{B}\xi \tag{12}$$

Where $\xi = \begin{pmatrix} \xi_1 \\ \xi_2 \end{pmatrix}$ and $B = \begin{pmatrix} 0 & -\frac{\varepsilon}{2}\left(\beta + \frac{1}{2}\right) \\ \frac{\varepsilon}{2}\left(\beta - \frac{1}{2}\right) & 0 \end{pmatrix}$ It is interesting to remark that the qualitative result by Floquet can now be expressed quantitatively as follows:

$$\bar{\mathbf{u}} = \begin{pmatrix} \bar{u_1} \\ \bar{u_2} \end{pmatrix} = \mathbf{P(t)e^{Bt}c} + \mathbf{O}(\varepsilon)$$

where $P(t) = \begin{pmatrix} \cos\omega t & \sin\omega t \\ -\omega\cos\omega t & \omega\cos\omega t \end{pmatrix}$,
$e^{Bt}\mathbf{c}$ the general solution of (12) with $\mathbf{c} \in \mathbb{R}^2$, an arbitrary constant vector and $\bar{u}$ an accurate $O(\varepsilon)$ approximation for $\mathbf{u} = \begin{pmatrix} u_1 \\ u_2 \end{pmatrix}$ if $\mathbf{u}(0) = \bar{\mathbf{u}}(\mathbf{0})$.
Special solutions of (12) are:

$$\begin{array}{lll} \bar{y_1} \equiv 0 & \text{and} & \beta = -\frac{1}{2} \\ \bar{y_2} \equiv 0 & \text{and} & \beta = \frac{1}{2} \end{array}$$

For the original equation (1):

$$\ddot{u} + (\delta + \varepsilon\cos 2t)u = 0$$

this implies that

$$\begin{array}{llll} \text{for} & \delta = 1 + \frac{1}{2}\varepsilon & , & \bar{u} = y_2(0)\sin t \\ \text{and for} & \delta = 1 - \frac{1}{2}\varepsilon & , & \bar{u} = y_1(0)\cos t \end{array}$$

It can be shown rigorously, but this will not be done here that for $\delta = 1 \pm \frac{1}{2}\varepsilon + O(\varepsilon^2)$ periodic solutions exist of which the first order approximations are given above by $\bar{u}$.
From (11) it also follows that for $\beta^2 > \frac{1}{4}$ the solutions of (11) are bounded and for $\beta^2 < \frac{1}{4}$ there are only unbounded solutions.
In the $\delta - \varepsilon$ parameter plane as sketched in fig. 1 the lines $\varepsilon = \pm 2(\delta - 1)$ represent the transition between stable and unstable solutions.

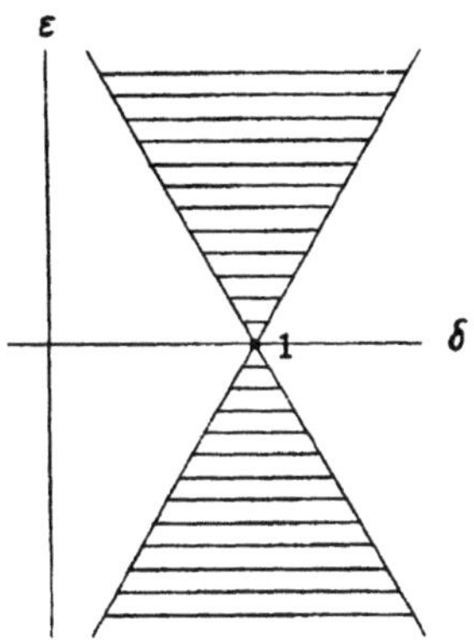

Figure 1. Shaded domains of instability around $\delta = 1$

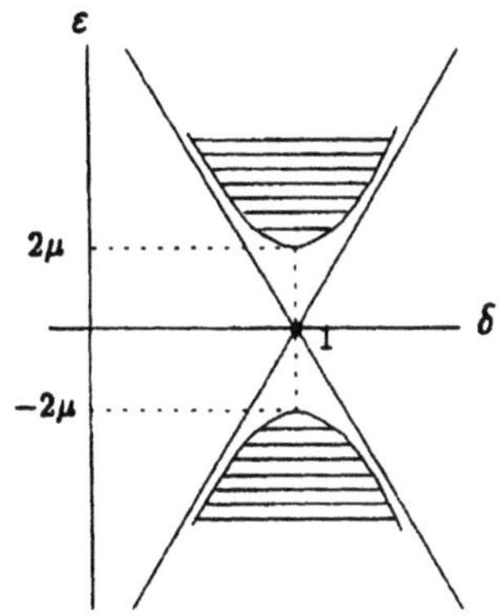

Figure 2. Shaded domains of instability around $\delta = 1$ for the Mathieu equations with damping

It is of interest to study additionally the Mathieu equation with small linear damping:

$$\ddot{u} + \mu\dot{u} + (\delta + \varepsilon \cos 2t)u = 0$$

where $\mu > 0$.
The computations can be carried out analogously to the case without damping ($\mu = o$). As transition curves one finds

$$\delta = 1 \pm \frac{1}{2}\sqrt{\varepsilon^2 - 4\mu^2}$$

where μ has been taken $O(\varepsilon)$.
Typical regions of instability are sketched in fig. 2. For more details one may consult [2].

4. The pendulum with varying length

The equation of motion for a pendulum with varying length reads:

$$m\ddot{x} = -mg\sin\varphi$$
$$x = \ell(t)\varphi,$$

where the variables are indicated in fig 3.

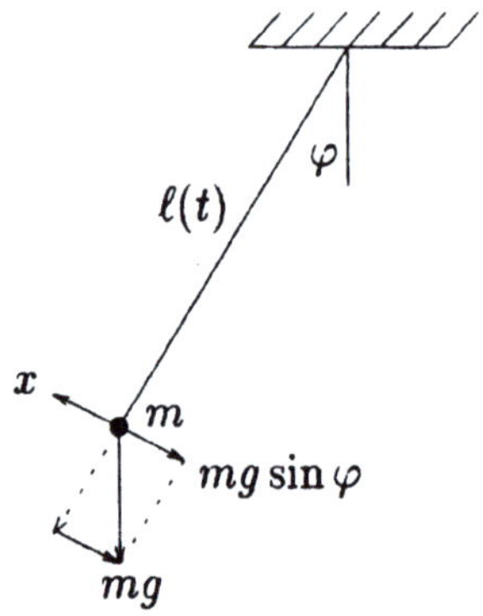

Figure 3: Sketch of the pendulum with varying length.

Consider small oscillations around the equilibrium state $\varphi = 0$ such that $\sin\varphi \approx \varphi$.
Then it follows:

$$\ddot{x} + gx/\ell(t) = 0$$

If we set $\ell(t) = \ell_0(1 - \varepsilon\cos\omega t)$ where $|\varepsilon| \ll 1$ we obtain

$$\ddot{x} + \omega_0^2(1 + \varepsilon\cos\omega t)x + O(\varepsilon^2) = 0$$

where $\omega_0^2 = g/\ell_0$.
If we now set $\omega_0 = 1$ and $\omega = 2$ we deduce from fig. 1 that for all $0 < \varepsilon \ll 1$ the solutions are unstable. The physical interpretation of this result is as follows. Let $\omega_0 = 1$ be the natural frequency of the pendulum (in absence of the parametric excitation i.e. $\varepsilon = 0$). Then, when one applies a parametric excitation with amplitude ε and frequency $\omega = 2$, that is twice the frequency of the natural frequency of the pendulum, the equilibrium position of the pendulum will become unstable. Applying this result to the real life situation with the person on the swing, one easily deduces that in order to get the swing into motion one has to move vertically with a frequency twice as much as the natural frequency of the swing.

5. The stretched string with variable length.

Consider a perfectly flexible string in a stretched situation, between two points $x = 0$ and $x = \pi$ of the x-axis. At $x = 0$ the string is fixed whereas at $x = \pi$ the string is excited horizontally, that is along the x-axis. We shall model the vibrations of this string in a vertical plane.
Let the unstretched string have a length $\ell < \pi$ and let $X \in [0, \ell]$ be coordinate of each material point P of the unstretched string.
The string is stretched uniformly so that the stretched length is π. The material point P with coordinate $X\mathbf{i}$ in the unstretched situation will have the coordinate $(1 + w_0)X\mathbf{i}$ in the stretched state, where $\mathbf{i}$ is the unit vector along the x-axis and $w_0 = \frac{\pi}{\ell} - 1$, the relative (time independent) strain.
Denote the dynamic displacement of point P with coordinate $(1 + w_0)X\mathbf{i}$ by $U(X,t)\mathbf{i}$ and $V(X,t)\mathbf{j}$, where $\mathbf{j}$ is the unit vector perpendicular to $\mathbf{i}$.
The position $\mathbf{R(X,t)}$ of P in the dynamic state can be written as:

$$\mathbf{R(X,t)} = \{(\mathbf{1} + \mathbf{w_0})\mathbf{X} + \mathbf{U(X,t)}\}\,\mathbf{i} + \mathbf{V(X,t)j}$$

The relative strain per unit length of the unstretched string is:

$$\left|\frac{\partial}{\partial X}\mathbf{R(X,t)}\right| - 1 = \sqrt{(1 + w_0 + U_X)^2 + V_X^2} - 1$$

Introduce the new coordinate x by:

$$x = (1 + w_0)X,$$

and let

$$\begin{aligned} U(X,t) &= U(x/(1 + w_0), t) = \tilde{u}(x,t), \\ V(X,t) &= U(x/(1 + w_0), t) = \tilde{v}(x,t), \end{aligned}$$

then the relative strain $w(x,t)$ per unit length becomes

$$w(x,t) = (1 + w_0)\sqrt{(1 + \tilde{u}_x)^2 + \tilde{v}_x^2} - 1. \tag{13}$$

The kinetic and potential energy densities are respectively given by:

$$T = \frac{1}{2}\rho(\tilde{u}_t^2 + \tilde{v}_t^2)\ , \quad V = \frac{1}{2}E\ w^2(x,t) \tag{14}$$

where ρ is the mass of the string per unit length and E is Young's modulus.
By assuming that $|\tilde{u}_x|$ and $|\tilde{v}_x|$ are small with respect to 1, the potential energy V may be approximated by its Taylor expansion V_4 up to terms of the fourth degree:

$$\begin{aligned} V_4 = \tfrac{1}{2}E\,\big[&(w_0^2 + 2(1 + w_0)w_0\tilde{u}_x + (1 + w_0)^2\tilde{u}_x^2 \\ &+ w_0(1 + w_0)\tilde{v}_x^2 + (1 + w_0)\tilde{u}_x\tilde{v}_x^2 - (1 + w_0)\tilde{u}_x^2\tilde{v}_x^2 + \tfrac{1}{4}(1 + w_0)\tilde{v}_x^4\big] \end{aligned} \tag{15}$$

By using T and V_4 in a variational principle one obtains the following equations of motion:

$$\begin{aligned}\rho\tilde{u}_{tt} &= E(1+w_0)^2\tilde{u}_{xx} + \tfrac{1}{2}E(1+w_0)\tfrac{\partial}{\partial x}\left\{\tilde{v}_x^2 - 2\tilde{u}_x\tilde{v}_x^2\right\}, \\ \rho\tilde{v}_{tt} &= w_0E(1+w_0)\tilde{v}_{xx} + \tfrac{1}{2}E(1+w_0)\tfrac{\partial}{\partial x}\left\{2\tilde{u}_x\tilde{v}_x - 2\tilde{u}_x^2\tilde{v}_x + \tilde{v}_x^3\right\}.\end{aligned} \tag{16}$$

It may be remarked that the equations (16) differ from similar ones given in [2],[3] where the non-linear terms in the right-hand sides are multiplied by a factor $1-w_0$ instead of $1+w_0$.
Moreover in [3], p. 862, a remark is made on this factor. From this remark one may deduce that the authors think that the case $w_0 = 1$ leads to a decoupling of the two equations. It may be clear that in (16) the case $w_0 = 1$ is of no special interest, which seems quite natural from physical point of view

Set now: $\frac{E}{\rho} = c_1^2$ and $\frac{Ew_0}{\rho} = c_2^2$ and replace t by $t/\sqrt{1+w_0}$.
Also damping forces $\tilde{\alpha}_1\tilde{u}_t$ respectively $\tilde{\alpha}_2\tilde{u}_t$ where $\tilde{\alpha}_1$ and $\tilde{\alpha}_2$ are positive parameters are taken into account:

$$\begin{aligned}\tilde{u}_{tt} - c_1^2(1+w_0)\tilde{u}_{xx} &= -\tilde{\alpha}_1\tilde{u}_t + \tfrac{1}{2}c_1^2\tfrac{\partial}{\partial x}\left\{\tilde{v}_x^2 - 2\tilde{u}_x\tilde{v}_x^2\right\} \\ \tilde{v}_{tt} - c_2^2\tilde{v}_{xx} &= -\tilde{\alpha}_2\tilde{v}_t + \tfrac{1}{2}c_1^2\tfrac{\partial}{\partial x}\left\{2\tilde{u}_x\tilde{v}_x - 2\tilde{u}_x^2\tilde{v}_x + \tilde{v}_x^3\right\}\end{aligned} \tag{17}$$

$$0 < x < \pi\,, \quad t > 0$$

$$\begin{aligned}\tilde{u}(t,0) &= 0\,, \quad \tilde{u}(t,\pi) = \tilde{\varepsilon}\sin(\tilde{\lambda}t + \varphi_0) \\ \tilde{v}(t,0) &= \tilde{v}(t,\pi) = 0\end{aligned}$$

$\tilde{\varepsilon}$, $\tilde{\lambda}$ and φ_0 are real parameters defining the parametric excitation. We consider (17) as a model problem of which the case $w_0 \ll 1$ will be studied in more detail. Note that as we are studying excitation in a stretched string an initial relative strain w_0 is needed. It easily follows that $w_0 = (c_2/c_1)^2$ where c_2 and c_1 are the speeds of the transversal respectively strain waves in the string.

Let $w_0 = (c_2/c_1)^2 = \varepsilon$, where ε is a small parameter, and introduce a new time τ defined by $\tau = c_2t$. The first equation of model problem (17) becomes:

$$\varepsilon\tilde{u}_{\tau\tau} - (1+\varepsilon)\tilde{u}_{xx} = \frac{1}{2}\frac{\partial}{\partial x}\left\{\tilde{v}_x^2 - 2\tilde{u}_x\tilde{v}_x^2\right\} - \varepsilon\frac{\tilde{\alpha}_1}{c_2}\tilde{u}_\tau \tag{18}$$

In a first order approximatoin, that is for $\varepsilon \to 0$ (18) reduces to:

$$-\tilde{u}_{xx} = \frac{1}{2}\frac{\partial}{\partial x}\left\{\tilde{v}_x^2 - 2\tilde{u}_x\tilde{v}_x^2\right\} \tag{19}$$

If we consider additionally vertical vibrations $\tilde{v}(t,x)$ of $O(\varepsilon)$ it follows from (19) that $\tilde{u}(t,x) = O(\varepsilon^2)$. This implies that the term $2\tilde{u}_x\tilde{v}_x^2$ can also be neglected in a first order approximation:

$$-\tilde{u}_{xx} = \tfrac{1}{2}\tfrac{\partial}{\partial x}\tilde{v}_x^2$$
$$\tilde{u}(t,0) = 0\ ;\quad \tilde{u}(t,\pi) = \tilde{\varepsilon}\sin(\lambda\tau + \varphi_0), \tag{20}$$

where $\lambda = \tilde{\lambda}/c_2$.
After integration of equation (20) twice and application of the boundary conditions the following expression for $\tilde{u}_x$ can be derived:

$$\tilde{u}_x(\tau,x) = -\frac{1}{2}\tilde{v}_x^2 + \frac{\tilde{\varepsilon}}{\pi}\sin(\lambda\tau + \varphi_0) + \frac{1}{2\pi}\int_0^\pi \tilde{v}_x^2 dx \tag{21}$$

With the new time varable τ the second equation of model problem (17) reads:

$$\varepsilon(\tilde{v}_{\tau\tau} - \tilde{v}_{xx}) = -\varepsilon\frac{\tilde{\alpha}_2}{\tilde{c}_2}\tilde{v}_\tau + \frac{1}{2}\frac{\partial}{\partial x}\left\{2\tilde{u}_x\tilde{v}_x - 2\tilde{u}_x^2\tilde{v}_x + \tilde{v}_x^3\right\} \tag{22}$$

Let now: $\tilde{v} = \varepsilon v$, $\tilde{u} = \varepsilon^2 u$ as assumed above and $\tilde{\alpha}_2/c_2 = \varepsilon\alpha_2$. Then after substitution of (21) in (22) the following equation is obtained:

$$v_{\tau\tau} - v_{xx} = -\varepsilon\alpha_2 v_\tau + \varepsilon\ v_{xx}\left\{\frac{1}{2\pi}\int_0^\pi v_x^2 dx + A\sin(\lambda\tau + \varphi_0)\right\} + O(\varepsilon^2), \tag{23}$$

where A is defined by: $\tilde{\varepsilon} = \varepsilon\pi A$ with the additional boundary conditions:

$$v(\tau,0) = v(\tau,\pi) = 0.$$

Note that in equation (23) $v(\tau,x)$ represents the vertical displacement of the string whereas in the same equation $A\sin(\lambda\tau+\varphi_0)$ represents the horizontal parametric excitation.
Equation (23) is a nonlinear partial-integro differential equation which has special exact solutions of the type

$$v(\tau,x) = q(\tau)\sin x \tag{24}$$

where $q(\tau)$ is a solution of the following ordinary differential equation:

$$\frac{d^2q}{d\tau^2} + q = \varepsilon\left(-\alpha_2\frac{dq}{d\tau} - \frac{1}{4}q^3 - qA\sin\lambda\tau\right). \tag{25}$$

Where for the sake of convenience φ_0 has been taken equal to zero.
Equation (25) is a nonlinear Mathieu equation with damping. Periodic solutions of (25) can be found by introducing amplitude and phase variables

$r(\tau)$ and $\varphi(\tau)$ defined by:

$$q(\tau) = r(\tau)\cos(\tau + \varphi(\tau))$$
$$\frac{dq}{d\tau} = - r(\tau)\sin(\tau + \varphi(\tau))$$

where $r(\tau)$ and $\varphi(\tau)$ are solutions of

$$\frac{dr}{d\tau} = \varepsilon\left(-\alpha_2 r\sin(\tau+\varphi) + \frac{1}{4}r^3\cos^3(\tau+\varphi) + Ar\cos(\tau+\varphi)\sin\lambda\tau\right)\sin(\tau+\varphi),$$
$$\frac{d\varphi}{d\tau} = \varepsilon\left(-\alpha_2 \sin(\tau+\varphi) + \frac{1}{4}r^2\cos^3(\tau+\varphi) + A\cos(\tau+\varphi)\sin\lambda\tau\right)\cos(\tau+\varphi).$$

After averaging with respect to τ one obtains the following approximating equations iff $\lambda = 2$:

$$\begin{aligned} \frac{d\bar{r}}{dt} &= \frac{1}{4}\varepsilon\bar{r}(-2\alpha_2 + A\cos 2\bar{\varphi}) \\ \frac{d\bar{\varphi}}{d\tau} &= \frac{1}{4}\varepsilon\left(\frac{3}{8}\bar{r}^2 - A\sin 2\bar{\varphi}\right) \end{aligned} \tag{26}$$

Critical points of system (26) follow from:

$$A\cos 2\bar{\varphi}_0 = 2\alpha_2 \quad \text{and} \quad A\sin 2\bar{\varphi}_0 = \frac{3}{8}\bar{r}_0^2 \tag{27}$$

or:

$$\frac{3}{8}\bar{r}_0^2 = \sqrt{A^2 - 4\alpha_2^2}.$$

It is clear that if $A > 2\alpha_2$, where A is the amplitude of the parametric exitation, time periodic solutions of (25) may exist and hence a time periodic solution of (23):

$$v(\tau, x) = \bar{r}_0\cos(\tau + \bar{\varphi}_0)\sin x$$

or

$$v(t, x) = \bar{r}_0\cos(c_2 t + \bar{\varphi}_0)\sin x, \tag{28}$$

where $\bar{r}_0$ and $\bar{\varphi}_0$ follow from (27) and the parametric exitation is given by

$$\tilde{\varepsilon}\sin\tilde{\lambda}t = \varepsilon\pi A\sin(2c_2 t). \tag{29}$$

It is of interest to remark that it can be shown rigorously that (28) represents an $O(\varepsilon)$ approximation to the exact time periodic solution of equation (23).

Note that the ratio between the frequency of the parametric excitation and the frequency of the transversally standing wave oscillation is 2 as can be understood on physical grounds. It may be clear that the fundamental mode oscillations as observed by Melde are represented by (28).

6. Some conclusions.

In this paper we have shown that two mechanical vibrating systems, the pendulum and the stretched string both with variable length display parametric excitation. In both cases the parametric excitation for the main resonance has been considered. The main resonance case applies to the case where the ratio between the frequency of the parametric excitation and the natural frequency of the mechanical system is 2. In case of the stretched string the natural frequency which was considered corresponds with the frequency of the fundamental mode. By using the averaging method transition curves in the neighbourhood of the main resonance were constructed. These transition curves agree locally well with the famous stability parameter diagram of the Mathieu equation. The advantage of the use of the averaging method is that also cases with damping and nonlinear perturbations can be investigated.
It is remarkable that the rather complicated model equation (23) could be reduced to a nonlinear Mathieu equation which on its turn could be analyzed by using the averaging method. Both in qualitative as in quantitative sense the pendulum and the stretched string have much in common, as far as its concerns the mechanism of unstability and the mechanism which causes finite amplitude oscillations. The last one has the character of nonlinear spring stiffness.

Acknowledgment

This paper was written during a stay in Marseille, France at the Ecole Supérieure de Mécanique de Marseille (ESM2) and the Laboratoire de Mécanique et d'Acoustique (LMA) of the Centre National de la Recherche Scientifique (CNRS). The author gratefully acknowledges the hospitality by M. Robert Pelissier, Mm. D. Habault and M. Paul J.T. Filippi. The author also expresses his gratitude to Mrs. Gemma van der Windt for the Latex word processing of the manuscript.

References

1. F. VERHULST, Nonlinear Differential Equations and Dynamical Systems, Second edition, Springer Universitext, 1996.
2. A.H. NAYFEH and D.T. MOOK, Nonlinear Oscillations, John Wiley, 1979.
3. P.M. MORSE and K. UNO INGARD, Theoretical Acoustics, Princeton Unversity Press, 1986.
4. F. MELDE, Über die Erregung stehender Wellen eines fadenförmigen Körpers, Pogg. Ann. d. Phys. Bd. III, pp. 193-215, 1860.

GENERATION AND SUPPRESSION OF RADIATION BY SOLITARY PULSES

HSUEH-CHIA CHANG,[†] EVGENY A. DEMEKHIN[§]
AND EVGENY KALAIDIN
Department of Chemical Engineering
University of Notre Dame
Notre Dame, IN 46556

Abstract. The description of the spatio-temporal dynamics of an extended active/dispersive medium can be simplified considerably if the dynamics is dominated by a fixed number of solitary pulses separated by radiation-free flat substrates. Radiation is generated in a noise-free environment when excess mass drains out of a pulse in the form of a spreading shelf and the radiation grows rapidly by feeding on the active substrate. However, this growth can be suppressed if the localized radiation packet is absorbed by a second pulse. It is shown that both the generation and suppression mechanisms can be quantitatively understood by analyzing the essential spectrum and a peculiar discrete spectrum of the equilibrium pulse called resonance pole. An explicit criterion is then developed and verified numerically for when the spatio-temporal dynamics of the generalized Kuramoto-Sivashinsky (gKS) equation is dominated by robust solitary pulses.

1. Introduction

There is now considerable experimental and numerical evidence that the spatio-temporal dynamics of many active/dissipative, excitable and dispersive media are dominated by localized solitary pulses (waves). A physical system of particular interest to us is the falling film [1- 4] whose pulses are separated by a wide span of substrate layer which is remarkably smooth

[†]Corresponding author, Email: Hsueh-Chia.Chang.2@nd.edu

[§]Permanent address: Department of Applied Mathematics, Kuban State Technological University, Krasnodar 350072, Russia eventhough it is strictly unstable. The average separation between pulses is also seen to increase downstream from the inlet such that the pulse density per unit length decreases. This coarsening stops at some position downstream as the average separation reaches an equilibrium value. The mechanism that drives this coarsening dynamics is that there are occasional "excited" pulses that travel faster than its front neighbors such that they eventually capture the front pulses to reduce the pulse density. Considerable experimental [5, 6] and numerical [3] efforts have been expended to scrutinize this coalescence event between a bigger and faster back pulse and a smaller and slower front pulse. Both experiments [5] and simulation [3] show that the back pulse actually loses mass to the back as it approaches the front one. The speed differential between the two pulses hence decreases as mass drains from the back pulse. In fact, if the drainage rate is too high or the separation is too large, coalescence will not occur [3]. If coalescence does occur, a new excited pulse is created from the two pulses which then attempts to capture the next pulse in front. The equilibrium is reached when all "excited" pulses have drained their excess mass without inducing further coalescence. We have statistically modeled the domino effect of successive coalescence during the coarsening stage and estimated the equilibrium when coalescence ceases [7].

Once equilibrium is reached, the pulses are nearly identical and their number remains the same but their spacing can be quite irregular and time-varying. Their large separation, however, allows the construction of a coherent structure theory [8, 9, 10, 11] that captures the position of every pulse with a low-order dynamical system whose dimension is equal to the number of equilibrated pulses. Such a theory describes how the front and back neighbors of a pulse shift its position by exciting the slow translation mode, which has a zero eigenvalue in the absence of the neighbors due to translational symmetry. The possibility of describing the complex spatio-temporal dynamics of an extended system like the falling film with a low-order dynamical system is hence very attractive and one wonders if other extended dynamics can likewise be simplified. Localized wave patterns have now been found in Faraday instability, shear boundary layers, Taylor-Couette flow, sheared films, reactive media etc. [2, 7, 11] and the dynamics of these extended-domain systems could be simplified if these coherent structures dominate over other patterns.

However, these are several stringent conditions that must be met before a coherent structure description is possible. Firstly, the pulses must be stable. Since the pulses approach the flat substrate in both directions and since the substrate is unstable in an active medium, the stability of a pulse must be defined locally with respect to a localized disturbance. However, a localized disturbance can also generate an expanding radiation packet that threatens to engulf a pulse unless it outruns or is outrunned by the expanding and growing radiative turbulence. This localized disturbance is typically introduced by the excess mass that drains out of an excited pulse in a relatively noise-free environment. Pego and Weinstein [12] have developed a convective stability theory with a cut-off weighting function to estimate whether a pulse can out run an expanding radiation sea for a purely dispersive medium. We have developed a simpler theory [13, 14] valid even for dissipative/active media based on an estimate of the envelope that bounds the unstable Fourier modes of the radiation packet. Connection between the two theories will be made more explicit here.

The second condition is that there cannot be annihilation of pulses due to coalescence and nucleation of new pulses from the radiation wavepackets. Both are determined by how the radiation is generated and how it is altered by the pulses. For the falling-film pulses, annihilation by coalescence is only possible if the mass draining out of an excited pulse is sufficiently slow such that the pulse has sufficient speed differential to capture its front neighbor. One hence needs to estimate the mass drainage rate and its correlation with the pulse speed. In this regard, extensive experiments on decaying pulses on a falling film [5, 6, 15, 3] have shown that, during the slow decay/drainage, the instantaneous speed scales linearly with respect to the instantaneous pulse amplitude (area) and both decay exponentially in time in a frame moving with the decaying pulse and linearly in space. A faithful theory must reproduce this observed decay dynamics. On the other hand, the mass that drains out of the decaying pulse generates radiation waves that can pick up more mass from the substrate medium and grow into new pulses. This generation dynamics typically involves a distinctive shelf behind the pulse that breaks up into radiation packets. While the original excited pulse survives the ordeal, the new pulses that grow out of the radiation clearly cannot be described by a dynamical system of fixed dimension.

The generation of new pulses from a localized wave packet can be suppressed if another pulse comes along and absorbs the growing radiation. However, since the radiation is mass-carrying, such pulse-radiation interaction can also drain mass into the absorbing pulse and excite it. The excited pulse, in turn, will drain and create more radiation behind it. However, provided that such radiation is attenuated by the absorption and is hence always swept up before they grow into bona-fide pulses, the finite-dimensional description remains accurate with radiation acting only as a noise source. Failing that, complex radiation/pulse exchanges take place and full-blown spatio-temporal chaos ensues. Such chaotic behavior may only be transient, however, if the average pulse separation can relax to a smaller value such that the radiation in between pulses has less time to grow and hence easier to sweep. Whether the pulses can absorb the radiating mass and whether it can retain it sufficiently long (to attenuate the radiation amplitude to the extent that new pulses will not form) then become key issues in the dynamics.

In this report, we develop a general theory for when pulse-dominated low-dimensional dynamics is appropriate for an extended active/dispersive medium by analyzing both the drainage mechanism for radiation generation that determines whether pulse annihilation by coalescence will occur and the radiation suppression mechanism that prevents pulse nucleation for the gKS (generalized Kuramoto-Sivashinsky or Kawahara) equation

$$\frac{\partial h}{\partial t} + 4h\frac{\partial h}{\partial x} + \frac{\partial^2 h}{\partial x^2} + \delta\frac{\partial^3 h}{\partial x^3} + \frac{\partial^4 h}{\partial x^4} = 0 \tag{1.1}$$

This equation is a generic long-wave evolution equation for mass-conserving gradient-flow systems [8, 16, 17] and it contains active, dissipative and dispersive terms. Moreover, it yields localized pulse solutions which propagate at constant speeds [8, 18]. When the dispersion term vanishes ($\delta = 0$), it reduces to the famed KS equation. We have established that the solitary pulses are unstable for $\delta < 0.137$ to radiation packets [13, 14]. As a result, turbulent fluctuation (KS chaos when δ vanishes) is the rule as the pulses are overwhelmed by radiation instead of absorbing it. For larger δ, however, a coherent structure theory becomes a possibility as the pulses stabilize and are radiation-absorbing [8, 18, 19]. Nevertheless, the issue of radiation generation and suppression remains unresolved. That the gKS pulses generate wavepackets by draining mass is seen in our simulation of Figs. 1 and 2 for $\delta = 1$ and $\delta = 5$. Excess mass is added to a gKS pulse to simulate an excited pulse after coalescence. The excess mass drains slowly out of the back of the pulse in a frame moving at the speed of the equilibrium pulse. As observed in earlier simulations of KdV equations [20], the exiting mass forms a shelf which later breaks up into oscillatory wavepackets (radiation). In Fig. 2, the excess mass (area) and speed of the draining pulse relative to the equilibrium are seen to decay exponentially in time with the same exponent and hence correlate linearly with each other. If there is an equilibrium pulse in front of the draining excited pulse, the latter's drainage rate and their separation determine whether the excited pulse will ever overtake the equilibrium pulse to induce coalescence. It is then important to decipher the decay dynamics of an excited pulse observed in Figs. 1 and 2. In Fig. 3, an equilibrium pulse at $\delta = 5$ is allowed to overrun a slower moving wavepacket. It is seen that the transmission through the pulse effectively flattens the wavepacket such that its amplitude is reduced considerably. This suppression does not allow the wavepacket to amplify under the destabilizing effect of the active medium and hence prevents nucleation of new pulses and destruction of old pulses. In contrast to the severe attenuation of the wavepacket, the equilibrium pulse is scarcely affected by the transmission. By constructing the essential and discrete spectra of an

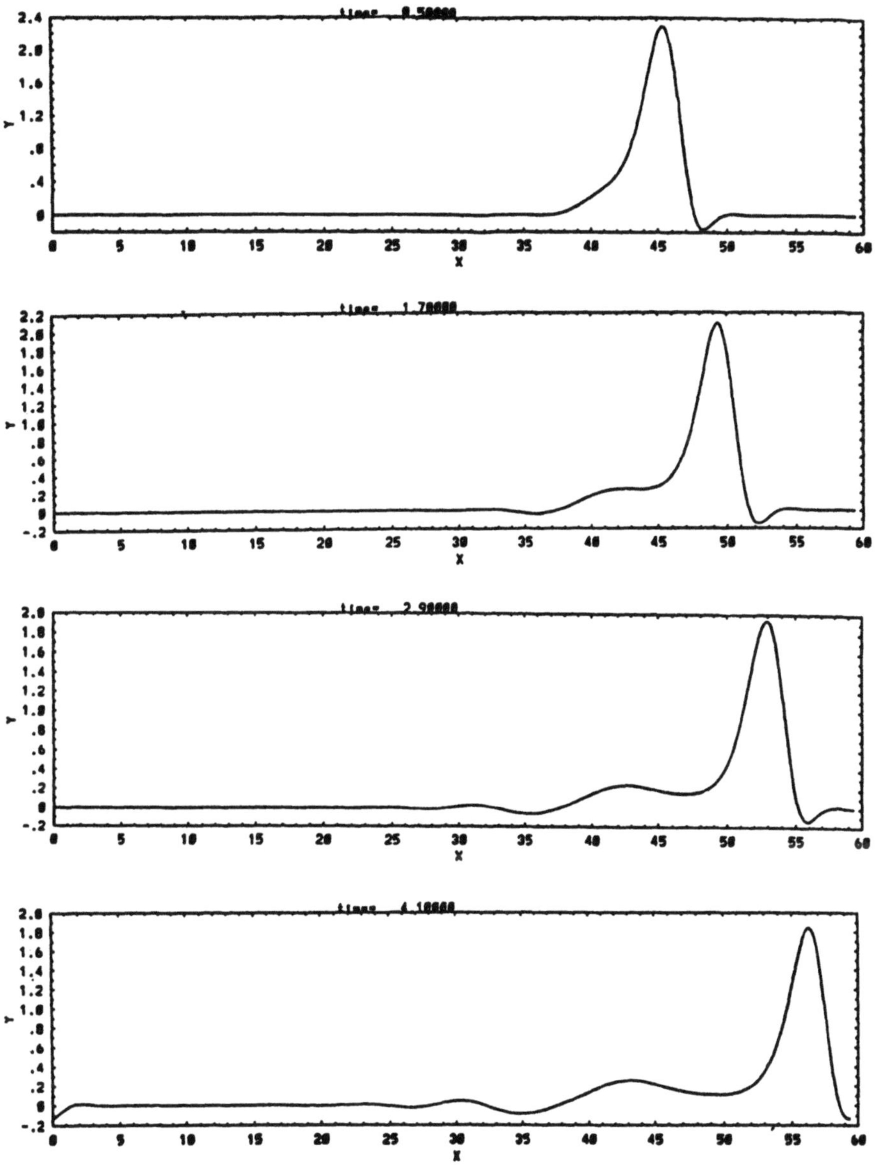

Figure 1. Drainage from an excited pulse when excess mass is added to anequilibrium pulse for $\delta = 1$. The draining mass in the back forms a shelf which breaks up into an oscillatory tail. The snapshots begin at $t = 0.5$ with $\Delta t = 1.2$ increments.

equilibrium pulse, we shall capture the radiation generation/suppression dynamics here and explain the correlation between the decaying speed and drainage dynamics of an excited pulse. Equilibrium states amenable to a coherent structure treatment are possible if one can find an average separation short enough to suppress radiation growth but long enough to prevent coalescence. While the technique is presented for pulses of the gradient-flow gKS equation, it should be extendable to any general system with coherent structures(fronts, spirals, defects etc.) including integrable systems that are difficult to analyze with exact inverse-scattering transforms.

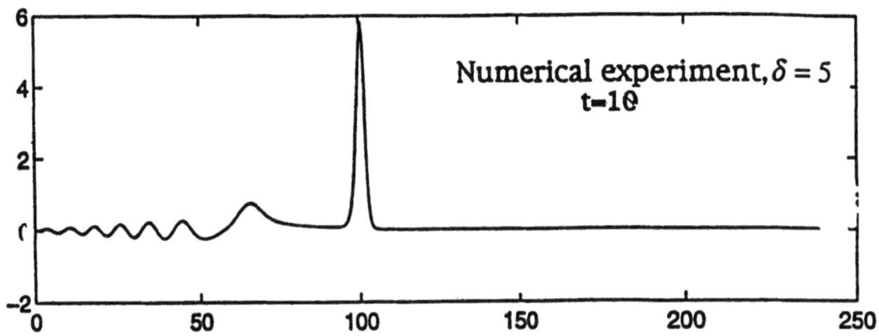

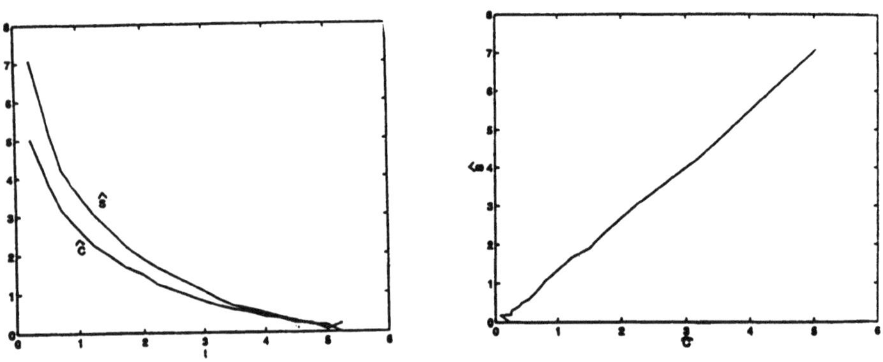

Figure 2. The excess speed $\hat{c}$ (deviation from the equilibrium pulse speed) and excess mass(area) $\hat{s}$ of the draining pulse both decrease exponentially in time and correlate linearly to each other.

2. Pulses and Symmetry

Transforming (1.1) to the moving frame shifting with the speed c of a pulse and integrating once, we obtain the defining equation for the equilibrium

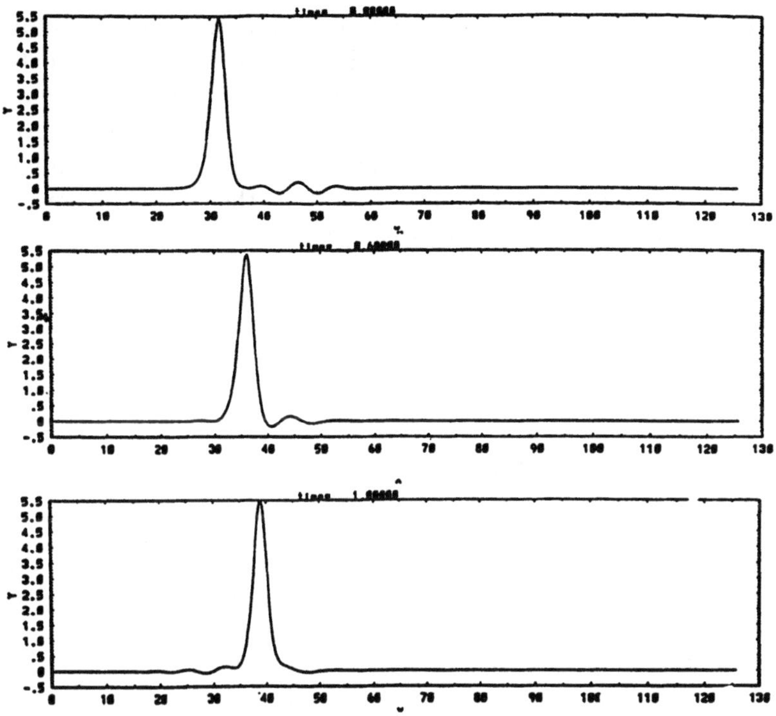

Figure 3. Suppression of localized wavepacket when it goes through an equilibrium pulse which is relatively unaffected by the radiation transmission.

pulses

$$h_{xxx} + \delta h_{xx} + h_x + 2h^2 - ch = q \tag{2.1}$$

$$h(x \to \pm\infty) = 0$$

where q is the constant "flow rate" in the moving frame since (1.1) in the moving frame is of the gradient form

$$\frac{\partial h}{\partial t} = -\frac{\partial q}{\partial x} \tag{2.2}$$

In free-surface flow like the falling film, this represents kinematic mass conservation where the change in the interfacial height h is proportional to the negative gradient of the flow rate q. The pulses are hence parameterized by both δ and q with c being the "nonlinear eigenvalue". The basic pulse family corresponds to one with the same zero flow rate carried by the zero substrate layer, $q = 0$. We shall show that, due to a Galilean symmetry,

pulses for all other flow rates can be determined from this reference family by a simple boost of both the speed and the substrate height.
We have exhaustively constructed all zero flow rate pulses whose solution branches can be represented by $c(\delta)$ [14, 18]. In a phase space formulation, these pulses correspond to homoclinic orbits connected to the origin. At small δ, the origin is a hyperbolic fixed point with a real positive eigenvalue and a complex pair with negative real parts. The usual Shilnikov bifurcations associated with homoclinic orbits of this type then occur and generate an infinite number of pulse families with different number of humps [18]. All families were found to be unstable, however, except for the single-hump pulse family which approach the symmetric KdV soliton at large δ:

$$\lim_{\delta\to\infty} c(\delta) = \frac{7}{5}\delta$$
$$\lim_{\delta\to\infty} h(x;\delta) = \frac{21\delta}{20 sech^2\sqrt{\frac{7}{20}}x} \tag{2.3}$$

Note, however, instead of an entire family of KdV solitons parameterized by the amplitude (or speed), a unique one-hump pulse exists at a given δ — the KdV limit at infinite δ is a singular limit.
We note that there are two symmetries associated with these gKS pulses. Due to the translational symmetry, (2.1) is invariant to a shift

$$x \to x + \xi \tag{2.4}$$

This symmetry then generates a one-parameter family from every pulse member at a given δ (and q). If we perturb this pulse slightly, it is reasonable to expect it to adjust slowly to a nearby pulse of this family, viz. to translate slowly. To leading order in the disturbance, we hence expect a zero eigenvalue in the discrete spectrum. More explicitly, since $h(x+\xi) \sim h(x) + \xi h_x(x)$ for a small shift, we expect ξ to be a constant with respect to time (zero eigenvalue) and h_x the eigenfunction of this zero mode

$$\mathcal{L}h_x = \left[\frac{d^4}{dx^4} + \delta\frac{d^3}{dx^3} + \frac{d^2}{dx^2} + \frac{d}{dx}(4h-c)\right] h_x = 0 \tag{2.5}$$

where $\mathcal{L}$ is the operator linearized about the pulse solution $h(x)$. That (2.5) is indeed the case can be easily demonstrated. It should also be noted that $< h_x >= \int_{-\infty}^{\infty} h_x dx = 0$ and this translation mode carries no mass. Higher order disturbance, say in the form of a nearby pulse, then induces slow dynamics in the solution mode $\xi(t)$ and one simply needs to project the perturbation onto the null space spanned by h_x to decipher the slow dynamics of $\xi(t)$.

Another symmetry associated with every pulse is the Galilean symmetry [10, 19, 2, 3]

$$\begin{aligned} h &\to h + \chi \\ c &\to c + 4\chi \\ q &\to q - c\chi - 2\chi^2 \end{aligned} \tag{2.6}$$

where the "boosted" pulse now has a different flow rate. Using (2.5), one can then generate pulses of all flow rate from the zero-flow rate pulse family. One expects another zero mode associated with the Galilean symmetry and, from the boost transformation (2.6), expects the eigenfunction to be a constant. In fact, it can be easily seen that this eigenfunction is a generalized one here it is

$$\mathcal{L}\left(\frac{1}{4}\right) = h_x \text{ or} \mathcal{L}^2\left(\frac{1}{4}\right) = 0 \tag{2.7}$$

Hence, this speed mode due to Galilean invariance appears to be a generalized zero mode. As a result, a constant boost in substrate thickness χ_0 remains constant but induces a linearly increasing shift in time, $\xi(t) = \xi_0 + 4\chi_0 t$,viz. a speed change of $4\chi_0$. The different flow rate under a boosted pulse has induced a change in its speed to produce this secular growth in time. Such secular behavior would presumably be arrested by higher order terms.

Pulses are seen to have different local substrates in simulations and experiments (observe the shelf on which the draining pulse sits in Fig. 1). Excited pulses formed after coalescence, in particular, seems to sit on a raised pedestal (shelf) [2, 3]. It was then thought that this speed mode χ is another slow modulation, imposed by symmetry, that is important in pulse dynamics. This belief is strengthened by the observation that the same zero mode is indeed excited during long-wave modulation of periodic waves [21]. For periodic waves,the modulation stability theory is constructed from the neutrally stable speed and translational modes which are both perioic over the wavelength in the short scale but the eigenfunctions for a pulse must cover an infinite domain. When we attempted to include this speed mode as a slow variable for pulses, we find that we cannot find the appropriate adjoint eigenfunction in $\mathcal{L}^2$ space to carry out the projection onto the position translation ξ and speed χ zero modes, viz. the generalized kernel of the adjoint of $\mathcal{L}$ is not two dimensional [3]. To remedy this, we had to adopt an ad hoc cut-off to permit the projection onto these modes. The ad hoc cut-off introduces a decay rate such that the generalized zero eigenvalue is shifted to a negative value.

In fact, the χ mode can never be excited in the infinite domain occupied by a localized pulse. A similar problem concerning the generalized speed mode $\frac{\partial h}{\partial c}$ of the KdV solution from a different symmetry was addressed by Pego and Weinstein [12]. They find that the projection can only be carried out in a weighted space and the generalized zero mode becomes a decaying mode known as a resonance pole. The weight then replaces the ad hoc cut-off we introduced earlier. In this report, we shall demonstrate that the essential spectrum serves as the true cut-off such that the shelf behind the draining pulse can be constructed with a proper analysis involving the essential spectrum. In essence, the mass-carrying speed mode of (2.7) is not a proper discrete mode in an infinite domain because of the infinite mass it carries. Instead, the mass is carried by the essential spectrum which is unstable. The instability is global but not pertinent to the localized pulse, however, when the mass-carrying unstable radiation is convected away from the pulse by dispersion. The local effect can be captured in a point-wise manner by shifting the essential spectrum to reveal a resonance pole which describes how mass carried by radiation drains out of pulse. Alternatively, the weight used by Pego and Weinstein can be used to focus on the local dynamics but the decay dynamics is then viewed through a less physical weighted projection. In both formulations, however, the continuum has conveniently manifested itself as a single discrete mode as in Landau damping in plasma physics. That the generalized speed mode χ is truly a decay mode for a gKS pulse has also been realized recently by Balmforth et al. [19] who assumed large periodicity without offering the decay rate. In contrast, the resonance pole concept can provide a specific estimate of the drainage rate. We shall use this approach to determine the dominant slow dynamics of the gKS pulse.

3. Pulse Spectra and Resonance Pole

The dynamics of small disturbances to a particular pulse is determined by the linear equation

$$\frac{\partial u}{\partial t} + \mathcal{L}u = 0 \tag{3.1}$$

and we shall restrict ourselves to bounded disturbances. As a result, the disturbances can be expanded in terms of the eigenfunctions of the following infinite domain eigenvalue problem [14]

$$\mathcal{L}\psi = \lambda\psi \qquad \psi \textit{ bounded as } x \to \pm\infty \tag{3.2}$$

Those eigenfunctions ψ_k that approach zero at infinite $|x|$ correspond to discrete eigenvalues λ_k and a simple integration of (3.2) shows that all discrete modes contain no mass, $<\psi_k> = \int_{-\infty}^{\infty} \psi_k(x)dx = 0$. This is a legacy of the gradient-flow mass conservation form. The eigenfunction $\psi(\lambda, x) = K(\alpha, x)e^{i\alpha x}$ which approach bounded oscillation in the infinities belong to the essential spectrum Γ defined by, for $\alpha \in \epsilon(0,\infty)$,

$$\lambda(\alpha) = i\alpha c + i\delta\alpha^3 + (\alpha^2 - \alpha^4) \tag{3.3}$$

Without loss of generality, the complex function $K(\alpha, x)$ is defined to approach unity at $x \to -\infty$ and the complex number $r(\alpha)e^{i\theta(\alpha)}$ at $x \to +\infty$ where $r(\alpha)$ is known as the transmission coefficient and $\theta(\alpha)$ the phase shift. Both $r(\alpha)$ and $\theta(\alpha)$ are real numbers. Due to the localized nature of the pulse, $K(\alpha)$ approaches its asymptotic values at $x \to \pm\infty$ only one pulse width from the pulse maximum, which can be conveniently used as the origin for x.

Construction of the discrete and essential spectra using Evans function has been detailed in our earlier report [14]. We find that the multi-hump pulses have at least an unstable discrete eigenvalue while only the one-hump family with the limiting behavior (2.3) has a stable discrete spectrum. We find that this single-hump family has exactly two discrete eigenvalues for positive δ - the translational position mode $\psi_1 = h_x$ with a simple zero eigenvalue and a stable mode ψ_2 with an eigenfunction similar to h_x. Both modes are shown in Fig. 4. The essential spectrum is also depicted in the same figure and it clearly protrudes into the right half of the complex plane. Any radiation wave packet is spanned by the essential spectrum. A particular eigenfunction of the essential spectrum is shown in Fig. 5 and it is clear that $|K| = |\psi|$ has reached its asymptotic values a pulse width σ (about 10 units for any δ) away from the pulse maximum. A Fourier mode within the radiation packet of wavenumber α will be attenuated by a factor of $1/r(a)$ as it travels through the pulse. As seen in Fig. 6, $r(\alpha)$ exhibits a sharp maximum for the maximum-growing Fourier modes around $\alpha \sim 1$. The attenuation also decreases for increasing δ such that the high-δ pulses do not attenuate passing wave packets at all. In fact, the *KdV* soliton has a transmission coefficient of unity! This suppression mechanism applies to both mass-carrying and zero-mean wavepacket disturbances. However, with repsect to the former disturbances, it suggests that mass carried by the radiation is not absorbed by the high-δ pulses and mass added to these pulses will quickly drain out. Since all the discrete modes do not carry mass, the mass draining from a pulse, previously associated with the speed mode of the Galilean symmetry, must be carried by the radiation continuum of the essential spectrum. However, the fact that the essential spectrum contains a band of unstable modes seems to suggest that it cannot describe a decaying pulse in drainage. Actually, the growing modes are convected away from the pulse while their cumulative drainage effect in the vicinity of the pulse can be conveniently represented by a single stable mode - the resonance pole. The radiation that drains out of the pulse, after appropriate attenuation, will accelerate and grow on the substrate. However, Pego and Weinstein [12] pointed out that if one places a weight that decays exponentially in

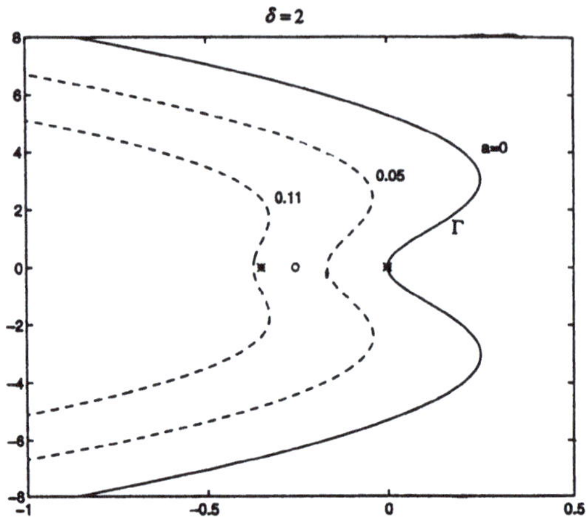

Figure 4. The typical spectra of a one-hump gKS pulse at $\delta = 2$ with a simple zero λ_1, corresponding to the translational mode, a stable discrete mode λ_2 and a discrete resonance pole $\lambda_\#$. The eigenvalues are represented by stars and the resonance pole by a circle. The essential spectrum Γ is shifted by the weight and pass both the resonance pole and the stable mode λ_2 to exchange their role.

the direction of negative x on the disturbance u, this distant growth can be filtered away and one could focus on the decay effect near the pulse. We hence use the e^{ax} weight of Pego and Weinstein [12] on the disturbance u in (3.1), $v = e^{ax}u$, and define a corresponding eigenvalue problem

$$\mathcal{L}_a\phi = e^{ax}\mathcal{L}(e^{-ax}\phi) = \lambda\phi \tag{3.4}$$

where $\phi = e^{ax}\psi$ is the weighted eigenfunction that spans the weighted disturbance $w(x,t)$. It is clear that the essential spectrum Γ_a of $\mathcal{L}_a$ is defined by shifting that of $\mathcal{L}$ in (3.3) by the transformation $\alpha \to \alpha + ia$.

$$\lambda_a = i(\alpha + ia)c + i\delta(\alpha + ia)^3 + [(\alpha + ia)^2 - (\alpha + ia)^4] \tag{3.5}$$

The net result is that the essential spectrum is shifted to the left in the complex plane as the real parts of λ decrease. This is demonstrated in Fig. 4. The fact that a particular value of a exists that can shift the entire essential spectrum to the left half plane implies that the pulse is stable. This is the essence of Pego and Weinstein's stability theory.

There is however, a limit to how deep into the stable left-half plane Γ_a can be shifted. The furthest position occurs when Γ_a reaches the saddle point defined by

$$\frac{d\lambda}{d\alpha}(\alpha_*) = 0 \tag{3.6}$$

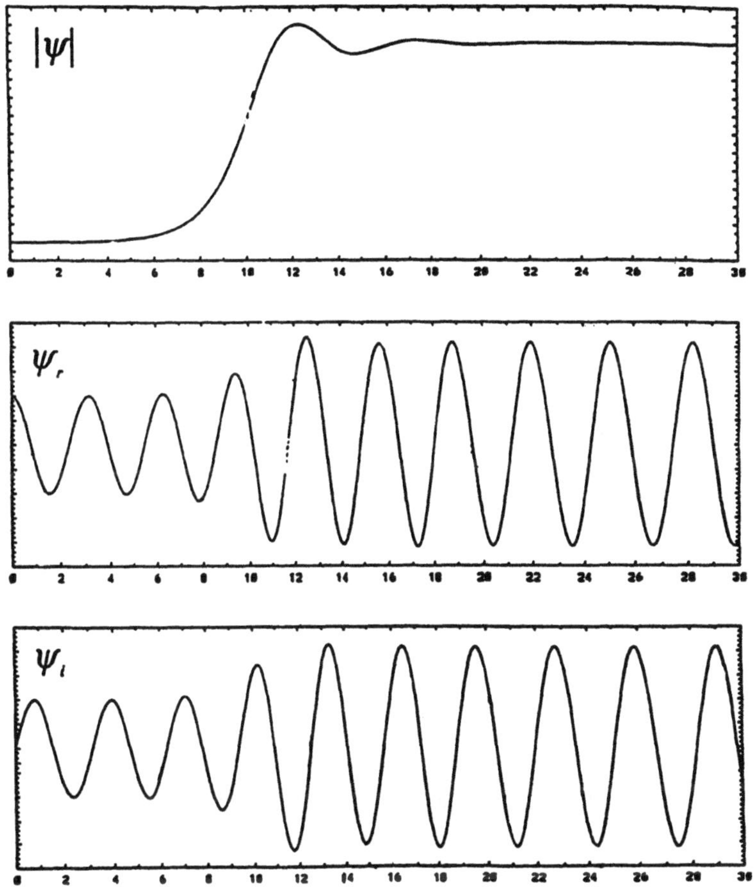

Figure 5. A typical complex eigenfunction ψ of the essential spectrum Γ at $\alpha = 2.0$ and $\delta = 0.5$. The modulus $|\psi|$ is unity as $x \to -\infty$ and $r(\alpha)$ at $x \to +\infty$.

where λ is given by (3.3). The complex saddle point α_* yields a complex growth rate

$$\lambda_* = \lambda(\alpha_*) \tag{3.7}$$

The shifted spectrum Γ_a cannot pass through λ_* such that a λ_* lying to the left of the unshifted spectrum Γ cannot lie on the right of the shifted spectrum Γ_a in the complex λ plane. Hence, if there exists a λ_* that lies to the left of Γ but is in the right half plane, there is no weight that can yield

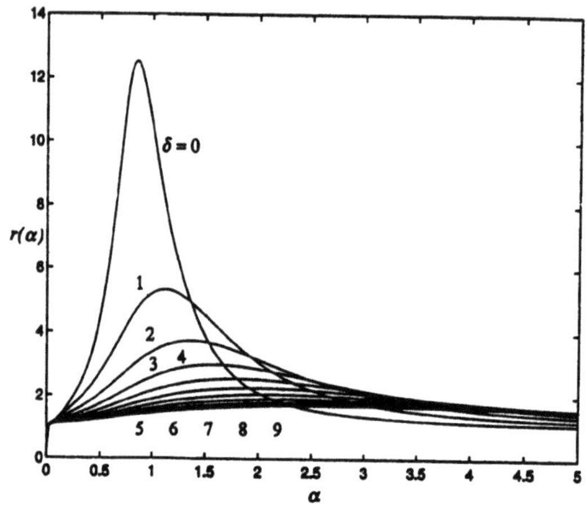

Figure 6. The transmission coefficient $r(\alpha)$ for single-hump pulses at various values of δ.

a stable Γ_a. In other words, one cannot find a weight to ensure convective stability for the pulse with the theory of Pego and Weinstein. However, we have shown earlier by steepest descent that the saddle-point wavenumber α_* corresponds to the dominant wavenumber along the ray traveled by the equilibrium pulse in the laboratory frame [13]. The condition $Re\{\lambda_*\} < 0$ is, in fact, the criterion we have derived for the pulse to escape unstable radiation. This then clarifies how Pego and Weinstein's stability theory for pulses in a purely dispersive medium can be connected to ours for an unstable medium. Our theory yields the result that only single-hump gKS pulses are stable and only for δ in excess of 0.17 [13].

It can be easily shown that, unlike the essential spectrum which is shifted by the weight, the discrete spectrum of $\mathcal{L}$ is also a discrete spectrum of $\mathcal{L}_a$ - with an important exception. The discrete eigenfunctions ψ_k of $\mathcal{L}$ decay to zero as $|x| \to \infty$ with an exponential rate determined by the roots α of the spatial characteristic polynomial

$$P(\alpha) = \lambda_k - \lambda(\alpha) \tag{3.8}$$

where λ_k is the discrete eigenvalue corresponding to ψ and $\lambda(\alpha)$ is given by the dispersion relationship (3.3). The roots with positive real part determine the decay rate of $x \to -\infty$ and ones with negative real parts the rate at $x \to \infty$. However, one can, in principle, also construct functions which satisfy $\mathcal{L}\psi_\# = \lambda_\# \psi_\#$ but do not decay to zero at one infinity. If, for example, there are two roots with negative real parts α_1 and α_2, the $\psi_\#$ can decay to zero at $+\infty$ with rate α_1 but grows exponentially at $-\infty$ with rate α_2. Since these modes do not decay to zero, they are not part of the discrete spectrum and are called resonance poles [23]. We have found only one real and negative resonance pole ($\lambda_\# < 0$) for all one-hump gKS pulses with positive δ [14], as shown in Figs. 4 and 7.

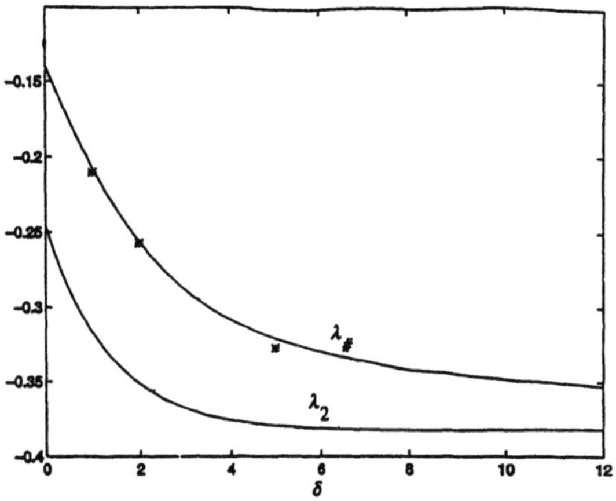

Figure 7. The resonance pole and stable eigenvalue for the single-hump pulse as a function of δ. The stars are the decay rates measured from simulations like in in Figs.1 and 2.

These resonance poles also satisfy the weighted operator $\mathcal{L}_a\phi_\# = \lambda_\#\phi_\#$ where $\phi_\# = \psi_\# e^{ax}$ but their decay properties can now change. Since the essential spectrum Γ_a for $\mathcal{L}_a$ is defined by $\lambda_a(\alpha)$ for α real, when there is a discrete eigenvalue on the essential spectrum Γ_a, one of the roots to the spatial characteristic polynomial (3.8) is purely imaginary. Consequently, as Γ_a is shifted across the resonance pole $\lambda_\#$, the negative root α_2 crosses the imaginary axis and its real part becomes positive. Hence, as Γ_a is shifted across the resonance pole, it becomes a bona fide discrete eigenvalue of $\mathcal{L}_a$. Conversely, if Γ_a is shifted across the discrete eigenvalue λ_2, is becomes a resonance pole to $\mathcal{L}_a$. The translational zero mode λ_1, as shown earlier [14], remains an eigenvalue as Γ_a passes by due to a hole in the Riemann surface. Other rules on the exchange between eigenvalues and resonance poles can be found in [14]. For every stable one-hump pulse, we are able to shift Γ_a such that $\lambda_\#$ becomes a discrete eigenvalue of $\mathcal{L}_a$ and λ_2 a resonance pole as seen in Fig. 4.

The presence of these resonance poles introduces a rare exception to the stability criterion $Re\{\lambda_*\} < 0$. If there are unstable resonance poles to the left of Γ, even if Γ_a can be shifted entirely to the left half plane, the uncovered resonance poles are unstable and can destabilize the pulse. While we found some rare examples of this instability [14], the resonance pole of the single-hump pulse is stable for positive δ and hence only represents decay dynamics. There are hence three possible mechanisms for a pulse to destabilize – unstable saddle points λ_*, unstable resonance poles $\lambda_\#$

(typically complex conjugates) and unstable discrete eigenvalues λ_k. Due to the exponential growth, $< \psi_\# >= \infty$ while its weighted counterpart $\phi_\#$ contains finite mass $< \phi_\# >=< e^{ax}\psi_\# >\neq 0$. As a result, the discrete spectrum of $\mathcal{L}_a$ is now mass-carrying. The shifted essential spectrum Γ_a is also mass-carrying but due to its far-left position in the complex plane, it represents a very fast transient. Hence, the asymptotic decay rate of the weighted disturbance $v(x,t)$ is governed solely by $\lambda_\#$. Since the original disturbance $u = e^{-ax}v$ is only slightly affected by the weight near the origin $x = 0$, the dynamics of the disturbance u around the pulse is also governed by $\lambda_\#$ - the local drainage rate is conveniently described by the resonance pole at large time. That the cumulative effect of the original unstable essential spectrum Γ on drainage can be described at large time by a stable discrete mode is remarkable. It requires significant dispersion to convect the radiation backwards and sufficient growth to counter this drainage to produce a finite decay rate. The KdV equation, for example, has no resonance pole.

Actually, the resonance pole concept can be introduced in a more physical manner without the weights of Pego and Weinstein. We can use the discrete and essential spectra $\mathcal{L}$ to expand any localized disturbance to the pulse. By localized we mean the disturbance must decay sufficiently rapidly from zero at a rate to be specified later. That the two spectra form a complete basis for localized disturbances is also known from standard spectral theory. We hence expand the disturbance $u(x,t)$ as

$$u(x,t) = \sum_{k=1}^{2} A_k \psi_k(x) e^{\lambda_k t} + \int_{-\infty+ia}^{+\infty+ia} A(\alpha) K(\alpha, x) e^{i\alpha x + \lambda_a(\alpha) t} d\alpha \qquad (3.9)$$

where A_k and $A(\alpha)$ are the coefficients of expansion for the discrete and essential spectral respectively. While the original expansion with the continuous eigenfunction would be carried out over the real wavenumber α, we can deform the contour of integration in the complex α plane by shifting it above the real line by a distance a. The expansion coefficient $A((\alpha)$ is also dependent on the shift but we shall not include this dependence in our notation for convenience. This shift in the contour of integration corresponds to the weight of Pego and Weinstein and shifts the essential spectrum $\lambda(\alpha)$ in (3.3) to that of λ_a in (3.5). In the complex λ space, this shift in the contour of integration then corresponds to a shift of Γ to Γ_a — shifting the essential spectrum to the left.

For $\delta > 0.137$, λ_* is in the left half of the complex λ plane, as we have shown earlier [13], and an a can always be found that shifts Γ_a to the left half plane as seen in Fig. 4. There is hence no pole in the term $\exp\{\lambda_a(\alpha)t\}$ that prevents the deformation of the contour of integration in (3.9). With a properly localized disturbance, there is also no singularity in the coefficient

$A(\alpha)$. Since $K(\alpha, x)$ blows up at $-\infty$ for the resonance pole, the only pole in the integral of (3.9) is the resonance pole. Hence, by the residue theorem, if one shifts Γ_a beyond $\lambda_\#$, the expansion (3.9) becomes

$$\begin{aligned} u(x,t) &= \sum_{k=1}^{2} A_k \psi_k(x) e^{\lambda_k t} + B\psi_\#(x) e^{\lambda_\# t} \\ &+ \int_{-\infty+ib}^{\infty+ib} A(\alpha) K(\alpha, x) e^{i\alpha x + \lambda_b(\alpha) t} d\alpha \end{aligned} \quad (3.10)$$

where b is such that Γ_b is to the left of $\lambda_\#$ and is completely in the left half plane. If Γ_b is shifted across λ_2 such that the latter is transformed into a resonance pole, the eigenfunction ψ_2 can be omitted in the above expansion.

Transformation of (3.9) to (3.10) by distorting the contour of integration simplifies the expansion drastically. Since Γ_b is more stable than $\lambda_\#$, the continuum in (3.10) decays much more rapidly than $\lambda_\#$ and can in fact be neglected at large time. If we likewise omit the very stable λ_2 mode, the large time dynamics can be approximated by

$$u(x,t) \sim A_1 \psi_1(x) e^{\lambda_1 t} + B\psi_\#(x) e^{\lambda_\# t} \quad (3.11)$$

That the dynamics of a continuum of modes, including unstable ones, can be described at large time by a single stable mode is extremely convenient! Note, however $\lambda_\#(x)$ blows up as $x \to -\infty$ and the decay dynamics is meaningless far from the origin. The local validity of (3.11) will be scrutinized in more detail in the next section but it is clear that the dominant dynamics of a stable gKS pulse is driven by two modes, a neutrally stable translation mode and a stable resonance pole mode. We also note that the stable continuum in the integral of (3.10) decay as e^{-bx} in the positive x direction and the resonance pole and discrete modes decay in a manner dictated by (3.8). Such decay rates of the eigenfunctions in the shifted expansion then stipulates that the disturbance must decay sufficiently rapidly in the $+x$ direction. We shall avoid a detailed estimate of the precise condition here. Nevertheless, we observe that (3.10) implies that a localized disturbance is expanded by a basis $\psi_\#$ which blows up at $-\infty$ but decays in time. Actually, the continuum in (3.10) also blows up negatively at $-\infty$ and essentially serves as a cutoff to truncate the growing tail of $\lambda_\#$. We shall demonstrate this and use the esential spectrum and the resonance pole to construct the shelf behind a draining pulse in the next section.

There is also a subtle connection between the pulse spectra of the gKS and the KdV equations. The transformation

$$h = w\delta/4$$

$$t = \tau/\delta \tag{3.12}$$

converts the gKS equation (1.1) to a perturbed form of the KdV equation,

$$w_\tau + ww_x + w_{xxx} + \delta^{-1}(w_{xx} + w_{xxxx}) = 0 \tag{3.13}$$

The solitary pulses of (3.13) are given by (2.3) in the limit of large δ and after proper transform. The same transform shows that the eigenvalues μ of pulses w^* in (3.13) to the operator

$$L_\delta = w^*\frac{\partial}{\partial x} + \frac{\partial w^*}{\partial x}\cdot + \frac{\partial^3}{\partial x^3} + \frac{1}{\delta}(\frac{\partial^2}{\partial x^2} + \frac{\partial^4}{\partial x^4}) \tag{3.14}$$

are related to λ by

$$\mu = \lambda/\delta \tag{3.15}$$

Hence, in the limit of large δ, the discrete eigenvalue λ_2 and the resonance pole $\lambda_\#$ correspond to eigenvalues of (3.14) that both collapse to the origin. This is true since, as seen in Fig. 7, both λ_2 and $\lambda_\#$ approach constant asymptotes at large δ. As is well known, the KdV operator $L_\infty = \lim_{\delta \to \infty} L_\delta$ has three zero modes [12],

$$\begin{aligned} L_\infty\frac{\partial w^*}{\partial x} &= 0 \\ L_\infty\frac{\partial w^*}{\partial c} &= -\frac{\partial w^*}{\partial x} \\ L_\infty 1 &= \frac{\partial w^*}{\partial x} \end{aligned} \tag{3.16}$$

where the first zero is the translational mode, the second generalized zero is a speed mode due to a symmetry of the KdV operator and the third one is again the improper Galilean mode. These three modes are the singular limits of λ_1, λ_2 and $\lambda_\#$ and the KdV soliton does not have a resonance pole. Nevertheless, since the last two modes of the KdV operator L_∞ are generalized zero modes, the eigenfunctions ψ_2 and $\psi_\#$ near the origin must approach ψ_1 to create the degeneracy.

In essence, δ and the shift b or the weight a break the degeneracy of (3.16) and "unfold" the generalized eigenvalues. As a result, for δ large, we expect both ψ_2 and $\psi_\#$ to be very close to $\psi_1 = h_x$ and that λ_2 and $\lambda_\#$ are small eigenvalues. This is clearly evident in Figs. 7 and 8 where $\psi_\#$ is compared to ψ_1. Although $\psi_\#$ blows up exponentially as $x \to -\infty$, its shape is very close to ψ_1 for x near zero and $x \to +\infty$. This is true even for moderately large δ of order unity.

This proximity to degeneracy allows us to approximate $\psi_\#$ near the origin by

$$\psi_\# \sim h_x + \lambda_\# f(x) \tag{3.17}$$

where $\lambda_\#$ is close to zero as seen in Fig. 7. Although (3.17) is strictly valid for large δ, we find it to be a good approximation, viz. $\psi_\#$ is close to h_x, even for small δ in our numerically constructed spectrum. The difference $f(x)$ also carries mass but blows up as $\psi_\#$ at $-\infty$. For positive x, (3.17) is actually valid even far from the origin as seen in Fig. 8. It is only at $-\infty$ where $\psi_\#$ blows up while h_x decays that the difference grows.

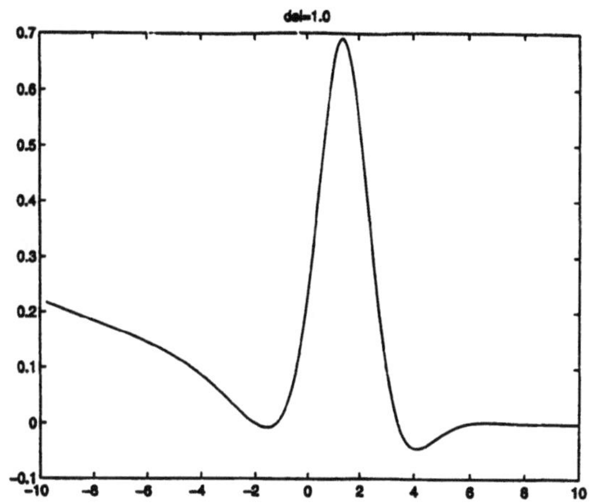

Figure 8a.

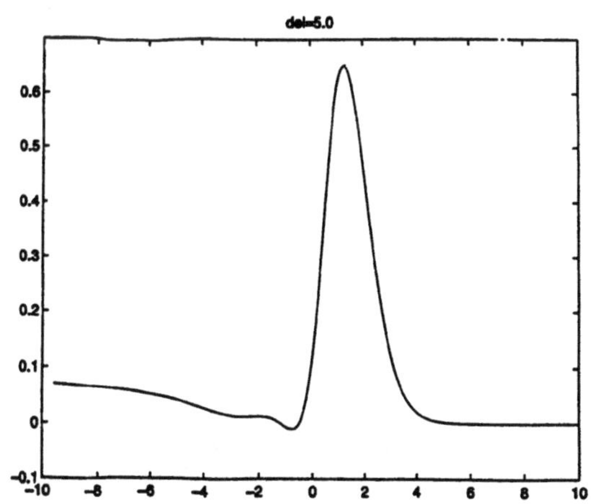

Figure 8b.

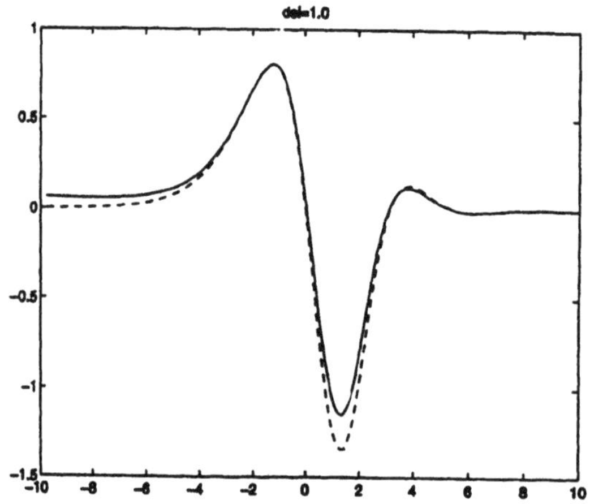

Figure 8c.

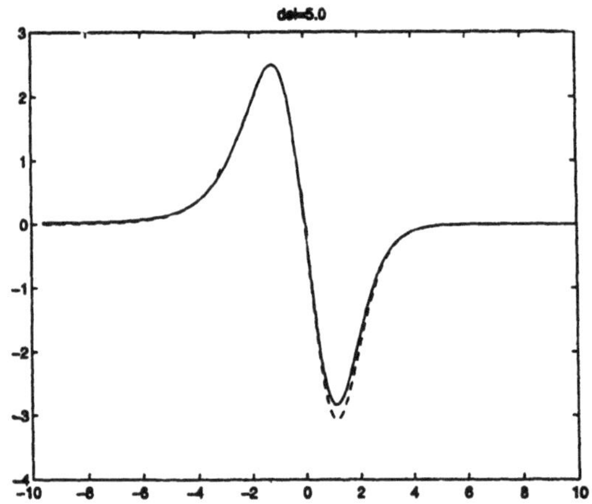

Figure 8d.

Figure 8. The local structure of draining excited pulses at $\delta = 1.0$ and 5.0 are shown in (a) and (b). A flat neck region is most obvious for $\delta = 5.0$ behind the back dimple. The draining mass grows exponentially into the shelf behind the neck. This tail region is driven by the resonance pole mode $\psi_{\#}$. In (c) and (d), we depict the discrete modes $\psi_{\#}$ and ψ_1 for $\delta = 1$ and 5. The resonance pole eigenfunction blows up exponentially as $x \to -\infty$ from the neck region behind the pulse but is identical to $\psi_1 = h_x$ near and in front of the pulse maximum at $x = 0$..

4. Drainage and Absorption

We shall first describe the drainage dynamics of Figs 1 and 2 with the weighted projection of Pego and Weinstein and then augment it with the more detailed expansion of (3.10) and (3.11). While the asymptotic drainage rate is immediately given by the resonance pole $\lambda_\#$, the property of $\lambda_\#$ in (3.17) also provides important information about the drainage process. At large time, the behavior of $u(x,t)$ near $x = 0$, as measured by $v(x,t) = e^{ax}u(x,t)$, can be estimated by expanding $v(x,t)$ in terms of the discrete spectrum of L_a, ϕ_1 and $\phi_\#$, and the essential spectrum Γ_a. The essential spectrum decays very rapidly and since ϕ_1 is a simple zero mode, its coefficient remains constant in time. Hence, at large time and near the origin

$$\begin{aligned} v(x,t) &\sim \xi_0\phi_1 + a_0\phi_\# \; e^{\lambda_\# t} \\ &= \xi_0\psi_1 e^{ax} + a_0 e^{ax}\psi_\# \; e^{\lambda_\# t} \\ &\sim \xi_0 h_x e^{ax} + a_0(h_x + \lambda_\# f(x))e^{\lambda_\# t}e^{ax} \end{aligned} \tag{4.1}$$

where a_0 is the initial value for the coefficient of expansion of the resonance pole mode. Transforming back to the original unweighted disturbance, we get

$$u(x,t) \sim \xi_0 h_x + a_0(h_x + \lambda_\# f(x))e^{\lambda_\# t} \tag{4.2}$$

which is simply (3.11) if one substitute $\lambda_1 = 0, \psi_1 = h_x$ and (3.17). Since h_x corresponds to a translation as seen in (2.4), the excited resonance pole mode causes a shift in the position ξ as well as drainage. This shift from the final equilibrium position ξ_0 is given by

$$\hat{\xi} = a_0 e^{\lambda_\# t}$$

Hence, the deviation speed decays as

$$\hat{c} = \frac{d}{dt}\hat{\xi} = a_0\lambda_\# e^{\lambda_\# t} \tag{4.3}$$

Although both the weighted projection and the complete expansion yields the same speed decay dynamics, they differ in the description of the amplitude decay of Fig. 2. The weighted disturbance $v(x,t)$ in (4.1) can be integrated immediately to yield

$$\begin{aligned} \frac{d<v>}{dt} &= \frac{d}{dt} < ue^{ax} > \\ &\sim a_0\lambda_\#^2 < f(x)e^{ax} > e^{\lambda_\# t} \end{aligned} \tag{4.4}$$

where $< fe^{ax} > \quad > 0$. Although $< v >=< ue^{ax} >$ is close to $< v >$, it nevertheless represents only a weighted projection of the disturbance $u(x,t)$ and not the true mass $< u(x,t) >$. The value $< v >$ is also dependent on the exponent a of the weighting function. One sees, however, from (4.4) that

$$< \hat{v} >=< v > (t)- < v > (\infty) \sim a_0 \lambda_{\#} < fe^{ax} > e^{\lambda_{\#} t} \qquad (4.5)$$

and the excess weighted mass $< \hat{v} >$ also decays exponentially in time as the speed (4.3). As a result,

$$\hat{v}/\hat{c} \sim < fe^{ax} > \qquad (4.6)$$

The integral on the right is bounded since f decays more rapidly than e^{ax} in the positive x direction and although $f(x)$ blows up to infinity in $-\infty$ as seen in (3.17), the weight e^{ax} suppresses this growth in the negative x direction as $\phi_{\#} = e^{ax}\psi_{\#}$ is bounded in both directions. One can approximate the deviation mass (area) $\hat{s} =< u >$ by $< v >$ but the tightness of the approximation is often dependent on the weight a.

It is quite inaccurate, on the otherhand, to approximate the measured deviation height $\hat{h}$ by $u(0,t)$ in (4.2). The mass-carrying resonance pole mode $\psi_{\#}$ resembles h_x near the origin (the maximum of the equilibrium pulse) as seen in (3.17). However, the maximum of h_x (and hence $\psi_{\#}$ near the origin) is away from the origin as seen in Fig. 8. As a result, when the disturbance is large, the maximum of the excited pulse lies not at the origin but shifts some place between the origin and the maximum of h_x as seen in Fig 8. For this reason, it is easier to measure the excess mass (area) $\hat{s}$ instead of the excess amplitude $\hat{h}$ of the excited pulse although the two are clearly proportional to each other because of the localized nature of $\psi_{\#}$ near the origin stipulates a constant width.

To obtain a better estimate of the true area, however, we must return to the full expansion of (3.10) and examine how the shifted essential spectrum cuts off the growing tail of the resonance pole function $\psi_{\#}$ to form a shelf. We shall ignore the translation mode ψ_1 here and write (3.10) as

$$u(x,t) = a_0 \psi_{\#} e^{\lambda_{\#} t} + \int_{-\infty+ib}^{\infty+ib} A(\alpha,x) e^{i\alpha x + \lambda_b t} d\alpha \qquad (4.7)$$

We shall focus on the region behind the pulse ($x < 0$) and set $K(\alpha,x)$ to its unit asymptotic value. We shall now seek an estimate of the Fourier coefficient $A(\alpha)$ for a given a_0 to investigate how the essential spectrum cuts off the exponential growth of $\psi_{\#} \sim e^{\eta x}$ in the negative x direction with $\eta < 0$ being the growth exponent of $\psi_{\#}(x)$ in $-\infty$. It is defined by (3.3),

$$\lambda_{\#} = c\eta - \eta^2 - \delta\eta^3 - \eta^4 \qquad (4.8)$$

As is evident in Fig. 8 and explained in (3.17), $\psi_{\#}(x)$ is close to $h_x(x)$ near the pulse. Far behind the pulse, $\psi_{\#}(x)$ grows exponentially by $e^{\eta x}$ and departs form $h_x(x)$. We determine the amplitude ϵ of this exponential tail by extrapolating it backwards to the point where $\psi_{\#}$ exhibits a minimum. This minimum (neck) of $\psi_{\#}$ can be chosen as the origin for x and its tail then behaves as $\epsilon e^{\eta x}$. This amplitude ϵ is a function of the pulse and hence a function of δ and must be estimated numerically. There is hence a neck to $\psi_{\#}$ which separates the pulse region where it can be approximated by $h_x(x)$ and the exponentially growing tail. This neck is evident in Fig. 8. The neck is several pulse widths to the left of the origin and the approximation $K(\alpha, x) = 1$ used in (4.7) can still be safely imposed for the neck region and behind it. In fact, $\psi_{\#}(x)$ in this exponentially - growing tail can be approximated by

$$\psi_{\#}(x) \sim \epsilon e^{\eta x} H(-x) \tag{4.9}$$

where we have shifted the origin $x = 0$ to the minimum of $\psi_{\#}$ for convenience. The results do not depend on the location of the origin.
Since the disturbance $u(x, t)$ vanishes in the tail region where (4.9) is valid, (4.7) can be written at time $t = 0$ as

$$a_0 \epsilon e^{\eta x} H(-x) + \int_{-\infty+ib}^{\infty+ib} A(\alpha, x) e^{i\alpha x} d\alpha = 0 \tag{4.10}$$

It is then clear that the localized initial disturbance can be decomposed as an exponential tail and a cut-off represented by the integral in (4.10) to ensure $u(x \to -\infty, 0)$ vanishes. This cut-off actually specifies $A(\alpha, x)$. A simple change of variable $\beta = \alpha - bi$ converts (4.10) to

$$a_0 \epsilon e^{\eta x} H(-x) + \int_{-\infty}^{\infty} A(\beta) e^{i\beta x - bx} d\beta = 0$$

or

$$-a_0 \epsilon e^{(\eta+b)x} H(-x) = \int_{-\infty}^{\infty} A(\beta) e^{i\beta x} d\beta \tag{4.11}$$

for $x < 0$. It is then clear that $A(\beta)$ is the Fourier coefficient of the function on the left hand side of (4.11). A simple Fourier integral then yields

$$A(\beta) = \frac{a_0 \epsilon i}{2\pi[\beta(\eta + b)]}$$

or

$$A(\alpha) = \frac{a_0 \epsilon i}{2\pi(\alpha + i\eta)} \tag{4.12}$$

Inserting this into (4.7), one can then resolve the tail evolution and, in particular, the mass spreading in the form of a thinning shelf with an ascillatory edge,

$$u(x, t) \sim a_0[\psi_{\#}(x) e^{\lambda_{\#} t} - \epsilon F(x, t)] \tag{4.13}$$

for $x \leq 0$. The cut-off function

$$F(x,t) = \frac{i}{2\pi}\int_{-\infty+ib}^{\infty+ib} \frac{e^{i\alpha x+\lambda_b(\alpha)t}}{\alpha+i\eta} d\alpha \tag{4.14}$$

is a positive cut-off function that blows up as $e^{\eta x}$ in $-\infty$ to cut-off the exponential tail of $\psi_\#$. Initially, the localized disturbance representing the drained mass from the pulse is in the form of a discontinuous shock (shelf) as seen in (4.9). However, the step function $H(-x)$ relaxes in time as seen in (4.14) as the shelf decreases in amplitude and its jump relaxes into the decaying oscillations seen in Figs. 1 and 2. The draining mass spreads over the substrate in the form of a thinning shelf with an edge relaxed by decaying oscillations.

The length $l_s(t)$ and the thickness (width) $w_s(t)$ of this shelf can also be estimated. Returning the origin to the position where the maximum of equilibrium pulse is located and denote x_* as the position of the neck of $\psi_\#(x)$ in Fig. 8. We also invoke approximation (3.17) for $x > x_*$. Initially, the shelf length $l_s(t)$ is zero and the shelf-edge is located at x_* or $x = 0$ in the previous coordinate of (4.9). The total excess mass (area) $\hat{s}$ at this instant is simply

$$\hat{s}_0 = < u > = a_0\lambda_\# \int_{x_*}^{\infty} f(x)dx = a_0 \parallel f \parallel \lambda_\# \tag{4.15}$$

where $f(x)$ is defined in (3.17). For larger time, the edge moves behind x_* to $x_* - l_s(t)$. Although the jump edge now relaxes to decaying oscillations, we shall approximate it still as a jump. From (3.17) and (4.13), it is clear that the excess mass contained in the pulse in front of x^* now decays as

$$\hat{s} = \hat{s}_0 e^{\lambda_\# t} = a_0\lambda_\# \parallel f \parallel e^{\lambda_\# t} \tag{4.16}$$

By mass conservation, this mass must be drained into the shelf between x_* and $x_* - l_s(t)$. Approximating this region by (4.9), the area of the shelf is then

$$\tilde{s} = \frac{a_0\epsilon}{\eta} e^{\lambda_\# t}\left(1 - e^{-\eta l_s(t)}\right) \tag{4.17}$$

Equating $\hat{s}$ to $\tilde{s}$, one obtains

$$l_s(t) = \frac{1}{\eta} ln \left\{1 + \frac{\eta \parallel f \parallel \lambda_\#}{\epsilon}\left(1 - e^{-\lambda_\# t}\right)\right\} \tag{4.18}$$

such that for $t << |\lambda_\#|^{-1}$

$$l_s(t) \sim |\lambda_\#^2 \parallel f \parallel /\epsilon| t \tag{4.19}$$

and for $t >> |\lambda_\#|^{-1}$

$$l_s(t) \sim |\lambda_\#/\eta|t \tag{4.20}$$

Hence, the length of the shelf increases linearly but with different rates at small and large times.

Similarly, by using (4.16) to determine how much mass has drained into the shelf, one concludes that the thickness $w_s(t)$ of the shelf remains relatively constant at $a_0\epsilon$ for $t << |\lambda_\#|^{-1}$ since the mass comes into the shelf at a linear rate while the shelf length also increases linearly. For $t >> |\lambda_\#|^{-1}$, however, the thickness decreases as t^{-1}

$$w_s(t) \sim a_0 \parallel f \parallel |\eta/\lambda_\#|t^{-1} \tag{4.21}$$

Comparing (4.16) to the weighted excess mass of (4.5), it is clear that we have replaced the weighted norm $< fe^{ax} >$ with the cut-off norm $\parallel f \parallel = \int_{x_*}^{\infty} f(x)dx$ which is independent of the weighting function e^{ax}. In real and numerical experiments, it is simple to measure the decrease in the area of the pulse captured by (4.16). One can also estimate the excess amplitude by defining a "virtual height" $\hat{h}$

$$\hat{h} = \hat{s}/\sigma \tag{4.22}$$

where σ is the pulse width. Both the excess area and the excess amplitude clearly decay exponentially in time with the same exponent $\lambda_\#$ as the speed in (4.3). Combining (4.3) with (4.16), one obtains

$$\hat{s}/\hat{c} = \parallel f \parallel \tag{4.23}$$

which verifies the linear correlation between excess speed and excess mass (or amplitude) observed in many experiments [5,6] and simulations [3,7] of decaying pulses.

We have carried out an extensive numerical study of how the gKS pulse drains fluid as seen in Figs 1 and 2. The excess area and speed are all observed to decay exponentially in time with exponent $\lambda_\#$ as seen in Fig. 7. The linear correlation between $\hat{c}$ and $\hat{s}$ predicted by (4.23) also fits the numerical data reasonably will, as seen in Fig. 9, despite some arbitrariness in the neck position x_*. To show how the essential spectrum cuts off the growing tail of the resonance pole eigenfunction $\psi_\#$, we have subtracted the solution $h(x,t)$, in our simulation of the gKS equation in Fig. 1, from the pulse solution to obtain the disturbance $u(x,t)$. At a particular time, we estimate a_0 and then use (4.9) to estimate the coefficients $A(\alpha)$ of the essential spectrum. In Fig. 10, we compare the reproduced signal according to (4.13) to the simulated disturbance evolution. It is clear that the full drainage evolution is faithfully reproduced. This decaying resonance pole

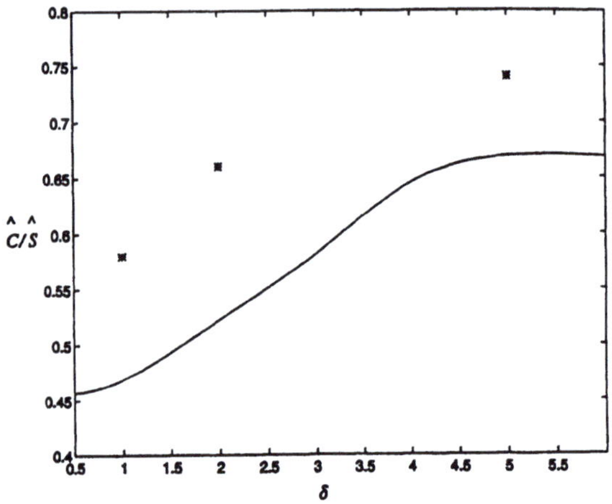

Figure 9. Comparison of the measured $\hat{c}/\hat{s}$ ratio of a draining pulse to the predicted one of (4.23) for different values of δ.

mode in the form of a shelf and the cut-off exercised by the essential spectrum as the shelf shifts to the left are also evident. A mass lump is simply spread over the substrate behind the pulse. For the small-time drainage depicted, the shelf thickness remains relative constant while its length $l_s(t)$ clearly increases linearly in time as predicted by (4.18). This finite-volume drainage mechanism, with one side of the draining shelf bounded by the growing resonance pole mode and the other radiation represented by the oscillating edge, is characteristic of draining pulses whose decay dynamics is described by a resonance pole.

The radiation emitted by the drainage shelf in Fig.10 will grow into pulses if a trailing pulse does not absorb it. Since this absorbing pulse will again emit the disturbance behind it, the radiation can only be suppressed if its amplitude is attenuated after each absorption inspite of the growth on the active substrate. The amplitude gained on the substrate is measured by the spatial growth rate [22] and it is very sensitive to the phase speed since the latter determines the time a particular Fourier mode spends on the substrate. As such, for a given δ, there is a critical separation l_c between two identical pulses beyond which radiation will not be suppressed in amplitude as it passes from the leading pulse to the trailing one. In this connection, we observe that, although the high δ pulses are less absorbing and suppresses the radiation less, as seen from the transmission coefficient in Fig. 6, the high dispersion on the substrate at large δ also implies that the radiation propagates faster and has less time to grow on the substrate. We shall show that the latter effect dominates and l_c actually increases with δ such that in the KdV limit l_c approaches infinity and the separation can become

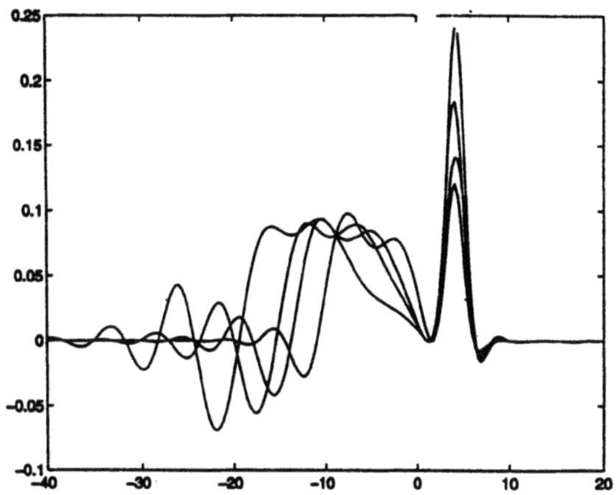

Figure 10a.

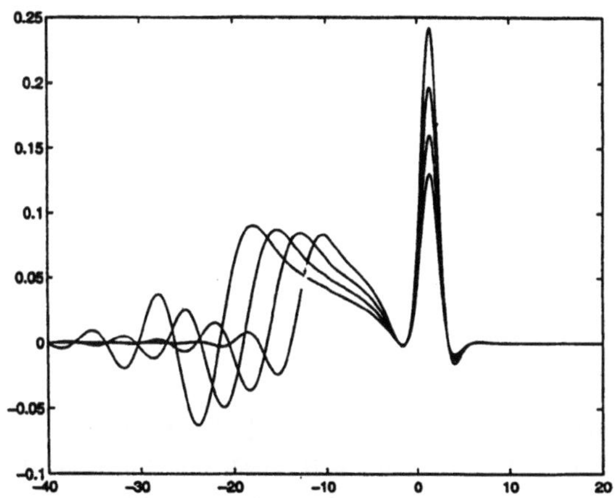

Figure 10b.

Figure 10. a) Simulated evolution of the disturbance $u(x,t)$ of a draining pulse at $\delta = 1.0$ for $t = 1.0, 2.0, 3.0$ and 4.0. The shelf retains its width and recedes roughly linearly in time. The deviation from perfectly linear spreading is due to the stable shifted essential spectrum, the stable ψ_2 mode and nonlinear effects. b) Analytical estimate using the coefficients of the resonance pole and the essential spectrum at time zero. Linear spreading is observed here.

infinitely large.

To render these arguments more explicit, we first note that the Fourier mode with frequency $\omega = \lambda_i(\alpha_r)$, where λ_i is the imaginary part of the eigenvalue in (3.3) for a real wavenumber α_r, emitted from the leading pulse will grow in amplitude by the amount $\exp[\alpha_i(l - \sigma)]$ over the length

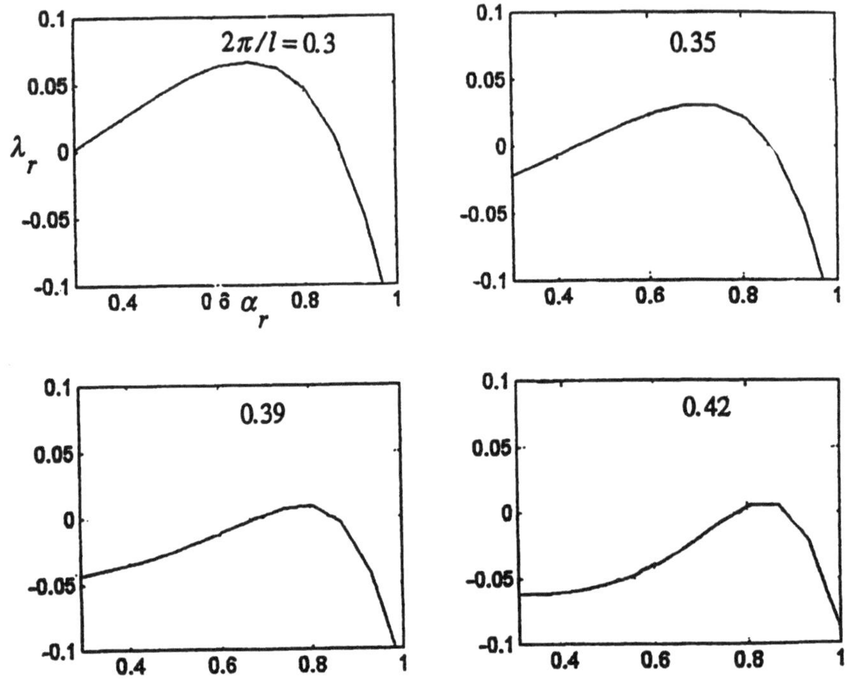

Figure 11. The temporal growth rate $\lambda_r(\alpha_r; l)$ of a spatially neutral Fourier mode with wavenumber α_r with the radiation packet for various values of $2\pi/l$. All modes are temporally stable for $2\pi/l > 2\pi/l_c = 0.40$ for $\delta = 0.13$ here.

of substrate $l-\sigma$ where α_i is the imaginary part of the complex wavenumber, l is the separation between the two pulses and σ the pulse width. The same mode, though, was attenuated in amplitude by a factor of $1/r(\alpha_r)$ when it passed through the front pulse. Hence, for the critical condition when the amplitude of the Fourier mode entering the trailing pulse is equal to its amplitude entering the front pulse,

$$\exp[\alpha_i(l-\sigma)] = r(\alpha_r) \tag{4.24}$$

or

$$\alpha_i = \frac{ln\ r(\alpha_r)}{l-\sigma} \tag{4.25}$$

which relate the real and imaginary parts of the complex wavenumber through the transmission coefficient of Fig. 6. However, the same Fourier mode with frequency λ_i and wavenumber α_r when it first leaves the leading pulse must also not grow temporally,

$$\lambda_r(\alpha_r, \alpha_i) < 0 \tag{4.26}$$

Since α_i is related to α_r by (4.25) for a mode that is spatially neutral, its temporal growth rate λ_r is a function only of α_r and l. In Fig. 11, we depict the computed temporal growth rate λ_r as a function of α_r for several functions of l. As is evident, all Fourier modes are spatially neutral and temporally stable for l in excess of $l_c = 2\pi/0.40$. For l less than l_c, all modes are spatially and temporally stable and any radiation mode will be suppressed by pulse absorption. Since unstable pulses cannot attenuate any growing radiation, we expect l_c to approach zero as δ approaches 0.137, the limit of stable pulses. For the purely dispersive KdV limit at infinite δ, we expect l_c to approach infinity as mentioned earlier since the substrate is not active.

In an earlier paper [18], we have considered the Floquet stability of periodic waves with wavenumber α of the gKS equation. In the limit of small α, these periodic waves resemble a periodic lattice of equally separated single-hump pulses. It was found then that if the wavelength exceeds a critical value (α less than critical value α_c), the periodic waves are unstable and the Floquet growth rate resembles the substrate growth rate of (3.3). It is clear that this α_c is simply $2\pi/l_c$ of the above absorption theory and in Fig. 12, we favorably compare the $2\pi/l_c$ determined from the transmission coefficient of the essential spectrum Γ to the Floquet result. The critical α_c goes to infinity at $\delta = 0.17$ and zero at infinite δ as expected. In Fig. 13, we show two sets of simulations where a small finite-mass disturbance is placed between the first two pulses of a periodic pulse trains whose separations are below and beyond l_c, respectively. It is clear that the radiation is swept clean by one and the other produces significant turbulence as new pulses are generated from the radiation. Although this transient seems to last forever, we suspect it will eventually die out and an irregularly spaced pulse train with an average pulse separation less than l_c will result. It is also possible that a sustained chaotic state, with continuous generation and annihilation of new pulses and radiation noise, is reached asymptotically.

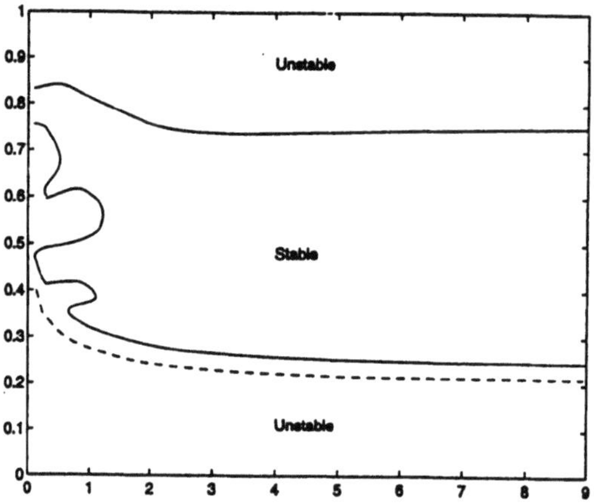

Figure 12. Comparison of $l_c(\delta)$ in dotted line to the results of Floquet theory for periodic pulse trains. The "holes" at low δ are due to subharmonic instabilities not considered here. The upper bound α_s for sideband stability from [18] is also shown.

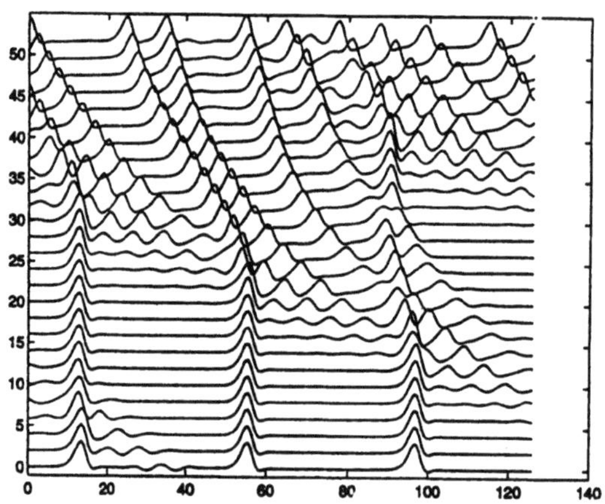

Figure 13a.

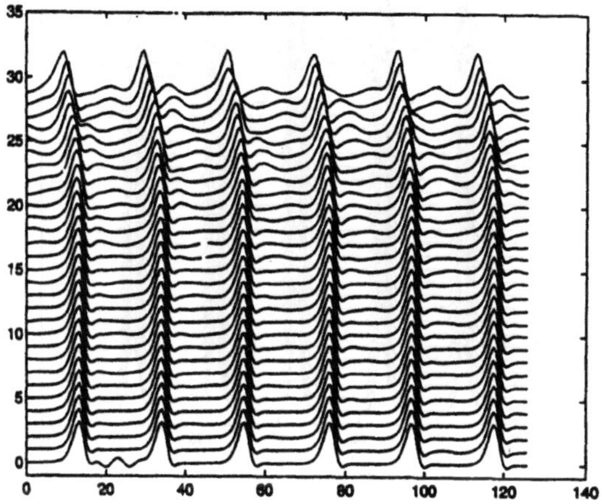

Figure 13b.

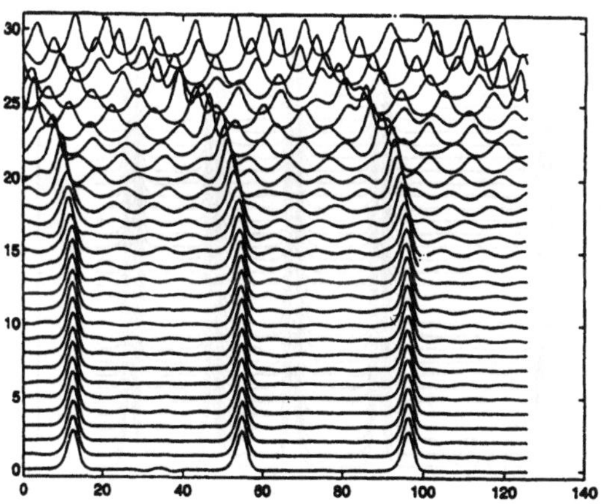

Figure 13c.

It should be emphasized that the suppression mechanism is not restricted to periodic pulse trains. The prior analysis can also be applied to an unevenly spaced train using (4.26) as a map to determine the growth / attenuation of the radiation packet as it passes through each pulse. We shall not study this in detail but simply show in Fig. 14 that a train with two spacing with

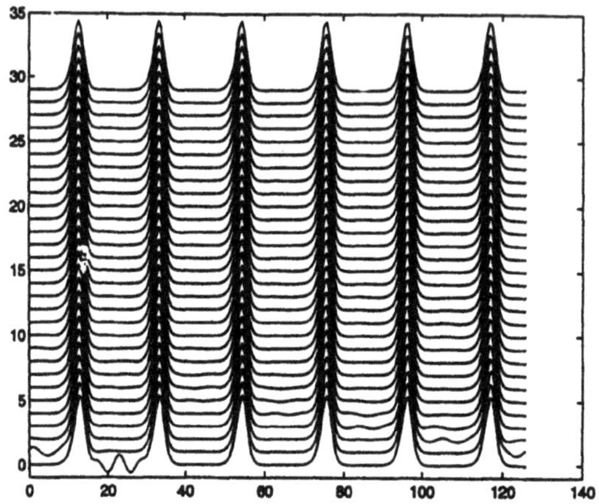

Figure 13d.

Figure 13. Simulations with periodic pulse trains with separation $l = 2\pi/\alpha$ larger and less than l_c. (a) $\delta = 1$ and $\alpha = 0.15$, (b) $\delta = 1$ and $\alpha = 0.3$, (c) $\delta = 5$ and $\alpha = 0.15$ and (d) $\delta = 5$ and $\alpha = 0.3$. From Fig. 12, the critical $\alpha_c = 2\pi/l_c$ is 0.27 for $\delta = 1$ and 0.2 for $\delta = 5$. Nucleation of new pulses is seen for unsuppressed localized disturbances initially between the first two pulses with $\alpha < \alpha_c$ and the number of pulses remains robust for $\alpha > \alpha_c$ although there is some pulse-pulse interaction that breaks the periodic lattice.

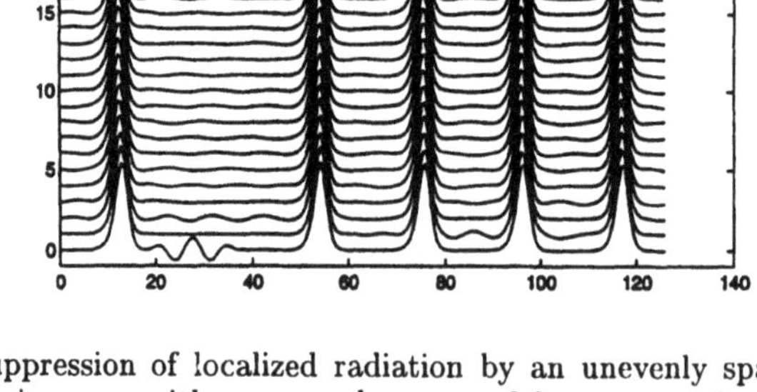

Figure 14. Suppression of localized radiation by an unevenly spaced pulse train at $\delta = 5$. Four spacings are with wavenumber $\alpha_1 = 0.3 > \alpha_c = 0.2$ while one spacing is with $\alpha_2 = 0.15 < \alpha_c$. The predominance of short spacing can still suppress the radiation growth.

$\alpha_1 > \alpha_c$ and $\alpha_2 < \alpha_c$ is still sufficient to suppress the radiation disturbance since there are more pulses with the shorter spacing $\alpha > \alpha_c$.

5. Summary and Discussion

We have clarified the drainage mechanism of an excited pulse with a distinctive spreading shelf that is a consequence of the cumulative local effect of the continuum of the entire stable and unstable essential spectrum, as represented by a single discrete stable resonance pole. One can use the dynamics (4.3) and (4.16) to determine the minimum separation l_{min} beyond which an excited pulse after coalescence cannot chase down its front neighbor to induce further coalescence, as we have done for the falling-film pulses in [3]. If successful, the final equilibrium pulse train after the transients seen in Fig. 13, which is without further pulse annihilation by coalescence or pulse generation from growing radiation, must have an average spacing that is between l_{min} and l_c. Unfortunately, unlike the falling-film pulses, the coalesence analysis of the gKS pulses is more involved and we have yet to derive l_{min}. Nevertheless, another lower bound on the separation was produced by our earlier Floquet theory [18] which shows that any local patch of periodic waves with wavenumber larger than $\alpha_s(\delta)$ is unstable due to a sideband instability also observed in falling-film waves [22]. The corresponding separation $l_s = 2\pi/\alpha_s$ is shorter than l_c and we can use l_c and l_s to bound the average separation of any equilibrium pulse train which no longer changes its number of pulses. In Fig. 12, we have plotted these two bands on equilibrium separation. Any non-chaotic asymptotic pulse train with a fixed number of pulses must have an average separation within this band. This selected equilibrium state will not have generation or destruction of pulses but the pulses need not lock up to form a periodic train. Coherent structure theory based on the translational mode of each pulse has shown considerable permanent fluctuation in the pulse separation eventhough the identical pulses are robust and the substrates are radiation free [9, 19]. The statistics of such spatio-temporal dynamics is easily decipherable.

Acknowledgment This work is supported in parts by an NSF grant and a NASA grant.

References

1. H.-C. Chang, *Ann. Rev. Fluid Mech.*, **26**, (1994) 123.
2. H.-C. Chang and E. A. Demekhin, *Adv. in Applied Mech.*, **32**, (1995) 1.
3. H.-C. Chang, E. A. Demekhin and E. N. Kalaidin, *J. Fluid Mech.*, **294**, (1995) 123.
4. H.-C. Chang, E. A. Demekhin and D. I. Kopelevich, *J. Fluid Mech.*, **270**, (1994) 251.
5. J. Liu and J. P. Gollub, *Phys. Fluids*, **6**, (1994) 1702.
6. R. R. Brock, *J. Hydraulic Div.*, **4**, (1969) 1401.
7. H.-C. Chang, E. A. Demekhin, E. Kalaidin and Y. Ye, *Phys. Rev. E* (to appear).
8. T. Kawahara and S. Toh, *Phys. Fluids*, **31**, (1988) 2103.

9. C. Elphick, E. Meron and E. A. Spiegel, *SIAM J. App. Math*, **50**, (1990) 490.
10. C. Elphick, G. R. Iereley, O. Regev and E. A. Spiegel, *Phys. Rev. A*, **44**, (1991) 1110.
11. N. J. Balmforth, *Ann. Rev. Fluid Mech.*, **27**, (1995) 335.
12. R. L. Pego and M. I. Weinstein, *Comm. in Math Phys.*, **164**, (1994) 305.
13. H.-C. Chang, E. A. Demekhin and D. I. Kopelevich, *Phys. Rev. Lett.*, **75**, (1995) 1747.
14. H.-C. Chang, E. A. Demekhin and D. I. Kopelevich, *Physica D*, **97**, (1996) 353.
15. S. V. Alekseenko, V. E. Nakoryakov and B. G. Pokusaev, *AIChE J*, **31**, (1985) 1446.
16. A. A. Nepomnhyashachy, *Physica D*, **86**,(1995) 90.
17. C. V. Conrado and T. Bohr, *Phys. Rev. Lett.*, **72**,(1994) 3522.
18. H.-C. Chang, E. A. Demekhin and D. I. Kopelevich, *Physica D*, **63**,(1993) 299.
19. N. J. Balmforth, G. R. Ierely and R. Worthing, (preprint).
20. W.L. Kath and N. F. Smyth, *Phys. Rev. E*, **51**,(1995), 661.
21. U. Frisch, S. S. Zhen and O. Thual, *J. Fluid Mech.*, **168**,(1986) 221.
22. M. Cheng and H.-C. Chang, *Phys. Fluids*, **7**,(1995) 34.
23. M. Reed and B. Simon, *Methods of Modern Mathematical Physics. IV Analysis of Operators*, New York, Academic Press (1978).

BUFFETING PROBLEMS MODELLING BY USE OF LIFTING SURFACE METHODS IN UNSTEADY AERODYNAMICS WITH SEPARATION

HARIJONO DJOJODIHARDJO

Department of Aeronautical Engineering
Institut Teknologi Bandung
Ganesha 10, Bandung 40132
Indonesia.

Abstract. The analytical approach to buffeting problem in subsonic flow is a challenging one and has been given attention by many researchers; no purely analytical approach has been satisfactorily capable of simulating the phenomenon. An attempt to develop a computational method for the calculation of the aerodynamic load for separated subsonic flow using kernel function technique in linearized aerodynamics has been made by the author and his associates, as an alternative to the more accurate but elaborate technique of direct numerical attack to the viscous flow problem. Two approaches were then made. First, to treat the buffeting problem as a stability one by solving the flutter equation. Second, treating buffeting problem using dynamic response approach. Comparison of the unsteady load calculation method with other existing results for non-separated flow cases indicated good agreements. For separated flow case, no other methods are available for comparison. However, flutter solution approach made by the present study indicated that the presence of the separation bubble tends to destabilize the system compared to the attached flow case. The analytical approach to the buffeting problem using dynamic response technique was made by assuming that there is no coupling between the motion induced unsteady loads and the independent-motion unsteady loads. The motion induced unsteady airloads is calculated using the present method while the buffeting excitation load was based on experimental data. Computational results for the bending moment of a wing experiencing buffeting with experimental results indicated good agreement but lower values for the onset of buffeting.

Nomenclature

A^+, A	-	source strength, doublet strength
$[A_m]$	-	generalized unsteady aerodynamicforce matrix
$\{B_n\}$	-	new state variable derive from Pade approximation
a_∞	-	freestream velocity of sound
$a_r(\gamma)$	-	series coefficient of pressure jump
$a_r^*(\eta)$	-	normalized $a_r(a_r^* C(\eta)/4s)$
b	-	semi-chord
c	-	mean chord
$c(\eta)$	-	sectional shord
$[C_{rs}]$	-	generalized damping matrix
$Cp(\xi,\eta)$	-	pressure coefficient
$\Delta Cp(\xi,\eta)$	-	pressure coefficient difference $(Cp_l - Cp_u)$
$d_q(\gamma)$	-	series coefficient of downwash jump
$d_q^*(\gamma)$	-	normalized $d_\eta(d_\eta^* c(\eta)/4s)$
$[F]$	-	state space matrix
f	-	frequency
g	-	damping factor
$h_r(X)$	-	pressure jump function
$h_q(X)$	-	downwash jump function
i	-	square root of -1
$l(k\|\eta-\eta'\|)$	-	modified Bessel function of 1 st order 1 st kind
k	-	$\omega s/U_\infty$, reduced frequency
$[K_r]$	-	generalized stiffness matrix
$K(k\|\eta-\eta'\|)$	-	modified Bessel function 1 st order 2 th kind
$K(\xi,\eta;\xi'\eta')$	-	Kernel function
$L(k\|\eta-\eta')$	-	modified Struve function of 1 st order
M	-	Mach number
$[M_r]$	-	generalized mass matrix
$\{q_r\}$	-	generalized coordinate vector
r	-	$[(x-x')^2+\beta^2(y-y')^2+\beta^2(z-z')^2]^{\frac{1}{2}}$
R	-	semi span, Laplace operator
t	-	time

$v(x,y,z,t)$	-	displacement of the wing surface
$u_r(x,y,z)$	-	the rth natural mode shape
U_∞	-	free stream velocity
$w(\xi,\eta)$	-	downwash
$\Delta w(\xi,\eta)$	-	downwash difference $(w_l - w_u)$
x	-	coordinate along flow direction
$\{X\}$	-	$[q_r.q_r.B]^T$
X	-	normalized coordinate in the flow direction
	-	$x - x_{le}(\eta)/c(\eta)$
y	-	coordinate in spanwise direction
x	-	coordinate perpendicular to xy plane
β_n	-	Lag term in Pade approximation
β^2	-	$1 - M^2$
$\phi(\xi,\eta)$	-	perturbation velocity potential
$\psi(\xi,\eta)$	-	acceleration potential
ω	-	frequency of oscillation
η	-	nondimensional y coordinate (y/x)
ξ	-	nondimensional x coordinate (x/x)
γ	-	$U_\infty t/s$, non dimensional time
ρ	-	M/β^2
m	-	mass per unit span
h	-	heaving mode direction
α	-	pitching mode direction (angle of attack)
S_m	-	static mass - moment
L	-	lift force per unit span
I_m	-	mass - moment of inertia
M_{ac}	-	moment aerodynamic at aerodynamic center
$M_{c/4}$	-	moment aerodynamic at quarter chord
$2cb = (0.5 + a)b$	-	distance from aerodynamic center to elastic axis
r_n	-	radius of gyration
m'_a, k'_a	-	real part of lift moment coeff.(heaving mode)
m''_a, k''_a	-	imaginary part of lift moment coeff.(heaving mode)
ρ	-	air density
$m'_b.k'_b$	-	real part of lift moment coefficient (pitching mode)
m''_b, k''_b	-	imaginary part of lift moment coefficient (pitching mode
X_c	-	axis of pitching oscillation
$H_o^{(2)}$	-	Hankel function zero order, second kind

$H_1^{(2)})$	-	Hankel function first order, second kind
$X_n = S_n/mb$	-	dimensionless static unbalance
$p = b/Us$	-	
$r = m/\pi\rho b^2$	-	
'	-	position of singularities
D	-	doublet
l	-	lower surface
m	-	non singular
s	-	singular
S	-	source
u	-	upper surface
p, j	-	indices in chordwise direction
v, n	-	indices in spanwise direction
le	-	leading edge
te	-	trailing edge
xt	-	second derivative with respect to x and t
xx, yy, zz	-	second derivative with respect to x, y, and z respectively

1. Introduction

The analytical approach to buffeting problem in subsonic flow is a challenging one and has been given attention by many researchers; no purely analytical approach has been satisfactorily capable of simulating the phenomenon. An attempt to develop a computational method for the calculation of the aerodynamic load for separated subsonic flow using kernel function technique in linearized aerodynamics has been made by the author and his associates, as an alternative to the more accurate but elaborate technique of direct numerical attack to the viscous flow problem. The analytical approach to buffeting problem associated with subsonic flow carried out here follows that of Foersching [1], Becker [2] and Zingel [3]. However, no purely analytical approach based on potential flow modelling can be used to satisfactorily tackle the buffeting phenomenon, since the complex phenomenon of buffeting originates from viscous flow interaction, as elaborated in great detail by Mabey [4]. All these predictions were based on semi-empirical method, whereby the unsteady aerodynamic forces associated with separation and attached part of the flow are obtained from windtunnel experiment and then the dynamic response of the structure is calculated by using analytical approach. Therefore, the attempt made here can only be regarded as a convenient way to approximate the modelling of buffeting phenomenon as a preliminary tool in design estimation, or in guiding experimentation, to arrive at improved semi-empirical approach. The analytical approach using Kernel function technique developed earlier by the author and his associates [5], [6], [7], [8], [9] will be utilized in the prediction of motion-induced unsteady airloads.

Using this technique, two approaches were then made. First, to treat the buffeting problem as a stability one by solving the flutter equation. Second, treating buffeting problem using dynamic response approach. Three dimensional cases will be considered.

Some idealization has been made to facilitate the analysis, such as the notion that the loci of the separation points are fixed along the span of the wing surface and does not move during the entire cycle of the oscillatory motion. Also, for convenience, the loci of the reattachment points along the span are assumed to coincide with the trailing edge.

Validation was made first in the unsteady aerodynamic loads calculation for the attached flow case by comparison with Laschka's technique [10] and Doublet Lattice Methods results using NASTRAN software. Validation for the flutter calculation were made by comparison to computational results using NASTRAN for the attached flow case.

Comparison of the unsteady load calculation method with other existing results for three-dimensional non-separated flow cases indicated good agreements. For separated flow case, no other methods are available for comparison. Parametric studies associated with flutter calculation for the separated flow case was made by changing the position of the separation points, using a hypothetical wing. The destabilizing trend exhibited due to the presence of the separation bubble as compared to the attached flow case will be used to asses the performance of the analysis.

2. Problem Formulation

Figure 1 depicts the wing structure configurations as modelled in the analysis. On the upper surface of the wing, a separation bubble appears. In the analytical approach to the buffeting problem, the pressure field in the separated region will be approximated by superposition of the mean and the fluctuating component. The mean pressure is assumed to be time-invariant during the entire cycle of unsteady or oscillatory motion. The fluctuating pressure is assumed to be harmonic and independent from the wing oscillation.

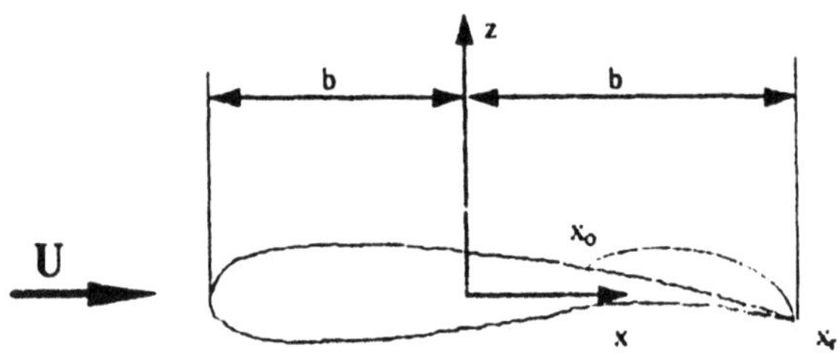

Figure 1a. Airfoil with separation bubble

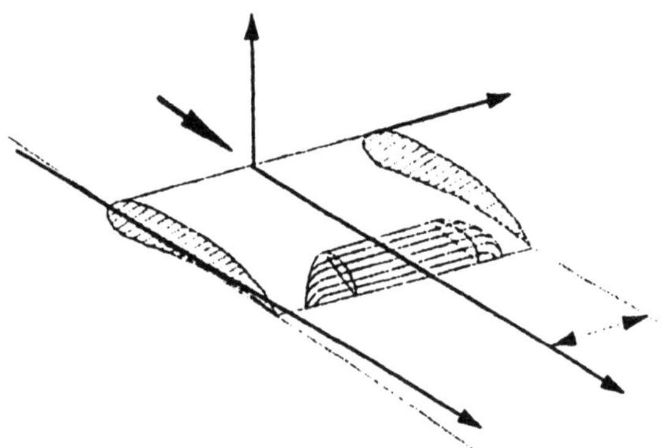

Figure 1b. Wing with separation bubble

With these assumptions, the aerodynamic force can be separated into two parts:

(a) Motion-induced aerodynamic forces associated with mean pressure value in the separated region;
(b) Motion-independent aerodynamic forces associated with pressure fluctuation in the separated region.

The corresponding functional diagram can be represented by an autonomous and force-vibration system, respectively, as depicted in figure 2.

Calculation of the unsteady aerodynamic forces F^A and excitation forces F^F can be carried out by linearized potential aerodynamics with additional pressure boundary condition in the separated region. In the three-dimensional problem worked out here, p-method is incorporated.

3. Governing Equations for Flutter Model

3.1. THREE DIMENSIONAL AUTONOMOUS FLUTTER EQUATION

Writing the deformation of the wing in terms of superposition of the natural mode shape times the generalized co-ordinates in the form:

$$u(x,y,z,t) = \sum_{r=1}^{R} q_r \bar{u}_r(x,y,z) \tag{1}$$

then the equation of motion for the wing structure for the autonomous flutter situation reads:

$$M_r\ddot{q}_r(t) + \sum_{r=1}^{R} C_{rs}\dot{q}_s(t) + k_r q_r(t) = 0.5\rho U^2 \sum_{s=1}^{R} A_{rs}q_s(t) \tag{2}$$

In matrix mode notation:

$$[M_r]\{\ddot{q}_r\} + [C_{rs}]\{\dot{q}_s\} + [k_r]\{q_r\} = 0.5\rho U^2[A_{rs}]\{q_s\} \tag{3}$$

For convenience, p-method will be utilized in solving the flutter equation. The motion-induced airloads can be approximated by using Pade approximation as follows:

$$[A_{rs}] \cong [A_{rs}]_0 + \frac{b}{U}s[A_{rs}]_1 + (\frac{b}{U}s)^2[A_{rs}]_2 + \sum_{l=1}^{R} \frac{[A_{rs}]_{(l+1)}\frac{b}{U}s}{(\beta_s)_l + \frac{b}{U}s} \tag{4}$$

A new state variable can be defined:

$$B_{s_l} = \frac{\frac{b}{U}s}{\beta_{s_l} + \frac{b}{U}s} q_s \tag{5}$$

The time derivative of the new state variable can be expressed as:

$$\dot{B}_{S_l} = \dot{q}_s - \frac{U}{b}\beta_{s_l} B_{s_l} \tag{6}$$

Writing equation (3) in state space form and substitution of (4)-(6) yields the flutter equation in state space formulation:

$$\{X\} = [F]\{X\} \tag{7}$$

where:

$$\{X\}^T = \{q_s, \dot{q}_s, (B_s)_1 (B_s)_2, ...(B_s)_N\} \tag{8}$$

$[F] =$

$$\begin{bmatrix} [0] & [1] & [0] & \cdots & [0] \\ [\overline{M}]^{-1}[-\overline{K}] & [\overline{M}]^{-1}[-\overline{C}] & [\overline{M}]^{-1}[\overline{A}_{rs}]_3 & \cdots & [\overline{M}]^{-1}[\overline{A}_{rs}]_N \\ [0] & [1] & -\frac{U}{b}[\beta_s]_1 & \cdots & [0] \\ \vdots & \vdots & \vdots & & \vdots \\ [0] & [1] & [0] & & -\frac{U}{b}[\beta_s]_{N-2} \end{bmatrix} \tag{9}$$

and where

$$\begin{aligned} [\overline{M}] &= [M] - .05\rho b^2 [A_{rs}]_2 \\ [\overline{C}] &= [C] - .05\rho U b [A_{rs}]_1 \\ [\overline{K}] &= [K] - .05\rho U^2 [A_{rs}]_0 \\ [\overline{A}_{rs}]_{(l+2)} &= .05\rho U^2 [A_{rs}]_{(l+2)} \end{aligned} \tag{10}$$

4. Calculation of Aerodynamic Loads

4.1. GOVERNING EQUATIONS

The equation governing the unsteady linearized compressible potential flow can be obtained by assuming small disturbances with respect to the free stream condition. Two equations in terms of perturbation velocity potential and acceleration potential will be available:

$$(1 - M^2)\phi_{xx} + \phi_{yy} + \phi_{zz} - \frac{12kM}{a_\infty}\phi_x + \frac{k^2}{a_\infty^2}\phi = 0 \tag{11}$$

$$(1 - M^2)\Psi_{xx} + \Psi_{yy} + \Psi_{zz} - \frac{12kM}{a_\infty}\Psi_x + \frac{k^2}{a_\infty^2}\Psi = 0 \tag{12}$$

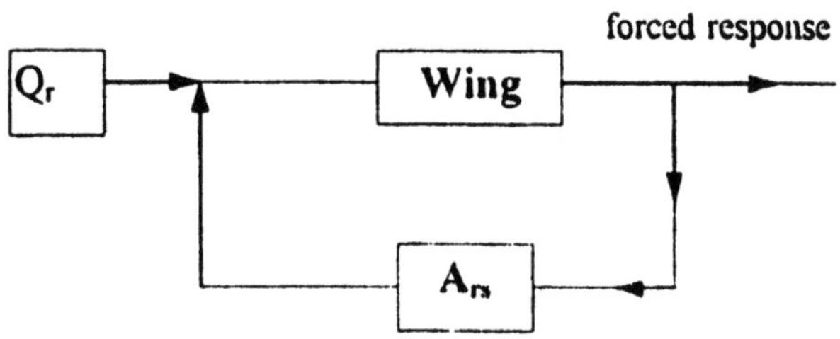

Figure 2a. Forced-vibration System

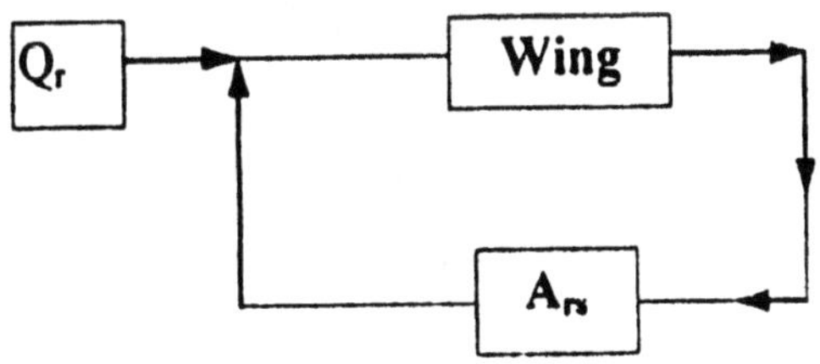

Figure 2b. Autonomous System

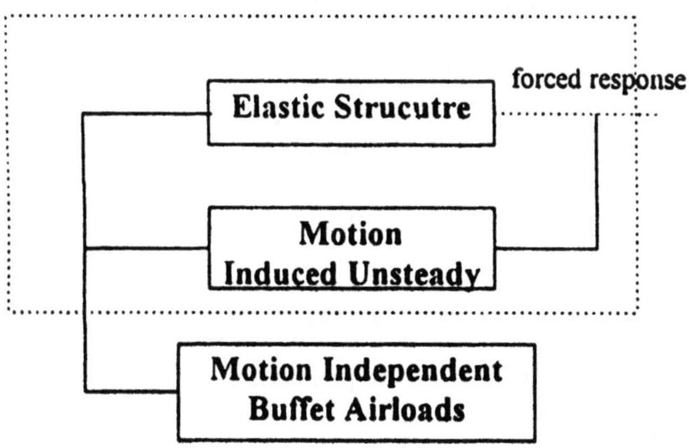

Figure 2c. Forced oscillation model buffeting

In these equations, ϕ and Ψ are complex quantities representing the amplitude of the acceleration and perturbation potential, respectively, indicating that in general there exists a phase lag between each of them with respect to the wing motion.

4.2. BOUNDARY CONDITIONS

4.2.1. *On the attached flow region*

The kinematical boundary condition of inviscid flow model dictates that the flow has to be tengential to the surface. This condition can be written as :

$$w(x,y,\tau) = \left[\frac{\partial}{\partial x} + \frac{\partial}{\partial \tau}\right] z(x,y,\tau) \tag{13}$$

where :

$$w(x,y,\tau) = \bar{\bar{w}}(x,y)e^{ik\tau t}$$

$$z(x,y,\tau) = \bar{\bar{z}}(x,y)e^{ik\tau} \tag{14}$$

Hence:

$$\bar{\bar{w}}(x,y) = \frac{\partial}{\partial x}\bar{z}(x,y) + ik\bar{z}(x,y) \tag{15}$$

where k is the reduced frequency and τ is the non-dimensional time.

4.2.2. *On the separated flow region*

The separation is considered only to take place on the upper side of the wing. On this region, some assumptions have to be made regarding the velocity and pressure of the fluid, based on experimental evidence. The true situation may be obtained using viscous flow approach, which is beyond the scope of this work.

It may be assumed that the velocity on the separated region is the same as that of the undisturbed flow. Alternatively, it may be assumed a priori that the pressure difference in the separation bubble is equal to same value γ

$$\bar{\bar{C}}p_u(x,y) = \gamma \tag{16}$$

4.3. SYSTEM OF INTEGRAL EQUATIONS

The system of integral equation for the problem is obtained from two sets of integral equations. The first set is the velocity equation with the kinematical boundary condition on the attached region and the other is the pressure equation with the prescribed pressure boundary condition on the separated region. The integral equations incorporated four kernel function. Major features of this development are reproduced below from references [5]-[9].

4.3.1. *Induced Velocity at Attached Flow Region*
The first kernel of the integral equation is associated with the induced velocity at the attached flow region. The velocity potential at this region induced by a doublet point is given by:

$$\Psi_D(x,y,z) = -\frac{1}{4\pi} A e^{i\mu k M(x-x')} \frac{\partial}{\partial z}\left(\frac{e^{-i\mu k r}}{r}\right) \tag{17}$$

Where

$$A = \frac{\Delta\overline{\overline{C}}p(x',y',z')}{2} \tag{18}$$

The downwash at the attached flow region induced by the doublet distribution over the entire surface is then given by:

$$\overline{w}_1(x,y) = -\frac{1}{8\pi}\int_{-s}^{s}\int_{x_{le}}^{x_{Le}} \Delta\overline{C}p(x',y')K_1(x,x';y,y')dx'\,dy' \tag{19}$$

Equation (19) is the usual integral equation of the lifting surface for attached flow. The kernel of the integral equation is given by Possio[11] and Kuessner[12] as

$$K_1(x,x';y,y') = e^{-ik(x-x')} \lim_{z\to o} \frac{\partial}{\partial z^2}\int_{-\infty}^{x-x'} \frac{e^{ik[\frac{x-x'}{\beta^2}-\mu r]}}{r} d(x-x') \tag{20}$$

4.3.2. *Induced Velocity at Separated Flow Region*
The contribution of the separated region, on which a source distribution is placed, to the first set of the integral equation, is derived using velocity potential formulation. The perturbation velocity potential induced by a source point is given by:

$$\phi_s(x,y,z) = -\frac{1}{4\pi} A^+ e^{i\mu k M(x-x')}\left(\frac{e^{-i\mu k r}}{r}\right) \tag{21}$$

where

$$\Delta\overline{w}(x',y',z') = -A^+ \tag{22}$$

The downwash induced by the source distribution on the upper surface where the flow is separated then becomes:

$$\overline{w}_l(\xi,\eta) = \int_{-l_s}^{l}\int_{\xi}^{\xi_r} \Delta\overline{w}(\xi',\eta')K_3(\xi,\eta;\xi',\eta')d\xi'd\eta' + \frac{\Delta\overline{w}}{2}\Big|_{\xi=\xi';\eta=\eta'} \tag{23}$$

The kernel function of equation (23) is:

$$K_3(\xi,\eta;\xi',\eta') = -\frac{1}{4\pi}\lim_{z\to 0}\beta^2 z e^{i\mu k M(\xi-\xi')}\frac{e^{-i\mu k r'}(-i\mu k r'-1)}{r'^3} \tag{24}$$

Figure 3a. Geometry of the hypothetical wing

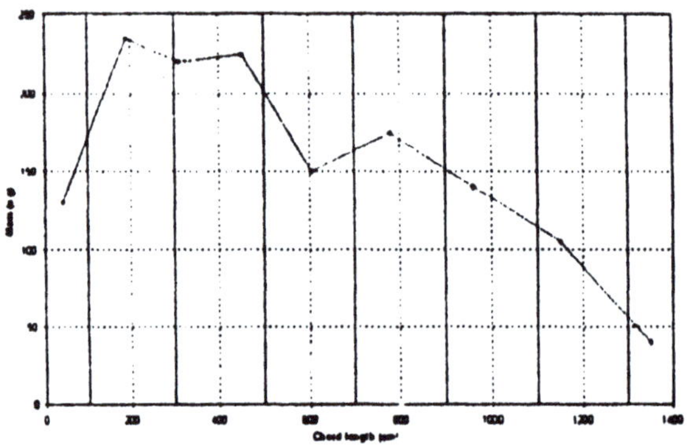

Figure 3b. Mass Distribution of the Hypothetical Wing

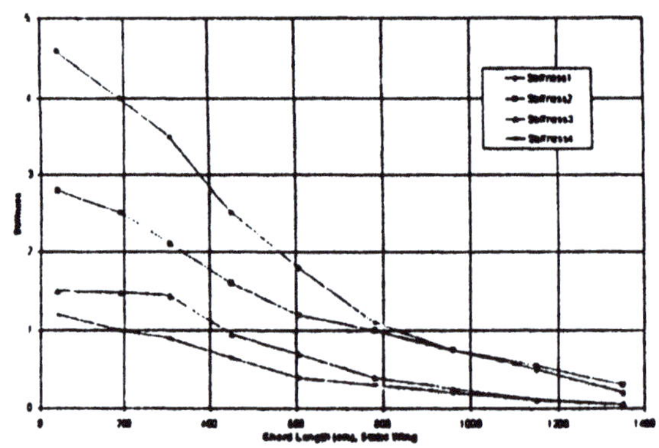

Figure 3c. Stiffness distribution of the hypothetical wing

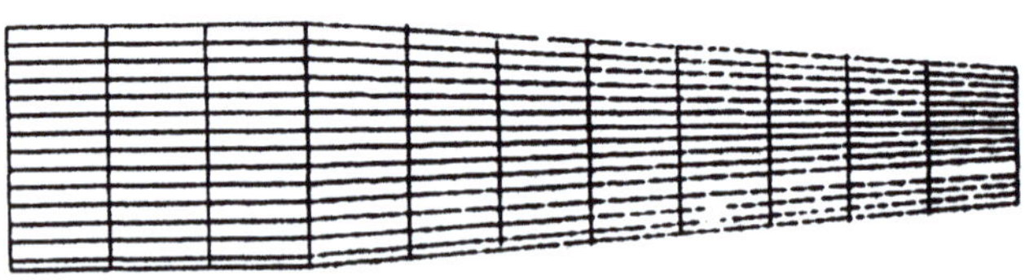

Figure 3d. Aerodinamic panneling of the hypothetical wing

4.3.3. *Induced Pressure at Attached Flow Region*

The third and fourth integral equations are pressure equations, with the pressure on the upper surface where the separation takes place serves as the boundary condition.

From the definition of acceleration potential and linearized Bernoulli equation, the pressure coefficient is given by:

$$\overline{C}p(x,y,z) = -2\psi_D(x,y,z) \tag{25}$$

The pressure coefficient on the upper surface of the wing is then:

$$\overline{C}p_u(\xi,\eta) = \int_{-l}^{l}\int_{\xi_{le}}^{\xi_{te}} \Delta\overline{C}p(\xi',\eta')K_2(\xi,\eta;\xi',\eta')d\xi' d\eta' - \frac{\Delta\overline{C}p}{2}|_{\xi=\xi';\eta=\eta'} \tag{26}$$

where the kernel function in equation (26) is given by:

$$K_2(\xi,\eta;\xi',\eta') = -\frac{1}{4\pi}\lim_{z\to 0}\beta^2 z e^{i\mu k[M(\xi-\xi')-r']}\frac{(-i\mu k r' - 1)}{r'^3} \tag{27}$$

4.3.4. *Induced Pressure at Separated Flow Region*

The kernel of the pressure integral equation is the contribution of the velocity potential source. The pressure coefficient is related to the velocity potential through the linearized Bernoulli equation, and is given by:

$$\overline{C}p(x,y,z) = -2\left(\frac{\partial\phi_s}{\partial\tau} + \frac{\partial\phi_s}{\partial x}\right) \tag{28}$$

The pressure on the upper surface is then given by:

$$\overline{C}p_u(\xi,\eta) = \int_{-l}^{l}\int_{\xi_x}^{\xi_r} \Delta\overline{w}(\xi',\eta')K_4(\xi,\eta;\xi'\eta')d\xi'\, d\eta' \tag{29}$$

Where the kernel is given:

$$K_4(\xi,\eta;\xi'\eta') = -\frac{1}{2\pi}e^{i\mu k[M(\xi-\xi')-r']}\left(\frac{ik}{\beta^2 r'} - \frac{(\xi-\xi')(i\mu k + 1/r')}{r'^2}\right) \tag{30}$$

Superposing all the elements contributing to the integral equation as elaborated in this section, the system of integral equtaion is found to be:

$$\overline{w}_1(\xi,\eta) = \int_{-l}^{l}\int_{\xi_{le}}^{\xi_{te}} \Delta\overline{C}_p pppp(\xi',\eta')K_1(\xi,\eta;\xi',\eta')d\xi' d\eta' + \frac{\Delta\overline{w}}{2}|_{\xi=\xi';\eta=\eta'} \tag{31}$$

$$\overline{C}p_u(\xi,\eta) = \int_{-l}^{l}\int_{\xi_{le}}^{\xi_{te}} \Delta\overline{w}(\xi',\eta')K_4(\xi,\eta;\xi',\eta')d\xi' d\eta' - \frac{\Delta\overline{C}p}{2}|_{\xi=\xi';\eta=\eta'} \tag{32}$$

4.4. SOLUTION OF THE INTEGRAL EQUATION

The unknown pressure and the downwash jump across the lifting surface will be represented by series expansion:

$$\Delta\overline{C}p(X,y) = \sum_{r=0}^{R} a_r(y)h_r(X) \tag{33}$$

Each function $h_r(X)$ should exhibit a well known singularity behaviour for $X \to 0$ of the type

$$\lim_{c\to 0} c^{-\frac{1}{2}}$$

The singularity models the suction force at the leading edge. At the trailing edge, where Kutta condition applies, the result of thin airfoil theory dictates that the pressure difference should go to zero like

$$\lim_{c\to 0} c^{\frac{1}{2}}$$

One set of functions which satisfy these physical characteristics in the chordwise direction is the one introduced by Glauert [10]:

$$h_r(X) = \frac{2}{\pi}\frac{\cos(r\phi) + \cos((r+1)\phi)}{\sin(\phi)} \tag{34}$$

Similarly, for the downwash jump across the lifting surface, the same series expansion is used:

$$\Delta\overline{w}(X,y) = \sum_{q=0}^{Q} d_q(y).h_q(X) \tag{35}$$

After substitution of the series expansion, the system of integral equations could be described in a set of linear simultaneous equations, and can conveniently be written in matrix notation:

$$\begin{bmatrix} g_1 & g_3 \\ g_2 & g_4 \end{bmatrix} \begin{Bmatrix} a' \\ d' \end{Bmatrix} = \begin{Bmatrix} \overline{w}_1 \\ \overline{C}p_u \end{Bmatrix} \tag{36}$$

Solving this set of linear algebraic equations, the coefficients of the series expansions are obtained. The detail of the evaluation of the chordwise and spanwise integration can be found in ref.[5][6].

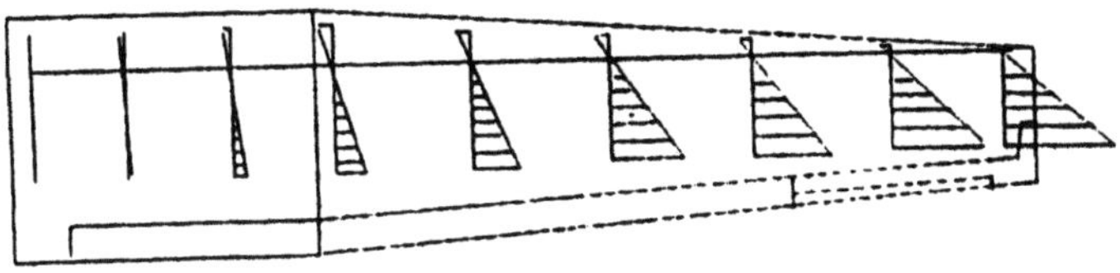

Figure 4a. The first mode shapes of hypothetical wing

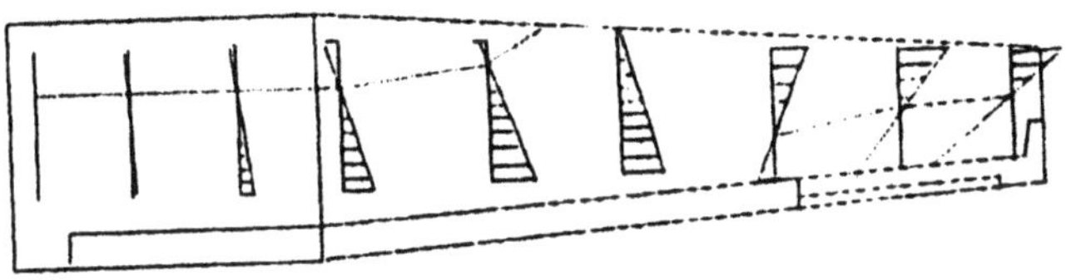

Figure 4b. The second mode shapes of hypothetical wing

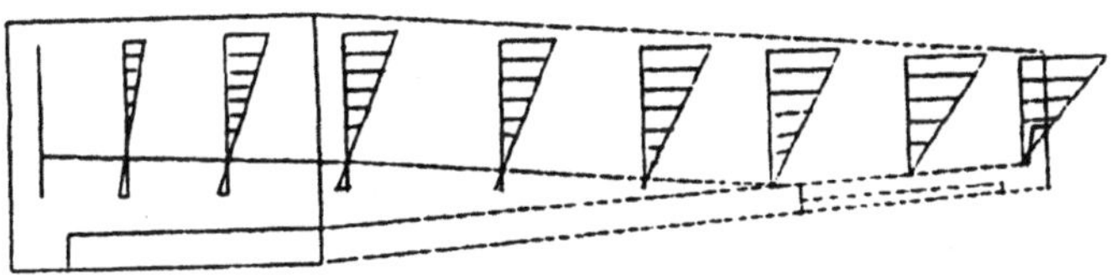

Figure 4c. The third mode shapes of hypothetical wing

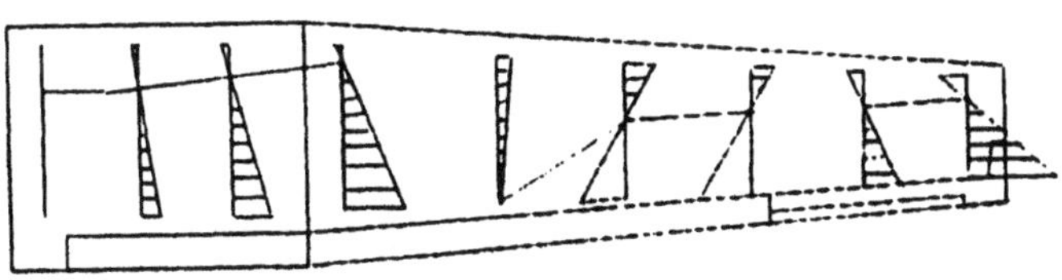

Figure 4d. The fourth mode shapes of hypothetical wing

4.5. PRESSURE COEFFICIENTS

The pressure coefficient consists of the attached and separated flow parts, which are calculated separately.

$$\Delta\overline{C}p(X,\eta) = \begin{cases} \Delta\overline{C}p_{att}(X,\eta), & for\ 0 < X < X_r\ and\ X_r < X < C(\eta) \\ \Delta\overline{C}ps_{sep}(X,\eta), & for\ X_s < X < X_r \end{cases} \tag{37}$$

The pressure difference on the attached flow region can directly be calculated from the series expansion as:

$$\Delta\overline{C}p_{att}(X,\eta) = \sum_{r=0}^{R} a_r(\eta)h_r(X) \tag{38}$$

The pressure difference at the separated flow region may be calculated using the following procedure.

$$\Delta\overline{C}p_{sep}(X,\eta) = \overline{C}p_l(X,\eta) - \overline{\overline{C}}p_u(X,\eta) \tag{39}$$

The lower pressure coefficient can simply be calculated from:

$$\overline{C}p_l(X,\eta) = \sum_{r=0}^{R} a_r(\eta)\frac{h_r(X)}{2} \tag{40}$$

while upper pressure coefficient can be calculated using the pressure integral equation:

$$\overline{C}p_u(X,\eta) = -\sum_{r=0}^{R} a_r(\eta)\frac{h_r(X)}{2} + \sum_{q=0}^{Q}\sum_{m=1}^{M} d_{qm}g_{4\ qm\ vp} \tag{41}$$

Substituting the two expressions for the upper and lower pressures in equation (39) there is obtained:

$$\overline{C}p_u(X,\eta) = \sum_{r=0}^{R} a_r(\eta)h_r(X) - \sum_{q=0}^{Q}\sum_{m=0}^{M} d_{qm}g_{4\ qm\ vp} \tag{42}$$

5. GOVERNING EQUATIONS FOR DYNAMIC RESPONSE MODEL

In a previous work [9], the author has approached the buffeting problem by the use of deterministic dynamic response approximation, and will not be reproduced here. In the following analysis, a stochastic approach will be presented, following closely the procedure outlined in [13]. In tackling the buffeting phenomena using dynamic response model, it will be assumed that there are no coupling between motion-induced unsteady airloads and motion-independent unsteady airloads. Then, following Foersching, the analytical model of the buffeting phenomena can be modelled as a forced vibration system depicted in Figure 2c.

In this model, the motion-independent unsteady airloads will act as the driving excitation forces which produce buffeting phenomenon, and the motion-induced unsteady airloads will give rise to aerodynamic damping that together with the structural damping will limit the dynamic aeroelastic response of the wing or the structure. In the present approach, flow situations modeled is such that the motion-independent buffeting forces can be approximated by using Powell and Bull's experimental results[14]. Since Powell and Bull's experiment is associated with turbulent flow, while in our case we are dealing with separated flow with buffeting, such modelling may only serve two purposes: to investigate the computational feasibility, and to utilize the procedure for parametric studies for practical applications.

A. **Equations of the Motion of the Wing:**

By looking into a infinitesimal slab along its span, the equation of motion of a flexible wing can be written as [11]

$$\frac{\partial^2}{\partial y^2}\left[EI\frac{\partial^2 u(y,t)}{\partial y^2} \right] = -m\ddot{u}(y,t) + F^A(t) + F^F(t) \tag{43}$$

$F^A(t)$ is motion-induced airload and $F^F(t)$ is motion-independent airloads.

The deflection of the structure can be expressed as:

$$u(P,t) = \sum_{r=1}^{n} u_r(P)q_r(t) \tag{44}$$

There fore, by substitution of equation (44), the equation of motion (43) can be recast into the form:

$$M_r\ddot{q}_r(t) + K_r q_r(t) - \sum_{r=1}^{n} A_{rs}q_s(t) = Q_r^F(t) \tag{45}$$

where K_r and M_r are the generalized stiffness and mass, respectively, of the structure, and the first one is given by:

$$\begin{aligned} K_r &= \int_0^l \frac{\partial^2}{\partial y^2}\left[EI\frac{\partial^2 u_r(y)}{\partial y^2} \right] u_r(y)dy \\ &= \omega_r \int_0^l u_r(y)m(y)u_r^T(y)dy = M_r\omega_r \end{aligned} \tag{46}$$

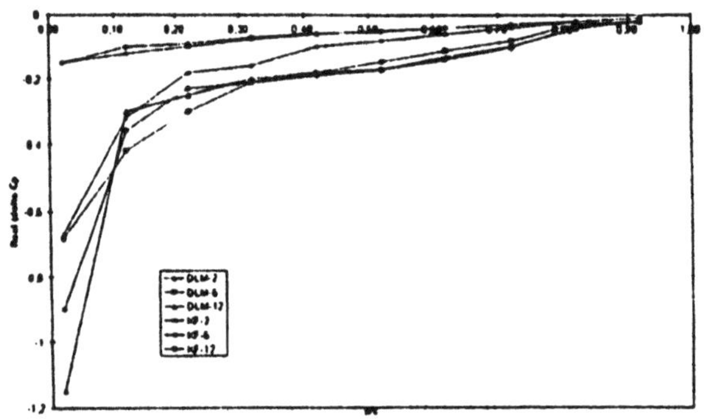

Figure 5a. Real Part ΔC_p Comparsion between Doublet Lattice Method and Kernel Function. Attached flow cases. First mode shapes, station 2,6, and 12, Mach Number 0.61 reduced frequency 0.001

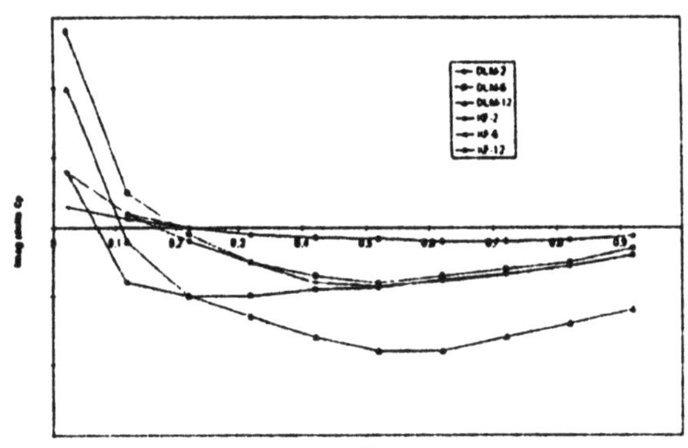

Figure 5b. Imaginary Part of ΔC_p Comparsion between Doublet Lattice Method and Kernel Function. Attached Flow case. First mode shapes, station 2,6, and 12. Mach number 0.61 reduced frequencyO.001

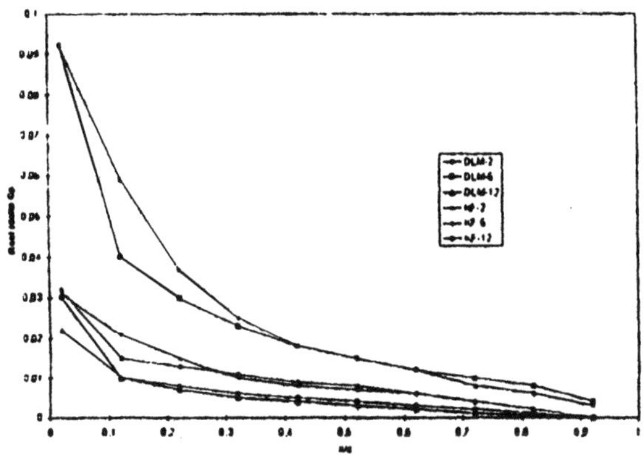

Figure 5c. Real Part ΔC_p Comparsion between Doublet Lattice Method and Kernel Function. Attached flow cases. Second mode shapes, station 2,6, and 12, Mach number 0.61 reduced frequency 0.001

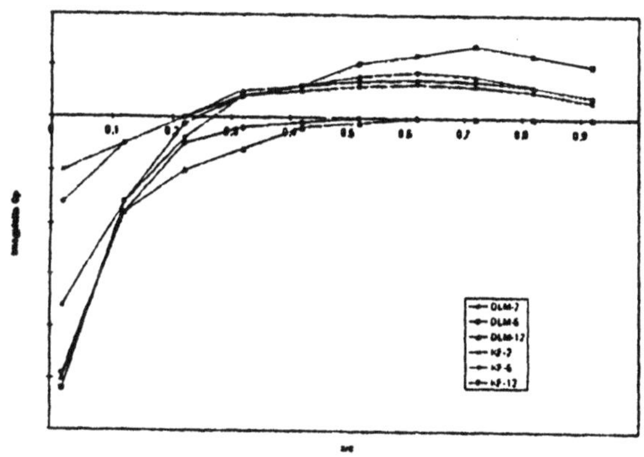

Figure 5d. Imaginary Part of ΔC_p Comparsion between Doublet Lattice Method and Kernel Function. Attached Flow cases. Second mode shapes, station 2,6 and 12 Mach number of 0.61 reduced frequency 0.001.

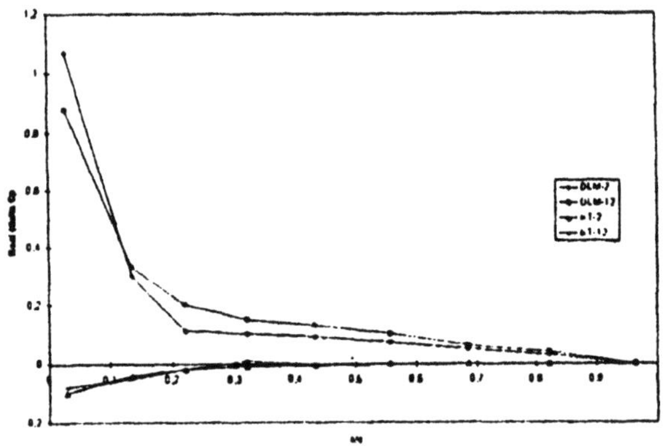

Figure 5e. Real Part of ΔC_p Comparsion between Doublet Lattice Method and Kernel Function. Attached flow cases. Third mode shapes, station 2,6, and 12, Mach umber 0.61 reduced frequency 0.001

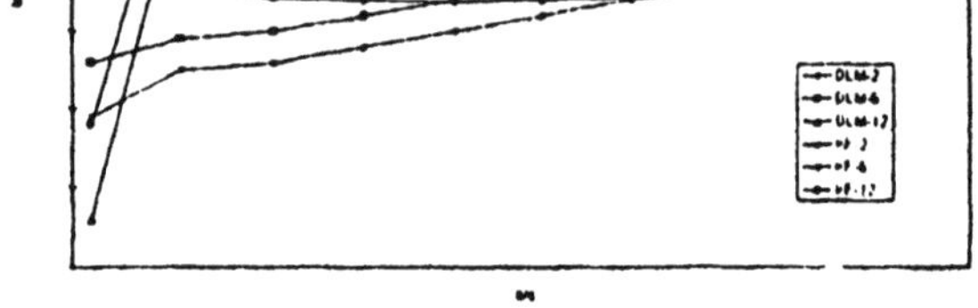

Figure 5f. Imaginary Part of ΔC_p Comparsion between Doublet Lattice Method and Kernel Function. Attached Flow cases. Third mode shapes, station 2,6 and 12 Mach number 0.61 reduced frequency 0.001.

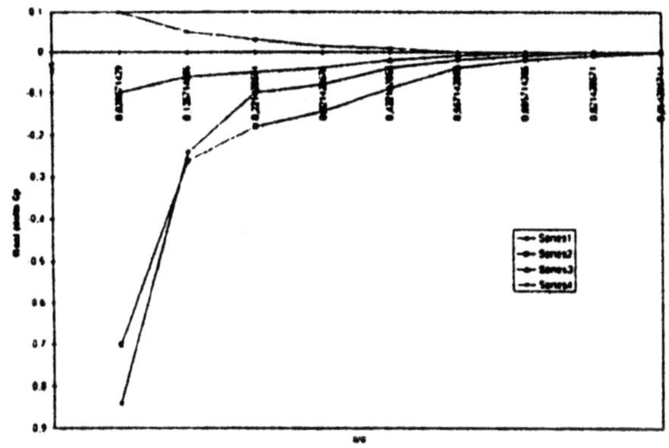

Figure 5g. Real Part of ΔC_p Comparsion between Doublet Lattice Method and Kernel Function. Attached flow cases. Fourth mode shapes, station 2,6, and 12, Mach umber 0.61 reduced frequency 0.001

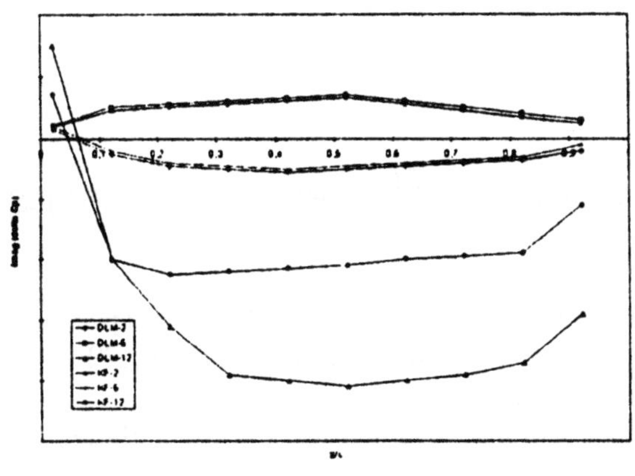

Figure 5h. Imaginary Part of ΔC_p Comparsion between Doublet Lattice Method and Kernel Function. Attached Flow cases. Fourth mode shapes, station 2,6 and 12 Mach number 0.61 reduced frequency 0.001.

The generalized unsteady aerodynamic forces due to the motion-induced unsteady pressure, $p_S^A(P,t)$, when the structure is vibrating in the s-th vibration mode, is given by:

$$A_{rs}(t) = \iint_{(s)} u_r(P) p_s^A(P,t) e_n dS \tag{47}$$

e_n is the unity vector normal to the surface.
The right hand side of equation (47) expresses the generalized aerodynamic forcing term associated with pressure fluctuation $p^F(x,t)$ due to buffeting excitation, which originates from unsteady flow separation, and can analogously be written as:

$$Q_r^F(t) = \iint_{(s)} \int_{l} p^F(P,t) u_r(P) e_n dS \tag{48}$$

$p^F(x,t)$ is the pressure fluctuation due to flow separation at the location $P(x,y,z)$ on the surface of the structure.
The unsteady pressure due to flow separation are stochastic quantities, which have to be treated by statistical methods (see for example, Bendat and Piersol [15]). Following Foersching and Zingel, a cross spectral density of the generalized excitation is formed in a manner analogous to the generalized excitation in equation (47) of the generalized unsteady airloads:

$$S_{Q_{rs}}(\omega) = \iint_{(s)} \iint_{(s)} u_r^T(P) S_p(P,P',\omega) u_s(P') dS\, dS' \tag{49}$$

where S_p is the diagonal matrix of the cross-spectral density of the fluctuating pressure at two different points (P,P').

The dynamic response problem is now formulated by means of the input-output relation:

$$S_u(P,\omega) = \sum_{r=1}^{n} \sum_{s=1}^{n} H_r^*(P,\omega) H_s^T(P,\omega) S_{Q_{rs}}(\omega) \tag{50}$$

In this expression, $S_u(P,\omega)$ is the diagonal matrix of the power spectral density of a deflection $u(x,t)$ of the structure considered at the location $P(x,y,z)$. $H_r(P,\omega)$ is the transfer function of the dynamic system for the r-th vibration mode, which can be determined from equation (45).

H^* is its conjugate form.
By assuming that both input and output are harmonics,

$$q_r(\omega) = \bar{\bar{q}}_r(\omega)e^{i\omega t}$$
$$Q_r(\omega) = \overline{Q}_r(\omega)e^{i\omega t}$$

Then by using the above equations equation (45) can be transformed into the stochastic form of the equation of motion in frequency domain:

$$M_r(\omega_r^2 - \omega^2)\overline{q}_r(\omega) - \sum_{s=1}^{n} A_{rs}\overline{q}_s(\omega) = \overline{Q}_r^F(\omega) \tag{51}$$

Solving the linear simultaneous equation (51), the transfer function relating the output variables modeshapes to the input stochastic buffeting forces can be determined as:

$$H_{u,r}(P,\omega) = \frac{u_r(P)\overline{q}_r(\omega)}{\overline{Q}_r^F(\omega)} \tag{52}$$

The bending moment at point P of the structure can be derived from simple beam theory:

$$M(P,t) = EI\frac{\partial^2}{\partial y^2}u(P,t) \tag{53}$$

Substitution of equation (53) into (52) yields the transfer function for the bending moment of the structure and is given by:

$$H_{M,r}(P,\omega) = \frac{[EIu_r''(P)]^2\overline{q}_r(\omega)}{\overline{Q}_r^F(\omega)} \tag{54}$$

where u'' indicates second spatial derivatives of $u(P)$. Neglecting the structural damping, the mean-square values of the bending moment can be formulated as:

$$\overline{M}_{P,r}^2 = \frac{[EIu_r''(P)]^2}{8\pi M_r^2\omega_r^3\gamma_{aero}}S_{Q,r}(\omega_r) \tag{55}$$

where the γ_{aero} is the aerodynamic damping factor.

B. Motion Induced Unsteady Airloads

The calculation of the motion induced unsteady airloads follows the procedure described in section 4.

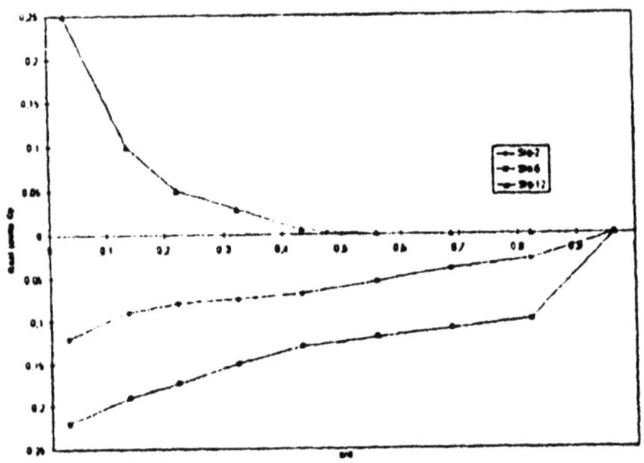

Figure 6a. Real part of C_{pu} Separated flow cases, $xs = 0.8c$. First mode shapes, station 2,6 and 12. Mach number 0.63, reduced frequency 0 001.

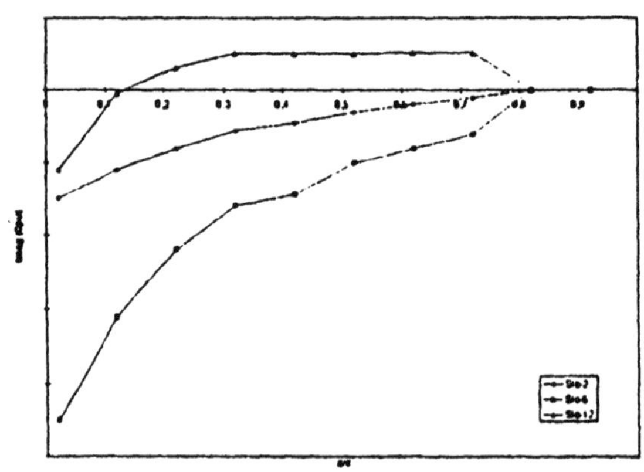

Figure 6b. Imaginary part of C_{pu} separated flow case, xs=0.8c. First mode shapes, station 2,6 and, 12. Mach number 0.63, reduced frequency 0.001

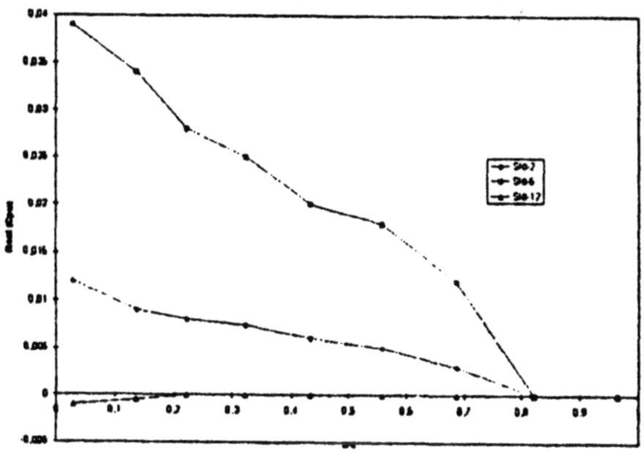

Figure 6c. Real part of C_{pu} Sparated flow case, xs=0.8c. Second mode shapes, station 2,6 and 12. Mach number 0.63, reduced frequency 0.001.

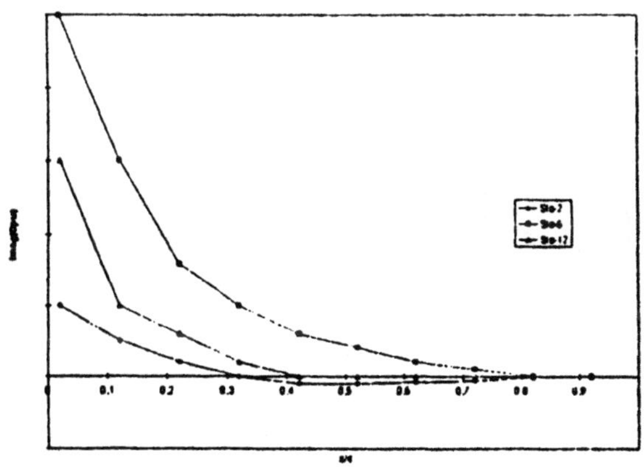

Figure 6d. Imaginary part of C_{pu} separated flow case, xs=0.8c. Second mode shapes, station 2,6 and 12. Mach number 0.63, reduced frequency 0.001

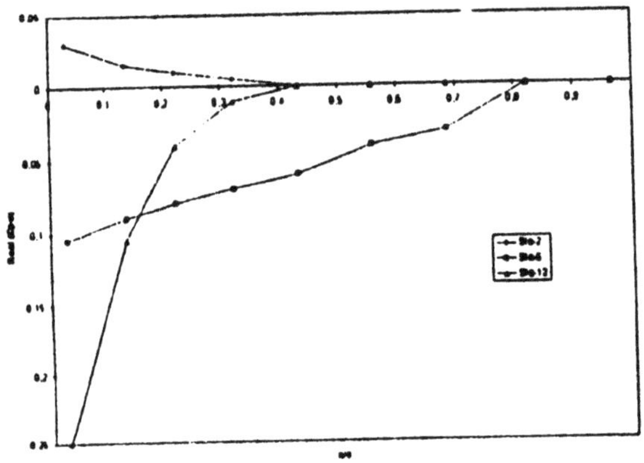

Figure 6e. Real part of C_{pu} Sparated flow case, xs=0.8c. Third mode shapes, station 2,6 and 12. Mach number 0.63, reduced frequency 0.001.

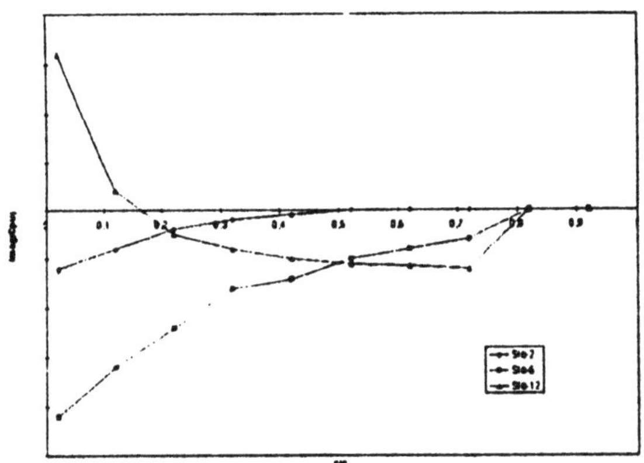

Figure 6f. Imaginary part of C_{pu} separated flow case, xs=0.8c. Third mode shapes, station 2,6 and 12. Mach number 0.63, reduced frequency 0.001

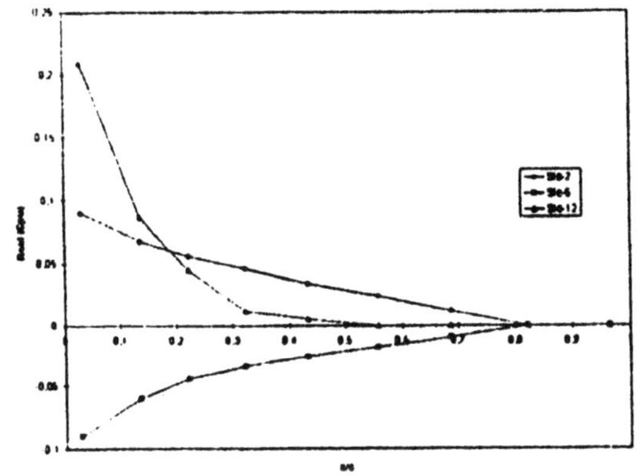

Figure 6g. Real part of C_{pu} Sparated flow case, xs=0.8c. Fourth mode shapes, station 2,6 and 12. Mach number 0.63, reduced frequency 0.001.

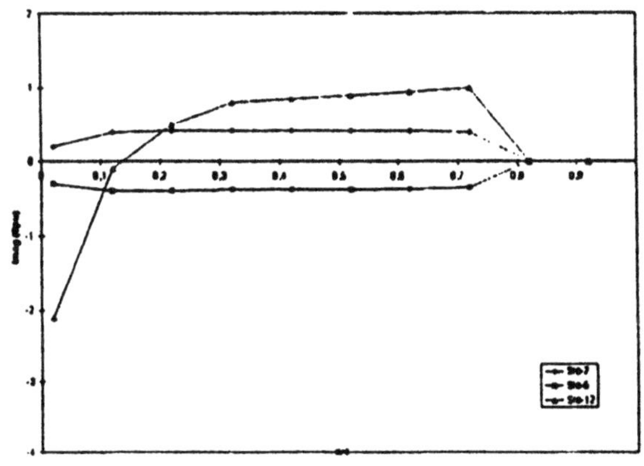

Figure 6h. Imaginary part C_{pu} separated flow case, xs=0.8c. Third mode shapes, station 2,6 and 12. Mach number 0.63, reduced frequency 0.001.

C. Motion-independent Unsteady Airloads

By assuming that the pressure fluctuations exhibit statistical regularity, then the magnitude of the motion-independent unsteady airloads can be written in statistical form. For this purpose, a planar surface sections are assumed and it is considered that the flow separation occurs in turbulent flow. Then the situation can be modelled using Bull and Powell's experiment.

Based on Bull and Powell's experiment, accepctance function reads:

$$J_{rs}(\omega) = \frac{1}{S^2 S_p(\omega)} \iint_{(s)} \iint_{(s)} u_r(P) S_p(P, P', \omega) u_s(P') dS dS' \tag{56}$$

where S and $S_p(P, P', \omega)$ are wing surface area and auto-correlation power spectral density of pressure fluctuations, respectively. Following Crocco's experiment [16], the cross-correlation power spectral density of the fluctuating pressure is given by:

$$S_p(P, P', \omega) = S_p(\omega) \exp\left[-\tfrac{\omega}{U_c}(0.1|\xi_1| + 0.55|\xi_2|) \right] \cos\left(\tfrac{\omega}{U_c}|\xi_1| \right) \tag{57}$$

ξ_1 and ξ_2 are distances between two points in the longitudinal and transversal directions to the flow, respectively, and Uc is the convective velocity which has often been assumed to be equal to about 70% of the free stream velocity.

The auto-correlation power spectral density at $w = w1$ can be approximated by [17]:

$$S_p(\omega) = 4K^2 q^2 c^3 \tag{58}$$

where q and c are dynamic pressure and chord, respectively. K is constant which is determined by experiment.

6. COMPUTATIONAL RESULTS AND DISCUSSION

A. Flutter Modelling Results

It should be mentioned that 2-dimensional results have been worked out in great detail [5]-[9] and provides confidence in proceeding to the present three-dimensional results.

Figure 3 shows the geometry, mass and stiffness distribution and aerodynamic panelling of the hypothetical wing. The wing was divided into 12 and 10 stations in the spanwise and chordwise direction, respectively.

The first four elastic modes was selected for flutter calculation. The mode shapes and natural frequencies are shown in Figure 4.
The unsteady pressure coefficients for each modes are shown in Figure 5 for the attached flow case. Comparison between the results obtained using Doublet Lattice Method incorporated in NASTRAN software and those obtained by the present method for each mode with reduced frequency 0.001 based on half-chord and M=0.63 shows good agreement. The largest discrepancies occur at the tip stations.
The pressure coefficients in the upper surface for the first four modes of the hypothetical wing in separated flow with separation point at 0.8 local chord and value equal to zero are shown in Figure 6.
The flutter calculation results for the hypothetical wing for the attached flow condition at sea level are shown in Figure 7; results obtained using NASTRAN software and present method are exhibited. The flutter velocity obtained by using NASTRAN software was 427 KTAS, while that obtained using the present method was 423 KTAS, both for the second mode.
For the separated flow case, the flutter diagram is shown in Figure 8. The flutter speed was found to be 150 KTAS and is associated with the first mode. Quantitative validity of this results needs to be verified in view of the modelling assumptions and other relevant factors.

B. **Dynamic Response Results**

In order to facilitate comparison with existing data, a rectangular wing structure used in Boyden and Johnson's experiment[17] was utilized. The computational result of the first natural bending mode of this model was 270 Hz, which is equal to Boyden and Johnson's.
The location of the separation points on our model is based on experimental results of Boyden and Johnson. Following this approach, approximation function for the location of the separation points were constructed as a function of angle of attack, and is given by:

$$\frac{x_s}{b} = \begin{cases} 1.0 & \text{if } \alpha_s < \alpha \\ 2\frac{(\alpha-\alpha_s)}{(\alpha_f-\alpha_s)} - 1.0 & \text{if } \alpha_s \leq \alpha \leq \alpha_f \\ -1.0 & \text{if } \alpha < \alpha_f \end{cases} \tag{59}$$

where α_s and α_f are the angle of attack associated with the onset of separation, and a fully separated situation, respectively, expressed in degrees. These are further modelled using the following empirical equation:

$$\alpha_s = (2.357)10^{-6} Re + 5.162 \tag{60a}$$
$$\alpha_f = (1.010)10^{-6} Re + 12.212 \tag{60b}$$

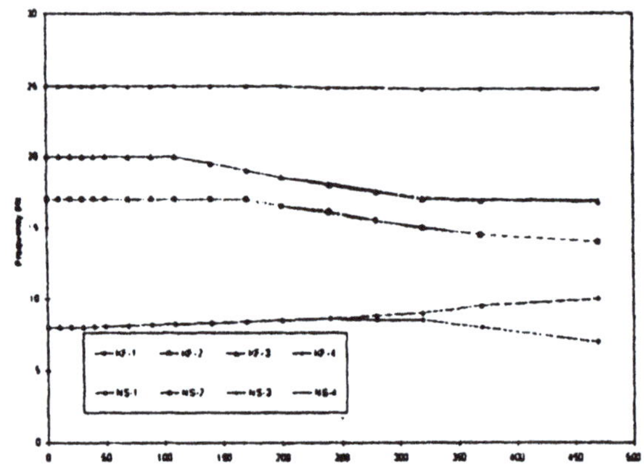

Figure 7a. Velocity-frequency diagram for hypothetical wing. Comparsion between NASTRAN and present work. Attached flow case.

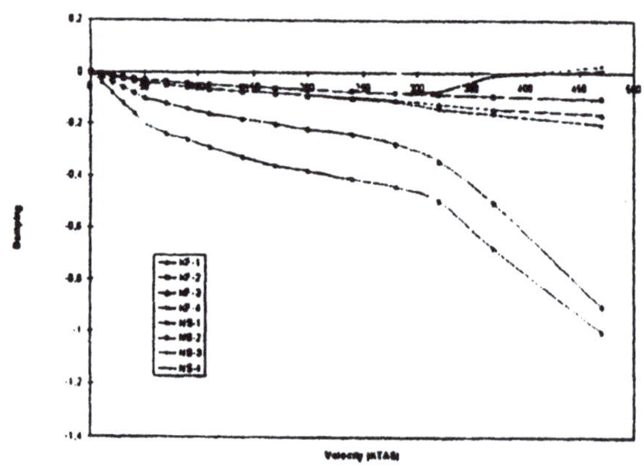

Figure 7b. Velocity-damping diagram for hypothetical wing. Comparsion between NASTRAN and present work. Attached flow case.

where Re is the Reynolds number. Figure 9 illustrates non-dimensional form of the chordwise co-ordinate of the separation points which are assumed to be located at a straight line along the span.

Figures 10 to 13 illustrate the real and imaginary parts of the motion-induced unsteady aerodynamic loads for various location of the separation points. The pressure distribution on the separated region of the upper surface was assumed to be zero. The motion-independent unsteady airload quantities expressed in terms of acceptance function for the first three modes are illustrated in Figure 14.

In the experimental results of Boyden and Johnson, the K value of equation (58) was found to be

$$K = \frac{\rho_\infty U_\infty}{\mu} \tag{61}$$

where, ρ, U_∞ and μ are air density, velocity, and viscosity of the undisturbed flow. In the present approach, this K value will be utilized as a first approximation. This value can later be corrected or updated by verifying the final results with available specific data. Following this approach, the root-mean-square bending moment coefficients, M_{rms}/qcS, for various values of Reynolds number and angle of attack can be found, and are given in Figure 15 to 18.

Compared to experimental results, the computational results give lower values of standard deviation of the bending moment. These results may be attributed to, among others, the fact that at the onset of buffeting, the pressure in the separation bubble is assumed to be zero. However, the qualitative agreement shown indicated that the method is capable to model the buffeting phenomena to some extent.

7. CONCLUSIONS AND FURTHER WORK

Two approaches have been attempted to model buffeting phenomena, utilizing the author's singularity method for modelling separated flow. First, to treat the buffeting problem as a stability one by solving the flutter equation. Second, treating buffeting problem using dynamic response approach. Three dimensional cases will be considered.

Validation was made in the unsteady aerodynamic loads calculation for the attached flow case by comparison with standard procedures in the literature. Validation for the flutter calculation were made by comparison to computational results using NASTRAN for the attached flow case. Comparison of the unsteady load calculation method with other existing results for three-dimensional non-separated flow cases indicated good agreements. For separated flow case, no other methods are available for comparison.

Parametric studies associated with flutter calculation for the separated flow case was made by changing the position of the separation points, using a hypothetical wing. The destabilizing trend exhibited due to the presence of the separation bubble as compared to the attached flow case will be used to assess the performance of the analysis.

For simplified cases, dynamic response behaviour that qualitatively agree with experimental data have been indicated. Computational results for the bending moment of a wing experiencing buffeting with experimental results indicated good agreement but lower values for the onset of buffeting.

These methods seem to indicate the plausibility of a systematic approach to buffeting modelling using analytical approach, which can be used for parametric studies for preliminary design purposes.

The use of other technique to calculate the aerodynamic forces, such as that developed in references [18] and [19], will be the subject of further work.

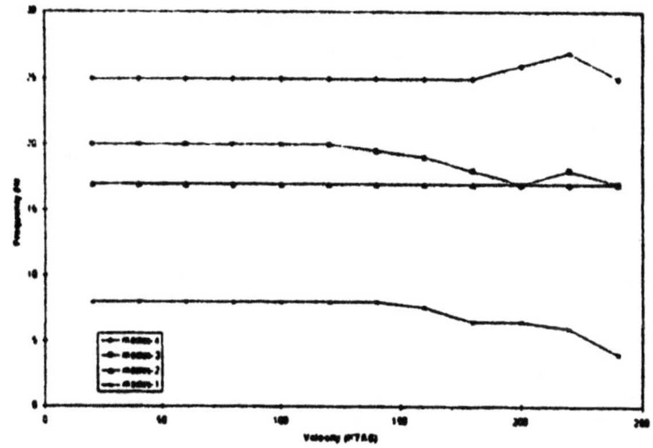

Figure 8a. Velocity-damping diagram for hypothetical wing. Sparated flow case, xs=0.8c.

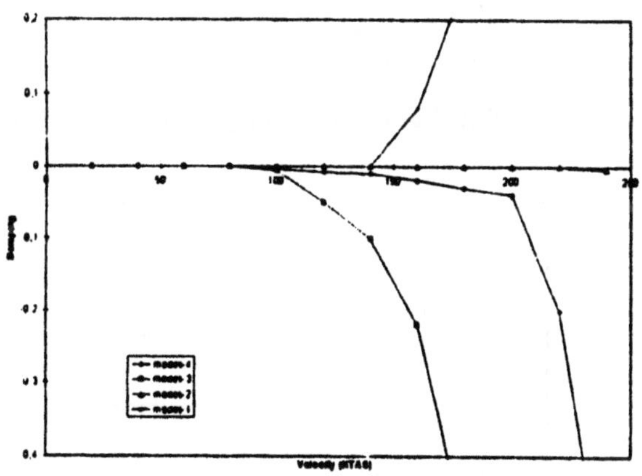

Figure 8b. Velocity-damping diagram for hypothetical wing. Sparated flow case xs=0.8.

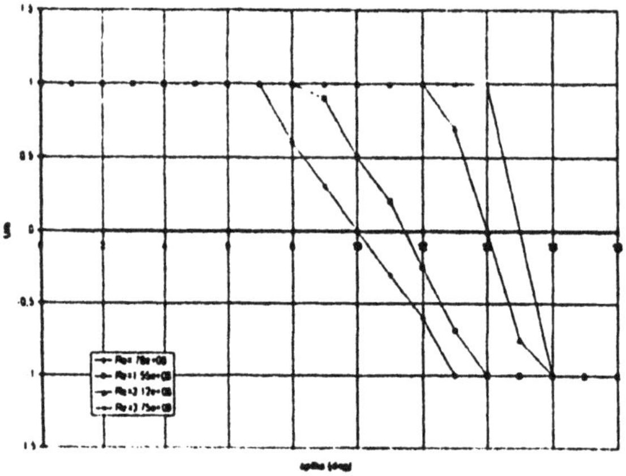

Figure 9. Sparated point model as a function of Reynolds number and angle of attack.

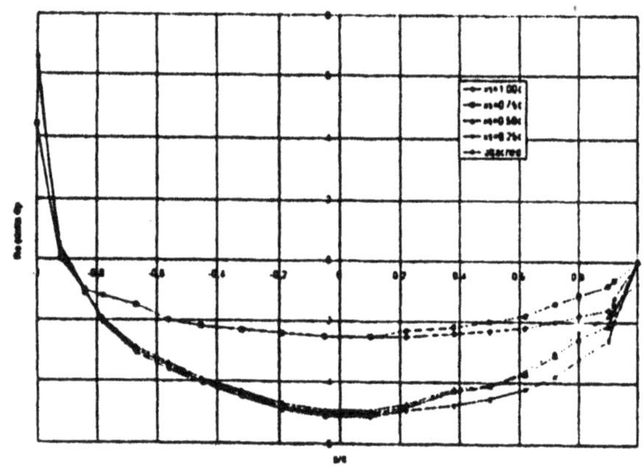

Figure 10. Real part presure distribution of bending mode, $y/i = 0.5, k = 1.67$, and $M = 0.3$.

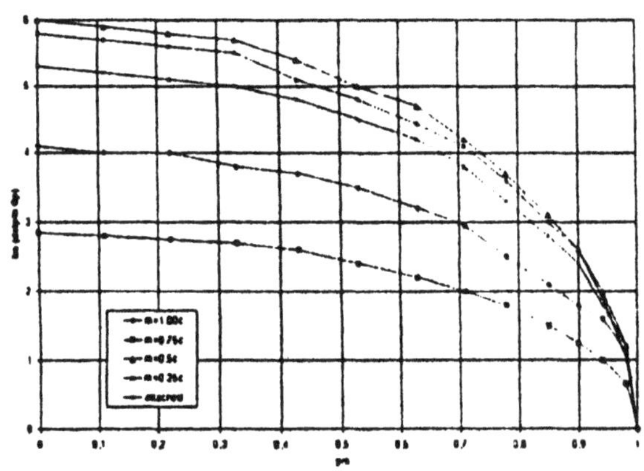

Figure 11. Imaginary part pressure distribution of bending mode, $x/c = 0.55, k = 1.67$, and $M = 0.3$.

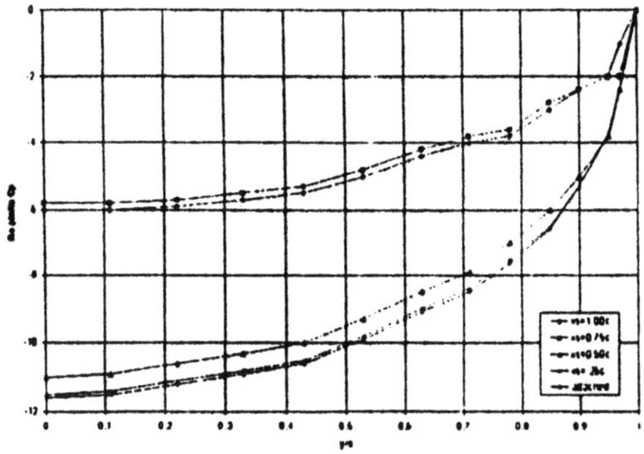

Figure 12. Real part pressure distribution of bending mode $x/c = 0.55, k = 1.67$, and $M = 0.3$.

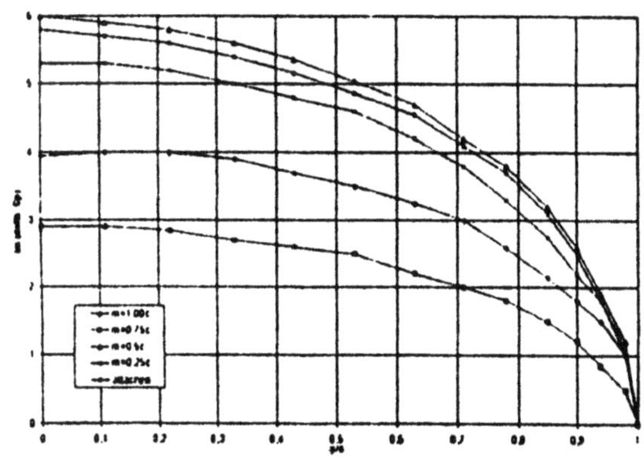

Figure 13. Imaginary part pressure distribution of bending mode, $x/c = 0.55, k = 1.67$ and $M = 0.3$.

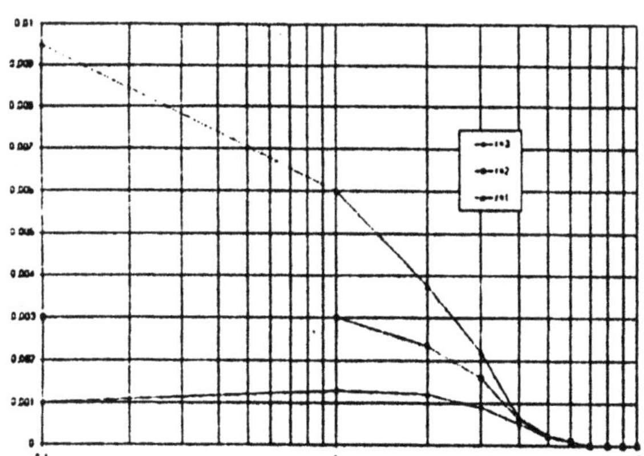

Figure 14. Acceptance function independent-motion airloads.

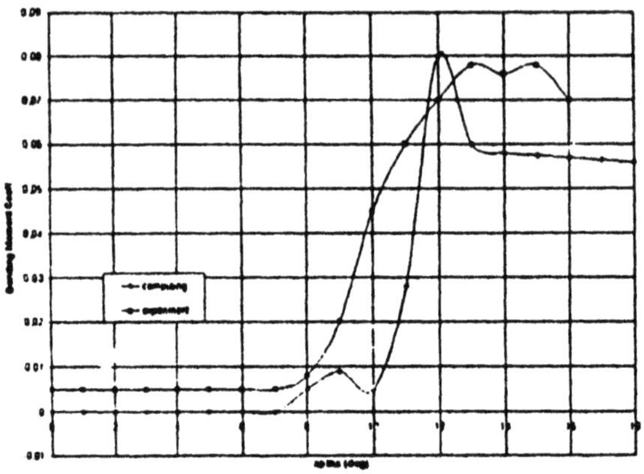

Figure 15. Bending Moment Coefficient for Re=0.78e+06.

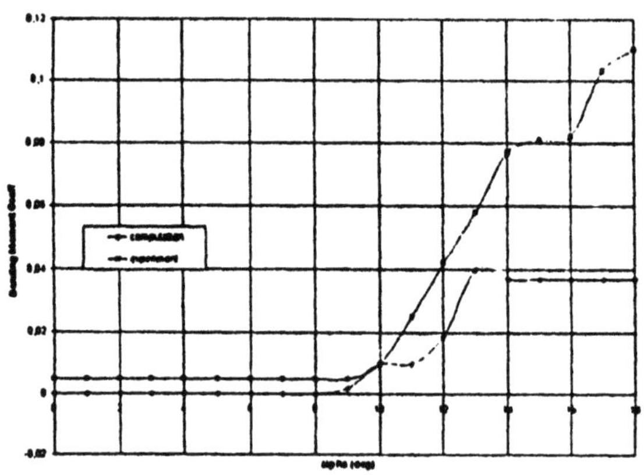

Figure 16. Bending moment coefficient for Re=1.55e+06.

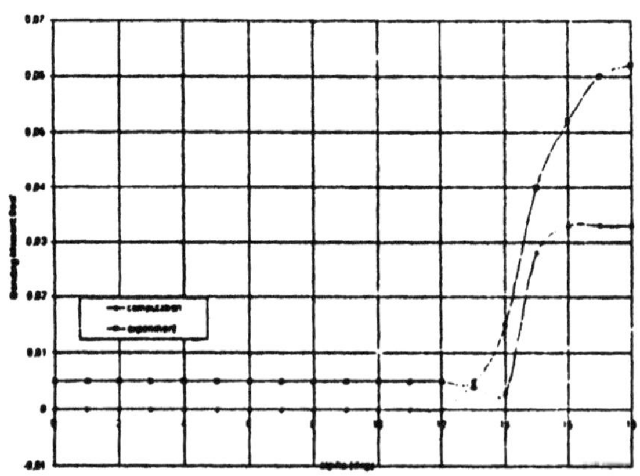

Figure 17. Bending moment coefficient for Re=3.12e+06.

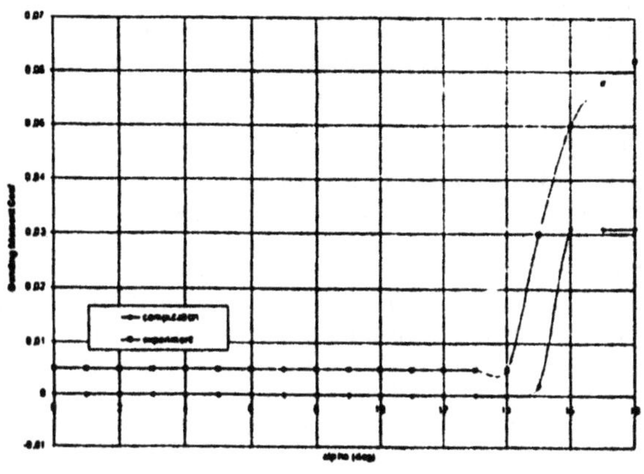

Figure 18. Bending moment coefficient for Re=3.75e+06.

8. ACKNOWLEDGEMENT

The author would like to acknowledge the contributions of many of his colleagues, among others Messrs Wijaya K. Sekar, Erwin Sulaeman, Ismoyo Haryanto and B. Bima Prananta, for their contributions in working out solutions to the problems discussed here. Mr. Thomas Aquino has assisted the author in completing this article.

References

1. Foersching,H., Aeroelastic Buffeting Prediction Techniques - A General Review, DFVLR, Goettingen, April 1981.
2. Becker, J. and Gravelle, A., Some Results of Experimental and Analytical Buffeting Investigations on A Delta Wing, International Symposium on Aeroelasticity, Aachen, April 1-3, 1985.
3. Zingel, H., Zur Bestimmung des aeroelastischen Verhaltens von Auftriebsystemen infolge von Stroemungsablosung, DFVLR, Goettingen, August 1986.
4. Mabey, D.G., Pressure Fluctuations Caused by Separated Bubble Flows at Subsonic Speeds, RAE TR 71160, August 1971.
5. Djojodihardjo,H., Sekar, W.K.,and Laschka, B., Calculation of Pressure Distribution on Two Dimensional Wing in Unsteady Subsonic Flow with Separation, International Forum on Aeroelasticity and Structural Dynamics, Aachen, Federal Republic of Germany, 1991.
6. Djojodihardjo, H., Singularity Methods in Unsteady Subsonic Aerodynamics with Separation, Franco-SEAMS Conference on The Application of Modern Mathematics in Engineering-related Problems, Institut Teknologi Bandung, June 22-24, 1992, appeared at Proceedings Institut Teknologi Bandung, Vol.25, No.1, 1993, Supplement.
7. Djojodihardjo,H., Sekar, W.K., and Prananta, B.B., Calculation of Three-Dimensional Unsteady Subsonic Flow with Separation Bubble using Singularity

Method, 18th Congress of the International Council of the Aeronautical Sciences (ICAS), Beijing, 1992
8. Djojodihardjo, H., and Sekar, W.K., Flutter Calculation of Two- and Three-Dimensional Subsonic Wing with Separation Bubble as Preliminary Estimation of Buffeting Problem, Forum International Aeroelasticite et Dynamique de Structures, Strassbourg, France, 1993.
9. Djojodihardjo, H., Sekar, W.K., and Sulaeman, E., Flutter and Dynamic Response Calculation of Unsteady Separated Flow Using Singularity Method for Preliminary Estimation of Buffeting Problem, Symposium on the Advances in Aerospace Sciences, in honour of Professor Holt Ashley, on the occasion of his 70th Birthday Stanford, California, 1993.
10. Laschka, B., Zur Theorie der Harminisch-schwingenden Tragenden Flaeche bei Unterschallanstroemung, Zeitschrift fuer Flugwissenschaften, Braunschweig, 1963
11. Bisplinghoff,R.L.,Ashley H. and Halfman R.L., Aeroelasticity, Addison Wesley Publishing Company, Reading, Massachusetts, 1955.
12. Kuessner, H.G., Algemeine Tragflaechentheorie, Luftfahrtforschung, No.17, 1940.
13. Haryanto,I., A Preliminary Study on Buffeting Problem Utilizing Dynamic Response Approach, 19th Congress of the International Council of the Aeronautical Sciences (ICAS),Anaheim, California, 1994.
14. Chyu, W.J., and Au-Yang, M.K., Random Response of Rectangular Panels to the Pressure Field Beneath A Turbulent Boundary Layer in Subsonic Flows, NASA TN D-6970, October, 1972.
15. Bendat, J.S. and Piersol, A.G., Random Data: Analysis and Measurement Procedure, John Wiley and Sons, Inc., Canada, 1971.
16. Bleviss, R.D., Flow-Induced Vibration, van Nostrand Reinhold Co., New York, 1977.
17. Boyden, R.P., and Johnson Jr.,W.G., Results of Buffet Tests in a Cryogenic Wind-Tunnel, NASA TM 84520, September, 1982.
18. Sulaeman,E. and Djojodihardjo, H., An Expansion Series for Solving Improper Cylindrical Functions Occuring in Subsonic, Supersonic and Transonic Kernel Formulations, in Romesh C.Batra(Ed), Contemporary Research in Engineering Science, in honour of Professor Satya N.Atluri, recipient of C.Eringen Medal, Springer,Berlin, 1995.
19. Djojodihardjo, H., Sulaeman,E., Aquino, T. and Sekar, W.K., A Discrete Element Approach For Oscillating Wing With Complicated Patterns Of Separated Flow Regions, presented at the 20th Congress of the International Council of the Aeronautical Sciences (ICAS), Sorrento, Italy, 1996

THE INFLUENCE OF AN EXTERNAL FORCE ON A SOLITARY WAVE

ROGER GRIMSHAW
Department of Mathematics, Monash University, Clayton, Victoria 3168, Australia.

Abstract. Using the forced Korteweg-de Vries equation as a generic model, we investigate the interaction of a solitary wave with an isolated external force of small amplitude, using both asymptotic analysis and numerical simulations. The analysis leads to a reduced dynamical system for the solitary wave amplitude and position, and this is analysed in detail. The theory predicts that the main regimes are passage, repulsion or trapping. These theoretical predictions are confirmed in our numerical simulations of the forced Korteweg-de Vries equation. Further, the effects of friction or variable force speed on these phenomoma will also be mentioned briefly.

1. Introduction

The forced Korteweg-de Vries (fKdV) equation is a canonical model for the description of the resonant generation and interaction of nonlinear waves with an external force in a weakly dispersive media (see, e.g. Grimshaw, 1990 and Grimshaw et al, 1994). The results from asymptotic approximations and numerical simulations show interesting examples of the nonlinear dynamics of solitary waves in the field of an external force. For instance, a solitary wave can be captured by the external force, either with or without oscillations near the forcing centre. In other circumstances, the solitary wave can pass through the force with a locally observable modulation in amplitude but often little permanent change. Again, the solitary wave can be reflected with a significant amplitude change. In this paper we will review these results, taking into account also situations when the external force has a variable speed, or when frictional processes maybe important.

For simplicity we will consider here the canonical form of the forced Korteweg-de Vries equation. Thus we consider

$$u_t + \Delta(t)\, u_x + \alpha\, u\, u_x + \beta\, u_{xxx} + \nu\, T\{u\} = f_x(x), \tag{1.1}$$

Here $\Delta(t)$ is the speed deviation between the force and an intrinsic linear long wave speed of the underlying physical system and is an important parameter. It may depend explicitly on the time t; situations where this may occur are described in Grimshaw et al (1996). $T\{u\}$ is an integro-differential operator describing dissipation; examples of where such a form of dissipation may arise are discussed by Grimshaw et al (1997). In the case of linear dissipation the operator T is given by (Grimshaw, 1983)

$$T\{u\} = \frac{1}{2\pi} \int_{-\infty}^{\infty} g(k)\, \hat{u}(k,t) \exp(ikx)\, dk, \tag{1.2}$$

and $\hat{u}(k,t)$ is the spatial Fourier spectrum of the wave field

$$u(k,t) = \int_{-\infty}^{\infty} u(x,t) \exp(-ikx)\, dx, \tag{1.3}$$

and $\nu g(k)$ is the decay rate of a linear wave. For instance, taking into account dissipation due to turbulence, we have $g(k) = k^2$ where here, and in the sequel, we shall assume that the dissipation coefficient $\nu > 0$. When the dissipation in the fluid is principally due to frictional drag at a rigid boundary, $g(k) = (-ik)^{1/2}$. More generally, we shall suppose that

$$g(k) \sim \mid k \mid^m, \tag{1.4}$$

with any positive m.

Here we will study the interaction between a single solitary wave and an isolated external force. We will assume that the external force has a small amplitude and that $\Delta(t)$ varies only slowly if at all, so that we can apply asymptotic methods. These methods are now well understood (see, for example, Grimshaw and Mitsudera, 1993). This asymptotic approach is described in Section 2 where we show that the result is a pair of ordinary differential equations describing the amplitude and position of the solitary wave. In Section 3 we obtain various solutions of this reduced system, exploiting various further approximations. These solutions show that there are three main regimes. These are *trapping*, when the solitary wave remains partially or totally trapped at the location of the external force; *repulsion*, when the solitary wave is reflected by the external force with a significant amplitude change; and *passage*, when the solitary wave passes through the external force with some amplitude modulation. Then in Section 4 we describe the results of some direct numerical simulations of the fKdV equation. The quantitative comparison between these numerical simulations and the asymptotic theory is good.

2. Asymptotic Model

Here we are concerned with the situation when the external force has a small amplitude, the friction coefficient ν is small, and the speed deviation Δ is slowly varying, so that f, ν and $d\Delta/dt$ are $0(\epsilon)$, where $0 < \epsilon << 1$, and the initial form of the wave is a solitary wave (i.e. a steady - state solution of (1.1) when $f = 0$ and $\nu = 0$). In this case we can use a perturbation analysis to describe the interaction between a given solitary wave and the weak external force. Following Grimshaw and Mitsudera (1993) and Grimshaw et al (1994) we seek the leading order solution in the following form, representing locally a solitary wave with parameters which are slowly varying functions of time. Thus

$$u_0 = a(T)\,\mathrm{sech}^2\{\gamma(T)\,\Phi\}, \tag{2.1a}$$

where

$$\Phi = x - \Psi(T), \tag{2.1b}$$

$$\Psi(T) = x_0 + \frac{1}{\epsilon}\int_0^T q(T')\,dT' \tag{2.1c}$$

and

$$T = \epsilon\, t. \tag{2.1d}$$

Then we seek an asymptotic expansion of the form

$$u = u_0 + \epsilon\, u_1 + \epsilon^2 u_2 + \ldots, \tag{2.2a}$$

$$q = q_0 + \epsilon\, q_1 + \epsilon^2 q_2 + \ldots. \tag{2.2b}$$

Substitution of these expressions into (1.1) confirms that u_0 is given by (2.1a-d) provided that

$$\alpha\, a = 12\,\beta\,\gamma^2 \tag{2.3a}$$

and

$$q_0 = \Delta(T) + \frac{1}{3}\alpha\, a \tag{2.3b}$$

In the next order of approximation (see Grimshaw et al, 1994) we obtain the equations for the solitary wave amplitude and position,

$$\frac{d}{dt}\int_{-\infty}^{\infty}\frac{u_0^2}{2}\,d\Phi = \int_{-\infty}^{\infty} u_0\,\frac{\partial f}{\partial x}\,d\Phi - \nu\int_{-\infty}^{\infty} u_0\,T\{u_0\}\,d\Phi, \tag{2.4a}$$

and

$$\frac{d\Psi}{dt} = \Delta + \frac{1}{3}\alpha a + \epsilon q_1. \tag{2.4b}$$

Here (2.4a) describes the leading order energy balance, and we recall that f and ν are $0(\epsilon)$. The speed correction q_1 is determined at second order, and is given by the next order energy balance,

$$\epsilon q_1 = \frac{\alpha}{24\beta\gamma^2}\int_{-\infty}^{\infty}(\tanh\gamma\Phi - \operatorname{sech}^2\gamma\Phi + \gamma\Phi\operatorname{sech}^2\gamma\Phi)\,(\frac{\partial f}{\partial\Phi} - \nu T\{u\})\,d\Phi, \tag{2.5}$$

In the general case the integrals in (2.4a) and (2.5) cannot be explicitly calculated. To demonstrate the main features of the interaction between the solitary wave and the external force we will focus on the case with broad forcing. Here the solitary wave acts like a delta-function and (2.4a) becomes

$$\frac{da}{dt} = 2\frac{\partial f}{\partial\Psi}(\Psi) - p_n a^n, \tag{2.6}$$

while the speed correction (2.5) becomes

$$\epsilon q_1 = -\frac{f(\Psi)}{a} + q_{n_1} a^{n_1}, \tag{2.7}$$

where p_n and q_n are positive constants and the parameters n and n_1 are given by the following expressions for linear dissipation in the form of (1.4),

$$n = 1 + \frac{m}{2}, \qquad n_1 = \frac{m - 1}{2}, \tag{2.8}$$

Equations (2.6) and (2.4b) with (2.7) are then used in the following sections to describe the interaction of a solitary wave with the external force.

3. Some solutions from the model

In order to keep the discussion focussed on the main results we shall consider only the simplest case when the speed deviation is a constant, and the friction term is omitted. The situation when these simplifying assumptions are not made are discused in detail in Grimshaw et al (1996) and Grimshaw et al (1997) respectively. Further for simplicity we initially omit the speed correction term (2.5); the case when this is included is discussed briefly below and in more detail in Grimshaw et al (1994). Thus the asymptotic model (i.e. (2.6) and (2.4b)) reduces to

$$\frac{da}{dt} = 2\frac{df}{d\Psi}(\Psi), \tag{3.1a}$$

$$\frac{d\Psi}{dt} = \Delta + \frac{1}{3}\alpha a \tag{3.1b}$$

This system was obtained by Warn and Brasnett (1983) and discussed in detail by Grimshaw et al (1994).

It is an integrable Hamiltonian system with Hamiltonian

$$H = \Delta a + \frac{1}{6}\alpha a^2 - 2f(\Psi), \tag{3.2}$$

Since H is an invariant, the orbits are defined by (3.2). To be explicit we consider the case when the external force is symmetric and isolated, and given by the representative function

$$f(x) = b \operatorname{sech}^2(x/l), \tag{3.3}$$

We then put

$$\Theta = \frac{\Psi}{l}, \quad \Gamma = \frac{a}{a_0},$$

$$\alpha a_0 = 3|\Delta|, \quad Q = \frac{4\alpha b}{3\Delta^2}. \tag{3.4}$$

With $\beta > 0$ it follws from (2.3a) that $\alpha a > 0$ and so also is $\Gamma > 0$ where Γ is the non-dimensional measure of the solitary wave amplitude. Also Q is the non-dimensional measure of the forcing amplitude where $Q > 0 (< 0)$ corresponds to a positive (negative) force. With the substitutions (3.4), (3.2) becomes

$$Q \operatorname{sech}^2\theta = [\operatorname{sign}(\Delta) + \Gamma]^2 + C. \tag{3.5}$$

Here the constant C is proportional to the Hamiltonian H.

The phase portraits give by (3.5) depend on the signs of of Q and Δ. First consider the case $\Delta > 0$ which is shown in Figures 1 and 2 for for $Q > 0$ and < 0 respectively. The main regime here is the *passage* of a fast-moving solitary wave through the external force. Note that here and elsewhere the behaviour as a becomes very small is not correctly predicted by the simplified model (3.1) since it is apparent from (2.7) that the omission of the speed correction term (2.5) is not valid in this case.

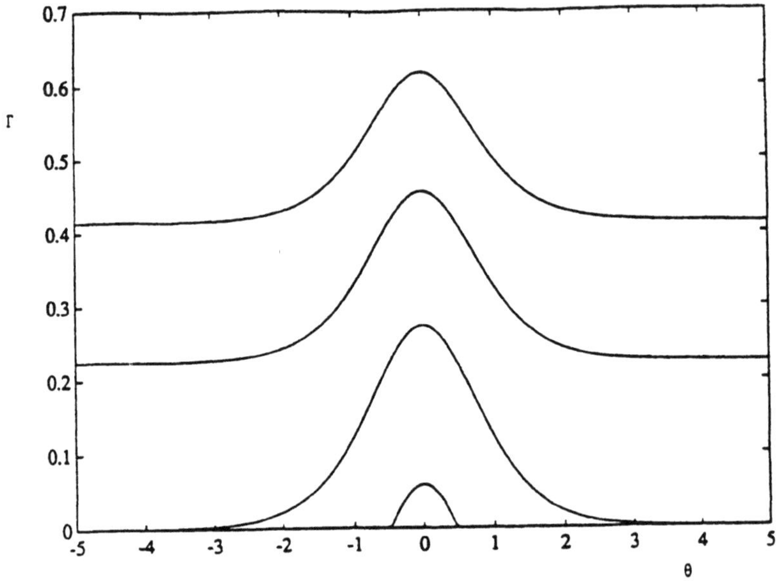

Figure 1. Phase portrait of system (3.1) based on the integral (3.5), for $\Delta = 0.36$ and $Q = 0.62$.

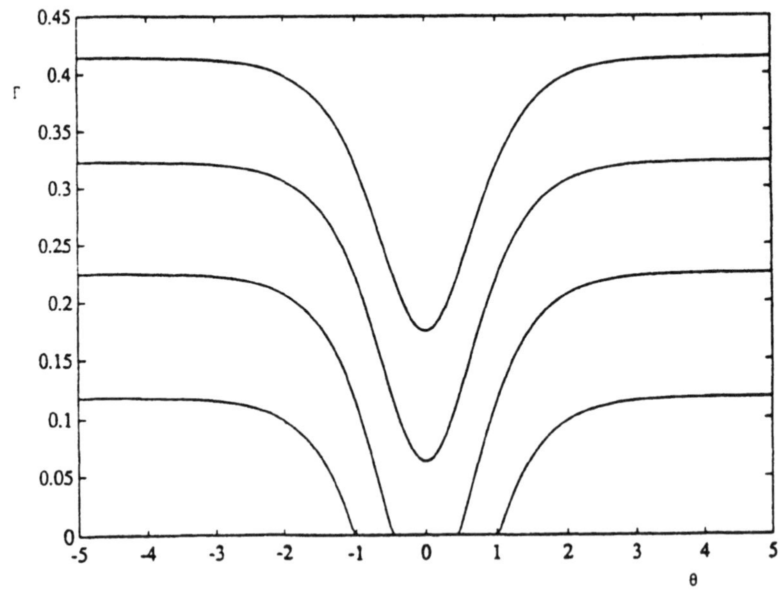

Figure 2. As for Fig. 1, but for $\Delta = 0.36$ and $Q = -0.02$.

Next we consider the opposite case $\Delta < 0$. Since it is now possible for the solitary wave speed to become zero or negative the dynamics is much richer. First, we note that there is an equilibrium state

$$\Gamma = 1 \quad \theta = 0. \tag{3.6}$$

This corresponds to a solitary wave captured by the external force and located precisely at the forcing location. The nature of the equilibrium state depends on the sign of Q. For $Q > 0$ we have a centre, and for $Q < 0$ we have a saddle point. The former case leads to a regime in which the solitary wave oscillates about the external force, which we call *trapping*, while the latter case leads to a regime in which the solitary waves reflect away from the equilibrium point and which we shall call *repulsion*. Typical phase portraits for each case are shown in Figures 3 and 4 respectively. Much more detail is contained in the work by Grimshaw et al (1994).

Next we include the speed correction term q_1 (i.e. (2.5)) so that the asymptotic model becomes

$$\frac{da}{dt} = 2\frac{df}{d\Psi}(\Psi), \tag{3.7a}$$

$$\frac{d\Psi}{dt} = \Delta + \frac{1}{3}\alpha a - \frac{f(\Psi)}{a}. \tag{3.7b}$$

Here of course (3.7a) is identical to (3.1a), but the inclusion of the speed correction term q_1 makes (3.7b) different from (3.1b). In general this term induces a dissipative aspect to the asymptotic model since this contains a term due to radiation emitted by the solitary wave. However, this term vanishes in the limit of broad forcing and hence the system (3.7) is also Hamiltonian with Hamiltonian

$$H = \frac{2}{3}\Delta\frac{a^2}{\gamma} + \frac{2}{15}\frac{\alpha a^3}{\gamma} - \frac{2a}{\gamma}f(\Psi). \tag{3.8}$$

Since H is an invariant, the orbits are defined by (3.8). Using the external force (3.3) and the nondimensionalisation (3.4) the orbits are given by

$$\frac{3}{4}Q\,\mathrm{sech}^2\theta = \mathrm{sign}(\Delta)\,\Gamma + \frac{3}{5}\Gamma^2 + C\,\Gamma^{-1/2}. \tag{3.9}$$

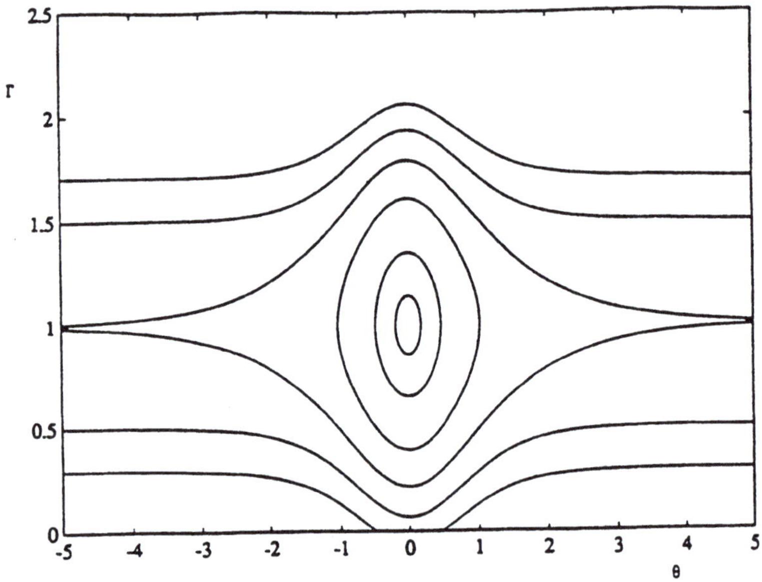

Figure 3. As for Fig. 1, but for $\Delta = -0.36$ and $Q = 0.62$

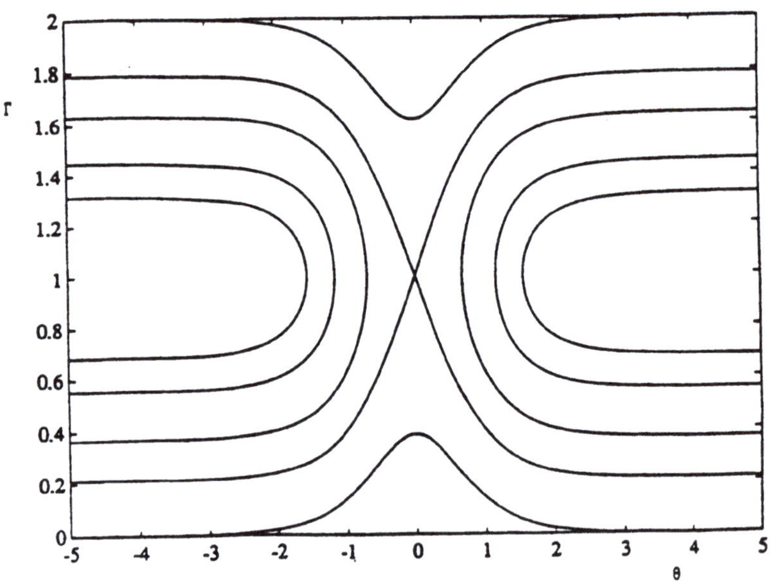

Figure 4. As for Fig. 1, but for $\Delta = -0.36$ and $Q = -0.62$

Here the constant C is proportional to H in (3.8). The main qualitative difference between the orbits defined by (3.9) and those discussed above and defined by (3.5) is for small amplitudes Γ. In particular we note that the equilibrium states are now given by

$$\operatorname{sign}(\Delta) + \Gamma - \frac{Q}{4\Gamma} = 0 \qquad \theta = 0. \tag{3.10}$$

We see that the effect of the speed correction term is to allow for the existence of an extra equilibrium state of small amplitude. Indeed, for $Q > 0$ there is now an equilibrium state which is a centre for all Δ, while for $Q < 0$ there are no solutions if $Q < -1$, and two solutions if $-1 < Q < 0$ where the larger solution is a saddle point and the smaller solution is a centre. In Figure 5 we plot the solitary wave amplitude as a function of Δ, and note the resemblance to the classical result for the forced periodic states of a nonlinear oscillator.

The phase portraits are shown in Grimshaw et al (1994) and the main regimes are again *passage*, *trapping* and *repulsion*. Further, when the assumption of broad forcing is relaxed, the speed correction term induces a dissipative effect and the asymptotic system is no longer Hamiltonian. In particular a centre become a stable or an unstable focus. In the latter case the trapping regime is only one of partial trapping.

4. Numerical solutions

In order to confirm the asymptotic theory we carried out a direct numerical simulation of the forced Korteweg-de Vries equation (1.1) for the main case of interest here when Δ is a constant and there is no dissipative term (i.e. $T\{u\} = 0$). Without loss of generality we put $\alpha = 6$ and $\beta = 1$. The numerical scheme was a pseudo-spectral scheme, with periodic boundary conditions. The external force was given by

$$f(x) = b \exp[-(x - x_0)^2 / l^2]. \tag{5.1}$$

Although this differs from the force (3.3) used in section 3, the phase portraits are very similar. The initial condition for the numerical simulations was a solitary wave (2.1a) located at the position x_{in} with amplitude a_{in}. We carried out numerical simulations for a wide range of parameters and found very good agreement with the asymptotic theory. Only a few representative cases are shown here.

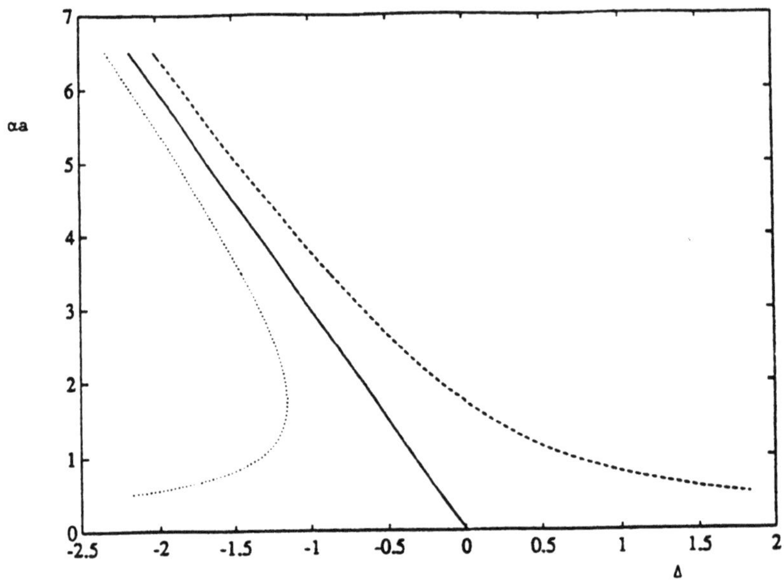

Figure 5. The equilibrium states for the solitary wave amplitude (actually $\alpha\ a$) as a function of $\Delta.(-)$ for $b > 0$ and $(\cdots)$ for $b < 0$, while the solid line is for $b = 0$.

First, in Figure 6 we show a case with $\Delta < 0$, which we see is a case of *partial trapping*. This case corresponds to the asymptotic model shown in Figure 3. There is very good agreement, after the effect of the speed correction term q_1 is taken into account, which converts the centre shown in Figure 3 to an unstable focus. Next, again with $\Delta < 0$, we show a case in Figure 7 of a *passage* regime. Again, this can be compared with the passage regime shown in Figure 3.

Then, in Figure 8, still with $\Delta < 0$, we show a case of *repulsion* in which a small-amplitude solitary wave is reflected from the external force as a large-amplitude solitary wave. This case corresponds very well to the asymptotic result shown in Figure 4. Finally, in Figure 9, again with $\Delta < 0$, we show another case of *repulsion* in which now a large-amplitude solitary wave is reflected from the external force as a small-amplitude solitary wave. This case corresponds also to the asymptotic result shown in Figure 4. Note that the contrast between these two cases is obtained by starting the solitary wave on opposite sides of the external force.

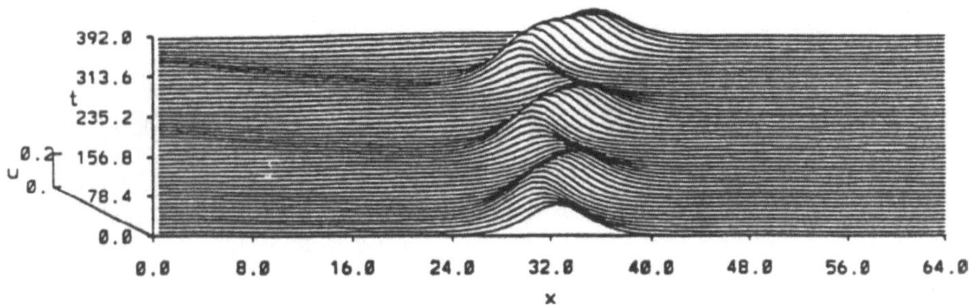

Figure 6. The numerical solution of the forced Korteweg-de Vries equation (1.1) for the external force (5.1) and $\Delta = -0.36, b = 0.01, l = 2.5, a_{in} = 0.185, x_0 = 31.5$ and $x_{in} = 32$.

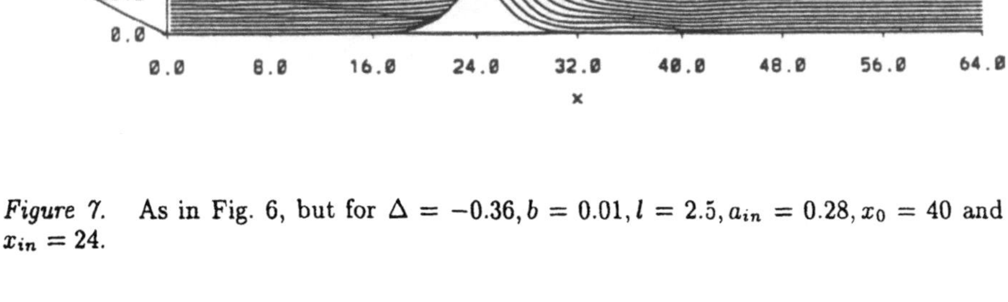

Figure 7. As in Fig. 6, but for $\Delta = -0.36, b = 0.01, l = 2.5, a_{in} = 0.28, x_0 = 40$ and $x_{in} = 24$.

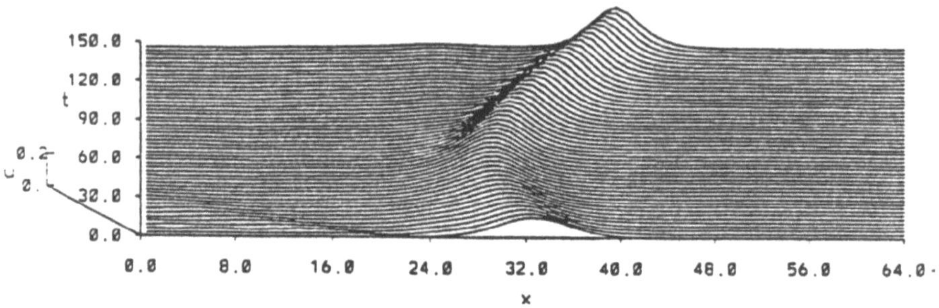

Figure 8. As in Fig. 6, but for $\Delta = -0.36, b = -0.01, l = 2.5, a_{in} = 0.125, x_0 = 24$ and $x_{in} = 32$.

Figure 9. As in Fig. 6, but for $\Delta = -0.36, b = -0.01, l = 2.5, a_{in} = 0.25, x_0 = 32$ and $x_{in} = 24$.

References

1. Grimshaw, R. 1983. Solitary waves in density stratified fluids. In "Nonlinear Deformation Waves", IUTAM Symposium, Tallinn, 1982, ed. U. Nigul and J. Engelbrecht, Springer, 431-447.
2. Grimshaw, R. 1990. Resonant flow over topography. In "Nonlinear Evolution Equations and Dynamical Systems", Proceedings of Fifth Workshop on Nonlinear Evolution Equations and Dynamical Systems, Kolymbari, Crete, 1989, ed. S.Carillo and O.Ragnisco, Research Reports in Physics, Springer, 209-211.
3. Grimshaw, R. and H. Mitsudera. 1993. Slowly-varying solitary wave solutions of the perturbed Korteweg - de Vries equation revisited. Stud. Appl. Math. 90, 75 - 86.
4. Grimshaw, R., E. Pelinovsky, and X. Tian. 1994. Interaction of a solitary wave with an external force. Physica D 77, 405 - 433.
5. Grimshaw, R., Pelinovsky, E. and Sakov, P. 1996. Interaction of a solitary wave with an external force moving with variable speed. Stud. Appl. Math., 97, 235-276.
6. Grimshaw, R., Pelinovsky, E. and Bezen, A. 1997. Hysterisis phenomena in the interaction of a damped solitary wave with an external force. Wave Motion, (to appear).
7. Warn T. and Brasnett B., 1983. The amplification and capture of atmospheric solitons by topography: a theory of the onset of regional blocking. J. Atmos. Sci. 1983, 40, 28-38.

AN ASYMPTOTIC METHOD IN SHIP HYDRODYNAMICS

A.J. HERMANS
Department of Applied Mathematics and Computer Science,
TU Delft, Mekelweg 4, 2628 CD Delft, The Netherlands

Abstract. In the field of diffraction of acoustic, electro-magnetic and elastic waves the ray method has proven to be a powerful tool to obtain reliable results for the behaviour of short waves. In ship hydrodynamics where the generation of short waves by a slowly sailing vessel and the diffraction of short surface waves by these vessels is of importance recently the ray method is applied successfully. In this presentation we present two cases, each with its own free surface condition, treated with this method.

In the first part we derive the asymptotic results for the wave pattern generated by a ship with a blunt bow and stern (for instance VLCC's). The dimensionless forward velocity, $Fn = U/(gL)^{1/2}$, of these ships is small. After an appropriate linearisation of the free surface condition is carried out, approximation for large wave numbers can be obtained. We derive equations for the ray pattern and for the wave ampitude. The wave excitation coefficients are obtained by the asymptotic evaluation of an approximate integral formulation. Results for the wave resistance are obtained by a ray tracing algorithm.

In the second part we derive an asymptotic theory for the diffraction of free surface waves by a very large floating flexible platform (floating airport) and the deflection of this platform. In the case of total reflection the evanescent mode will be shown. Also the deflection due to a landing airplane and its take off is considered.

1. Introduction

In naval hydrodynamics a well known short wave phenomena can be observed in the case of a blunt ship sailing at low speed, or a duck swimming

at the surface of a pond. The wave pattern is stationary with respect to the moving body. The resistance acting on the body due to this generated wave pattern can be a substantial part of the total resistance. It may be of the same order of magnitude as the viscous resistance. In 1898 Michell already derived a formula for the wave resistance for thin ships. His formulae are a good approximation except in the low speed range. Generally his result becomes unrealible if the Froude number becomes small. For the very large crude carriers (VLCC's) the Froude number is small and the predictions are not accurate. Compared to the length scales the waves are short. In the next section we derive a proper free surface condition and apply short wave approximations to obtain the wave resistance.
In section 3 the short wave theory is applied to determine the deflection of a floating platfor.
The method we use is the same for both examples. The two problems have their own free surface condition that lead to specific difficulties in the application of the problem. The method we use is well known in for instance the diffraction of short acoustic waves, therefore we recapitulate the main features of the method if it is applied to the Helmholtz equation in three dimensions,

$$\Delta u + k^2 n^2(\vec{x}) u = 0,$$

for large values of the wave number k. This equation describes the harmonic solutions $v(\vec{x}, t) = u(\vec{x}) e^{-i\omega t}$ of the wave equation for an inhomogeneous medium.

$$\Delta v - \frac{1}{c^2(\vec{x})} v_{tt} = 0$$

with the wave number $k = \frac{\omega}{c_0}$, where $c_0$is a constant reference speed and the index of refraction defined as $n(\vec{x}) = \frac{c_0}{c(\vec{x})}$.

In anology with the plane wave solution (for constant n)

$$u(\vec{x}) = z(\vec{k}) e^{in\vec{k}\cdot\vec{x}}$$

we seek, for large values of the propagation vector $\vec{k}$ with length k, solutions of the form

$$u(\vec{x}) = z(\vec{x}, k) e^{iks(\vec{x})}$$

If we insert this in the Helmholz equation we obtain

$$-k^2[\nabla s \cdot \nabla s - n^2] z + 2ik \nabla s \cdot \nabla z + ikz \Delta s + \Delta z = 0 \tag{1}$$

For large values of k we assume that $z(\vec{x}, k)$ can be expanded in inverse powers of k. It is convenient to write the expansion in terms of (ik) in the form

$$z(\vec{x},k) \sim \sum_{m=0}^{n} z_m(\vec{x})(ik)^{-m} + o(k^{-n}).$$

From equation (1) it follows that the coefficient of each power of k must be zero. We then obtain

$$[\nabla s \cdot \nabla s - n^2]z_0 = 0, \tag{2}$$

Because $z_0 \neq 0$ this leads to the eiconal equation for s

$$\nabla s \cdot \nabla s = n^2(\vec{x}) \tag{3}$$

and for $m = 0, 1, 2, \cdots$, the vanishing of the coefficients implies

$$2\nabla s \cdot \nabla z_0 + z_0 \Delta s = 0 \tag{4}$$

and for $m = 1, 2, \cdots$

$$2\nabla s \cdot \nabla z_m + z_m \Delta s = -\Delta z_{m-1}. \tag{5}$$

These equations are called the transport equations. In the case of acoustic diffraction one has to prescribe appropriate boundary conditions for the phase function $s(\vec{x})$ and the amplitude functions $z_m(\vec{x})$. For the diffraction of short waves by for instance a smooth interface, that is illuminated completely, these conditions follow from the exact boundary conditions. For the diffraction at an illuminated corner the excitation coefficients are determined by solving a canonical problem exactly.
In our first example in the field of water waves determination of excitation coefficients are of importance. In the second one an evenescent mode becomes important.

2. Nonlinear ship waves at low Froude number

It is well known in slow-ship theory that the usual linearized free-surface condition cannot be applied when obstacles are considered of which the beam to length ratio is not sufficiently small. In the free surface condition several terms that at neglected in the usual slender case have considerable effect. In the region of low fluid velocities extra terms have to be taken into account. In this section we first derive the appropriate free surface condition, then we consider the near field to construct the wave excitation mechanism and then we apply the ray method in the far field to describe the wave pattern and its amplitude. Finally we give some results for the wave resistance.

2.1. NEAR-FIELD FORMULATION

We consider a fluid that is nonviscous and incompressible with a constant density, while at the free surface we have the atmospheric pressure. The coordinate system is fixed to the ship (see Figure 1) and the incoming velocity field is $(U, 0, 0)$. We introduce the velocity potential $\Phi(\vec{x})$

$$\vec{V}(\vec{x}) = \nabla\Phi(\vec{x}) \tag{6}$$

which obeys the law of continuity, hence

$$\Delta\Phi = 0 \qquad \forall\vec{x} \text{ outside the ship and } z < \overline{\zeta}(x, y) \tag{7}$$

On the free surface $z = \overline{\zeta}(x, y)$ we have both the kinematic and the dynamic condition

$$\begin{aligned} 0 &= \Phi_x\overline{\zeta}_x + \Phi_y\overline{\zeta}_y - \Phi_z \\ \frac{1}{2}U^2 &= g\overline{\zeta} + \frac{1}{2}(\Phi_x^2 + \Phi_y^2 + \Phi_z^2) \end{aligned} \tag{8}$$

On the hull of the ship, the kinematic condition holds

$$\nabla\Phi \cdot n = 0 \qquad \text{on} \quad y = f(x, z) \tag{9}$$

In our example the bow and the stern are wedge-shaped with finite angles and the Froude number $Fn = U/(gL)^{1/2}$, where L denotes the length of the ship, is small. A formal low Froude number expansion can be made. However, it can be shown that no solutions with a wavy character downsteram can be found. Nevertheless, the lowest-order term leads to a good approximation of the fluid velocity near the hull. This so-called double body solution Φ_r is used as a first step to construct the wave solution.

In the layer near the free surface we assume that the potential has the form

$$\Phi(\vec{x}) = \Phi_r(\vec{x}) + \phi(\vec{x}) \tag{10}$$

The wave potential ϕ is a solution of the Laplace equation for $z < \overline{\zeta}(x, y) = \tilde{\zeta}_r(x, y) + \zeta(x, y)$, where $\zeta_r(x, y)$ is the surface elevation due to the double body potential

$$\zeta_r(x, y) = \frac{1}{2g}[U^2 - \Phi_{rx}^2(x, y, 0) - \Phi_{ry}^2(x, y, 0)] \tag{11}$$

We introduce a new coordinate system $(x, y, z) := (x, y, z - \zeta_r(x, y))$ and the first approximation to the potential function $\phi(\vec{x})$ is, for small values of Fn, a solution of the Laplace equation. The free surface condition becomes

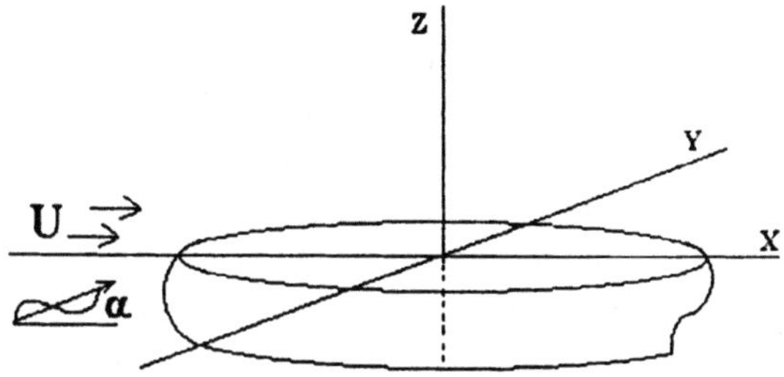

Figure 1. Coordinate system.

$$\phi_y + \frac{1}{g}[\Phi_{rx}^2\phi_{xx} + 2\Phi_{rx}\Phi_{ry}\phi_{xy} + \Phi_{ry}^2\phi_{yy}]$$
$$+ (3\Phi_{rx}\Phi_{rxx} + 2\Phi_{ry}\Phi_{rxy} + \Phi_{rx}\Phi_{ryy})\phi_x +$$
$$+ (3\Phi_{ry}\Phi_{ryy} + 2\Phi_{rx}\Phi_{rxy} + \Phi_{ry}\Phi_{rxx})\phi_y = \mathcal{D}(x,y) \qquad \text{at } z = 0 \quad (12)$$

where the function $\mathcal{D}(x,y)$ only depends on Φ_r

$$\mathcal{D}(x,y) = \frac{\partial}{\partial x}[\Phi_{rx}(x,y,0)\zeta_r] + \frac{\partial}{\partial y}[\Phi_{ry}(x,y,0)\zeta_r] \qquad (13)$$

The wave potential does not fulfill the condition on the hull. A second order outer solution is needed to correct this error of lower order. The function $\mathcal{D}(x,y)$ can be seen as a pressure distribution over the plane $z = 0$. Consequently the wave potential can be expressed as a source distribution over the free surface

$$c\phi(\vec{x}) = \int_{S_f}\int \mathcal{D}(\xi,\eta)G(\vec{x};\xi,\eta,0)\ \mathrm{d}\xi\ \mathrm{d}\eta \qquad (14)$$

with

$$c = \lim_{\epsilon \to 0} \int_{\partial B_\epsilon \cap D} \frac{\partial G}{\partial n_\epsilon}\ \mathrm{d}s_\epsilon$$

where ∂B_ϵ denotes the boundary of a ball of radius ϵ and having the center at $\vec{x}$ and D denotes the fluid domain.

The source potential is a solution of the homogeneous problem and behaves like

$$G(x,y,z;\xi,\eta,\zeta) \approx \frac{1}{[(x-\xi)^2 + (y-\eta)^2 + (z-\zeta)^2]^{\frac{1}{2}}} \qquad (15)$$

It will be clear that the source function cannot be constructed exectly. By means of a multiple scale approach an approximate solution can be obtained near the ship hull. We cosider the case where the wave number $k = g/U^2$ is large. The length scale for the variations of the waves is $1/k$. Therefore we introduce a new set of independent variables $\tilde{\vec{x}} = k\vec{x}$ and consider the potential to be a function of $\tilde{\vec{x}}$ and $\vec{x}$ independently. The solution obtained this way is valid only in the vicinity of the ship. At finite distance it becomes invalid due to secular terms in the expansion. This is in accordance with the Poincaré-Lighthill-Kuo (PLK) method. After some manipulations as can be found in Brandsma [4] and Baba [1, 2] we obtain for the wave elevation

$$\tilde{\zeta}(x,y) = \Re \frac{-k}{2\pi U} \int_{s_f}\int \mathcal{D}(\xi,\eta) \int_{\beta-\pi/2}^{\beta+\pi/2} \frac{\exp(ikk_0\omega)}{q_3\cos^3(\theta-\beta)}\, \mathrm{d}\xi\, \mathrm{d}\eta\, \mathrm{d}\theta \tag{16}$$

where k_0 and ω are given by

$$k_0(x,y;\theta) = [q(x,y)\cos(\theta-\beta(x,y))]^{-2}$$

and

$$\omega(x,y;\xi,\eta;\theta) = (x-\xi)\cos\theta + (y-\eta)\sin\theta$$

where q is the local double body velocity and β is its angle with respect to the x-axis. This is the classical result as can be found in Wehausen and Laitone if one considers the undisturbed flow instead of the double body flow. for large values of k the leading term in the integral becomes an integral along the waterline C denoted by $(x_0(s), y_0(s))$.

$$\tilde{\zeta}(x,y) \approx \Re \frac{1}{2\pi i U} \int_C \int_{\beta-\pi/2}^{\beta+\pi/2} A \exp(ikS)\, \mathrm{d}\theta\, \mathrm{d}s \tag{17}$$

with

$$A(x,y;\theta,s) = \frac{(n_x\cos\theta + n_y\sin\theta)\mathcal{D}(x_0(s),y_0(s))}{q\cos(\theta-\beta)}$$

and

$$S(x,y;\theta,s) = \frac{(x-x_0(s))\cos\theta + (y-y_0(s))\sin\theta}{q^2\cos^2(\theta-\beta)}$$

For a specific hull form this expression can be evaluated further, by means of partial integration, to furnish the excitation coefficients at the bow and the stern.

2.2. FAR-FIELD FORMULATION

To compute the wave pattern and the wave resistance, the continuation of the local solution into the far field must be considered. A powerful tool is the ray method. For large wave number k, the ray method suggests the expansion

$$\phi(\vec{x}) = \Re\left[\left(\frac{U}{ik}a_0(\vec{x}) + \frac{U}{(ik)^2 L}a_1(\vec{x}) + \cdots\right)\exp\left(ikS(\vec{x})\right)\right] \tag{18}$$

where S is the phase function and the a_i's are dimensionless amplitude functions. (The length scale L can be chosen to be the length of the ship) We substitute this form in the Laplace equation and in (12) and take equal powers of k equal to zero. We obtain the following equations for S and a_0:

$$\begin{aligned} \nabla S \cdot \nabla S = 0 \qquad & \text{for } z < 0 \\ \frac{\partial S}{\partial z} + i(MS)^2 = 0 \qquad & \text{at } z = 0 \end{aligned} \tag{19}$$

and

$$2\nabla S \cdot \nabla a_0 + a_0 \Delta S = 0 \quad \text{for } z < 0 \tag{20}$$
$$-i\frac{\partial a_0}{\partial z} + 2(MS)(Ma_0) + \left(M^2 S - \nabla\overline{\zeta}_r \cdot \nabla S - \frac{\Phi_{rzz}}{U}MS\right)a_0 = 0 \quad \text{at } z = 0$$

where M is the differential operator

$$M = \frac{1}{U}\left(\Phi_{rx}(x,y,0)\frac{\partial}{\partial x} + \Phi_{ry}(x,y,0)\frac{\partial}{\partial y}\right)$$

and

$$\overline{\zeta}_r(x,y) = \frac{g}{U^2}\zeta_r(x,y) = \frac{1}{2U^2}(U^2 - \Phi_{rx}^2 - \Phi_{ry}^2)$$

If the z-derivative in (19) is eliminated, we obtain a nonlinear equation for S in the plane $z = 0$:

$$(MS)^4 - \left(\frac{\partial S}{\partial x}\right)^2 - \left(\frac{\partial S}{\partial y}\right)^2 = 0 \quad \text{at } z = 0 \tag{21}$$

which is called the eikonal equation. In order to solve this equation the method of characteristics is used.
With $\vec{p} = (p,q) = \nabla S$ and $\vec{u} = \frac{1}{U}(\Phi_{rx}, \Phi_{ry})$, equation (21) is written in the standard form

$$\mathcal{F}(x,y,S,p,q) := (\vec{u}\cdot\vec{p})^4 - (\vec{p}\cdot\vec{p}) = 0 \tag{22}$$

In this case $\mathcal{F}$ does not depend on S explicitly. The *Charpit-Lagrange* relations lead to the following system of equations which has to be solved numerically in order to find the characteristic curves.

$$\begin{aligned}
\frac{\mathrm{d}x}{\mathrm{d}\sigma} &= \mathcal{F}_p = 4(\vec{u}\cdot\vec{p})^3 u - 2p,\\
\frac{\mathrm{d}y}{\mathrm{d}\sigma} &= \mathcal{F}_q = 4(\vec{u}\cdot\vec{p})^3 v - 2q,\\
\frac{\mathrm{d}p}{\mathrm{d}\sigma} &= -\mathcal{F}_x = -4(\vec{u}\cdot\vec{p})^3(\vec{u}_x\cdot\vec{p}),\\
\frac{\mathrm{d}q}{\mathrm{d}\sigma} &= -\mathcal{F}_y = -4(\vec{u}\cdot\vec{p})^3(\vec{u}_x\cdot\vec{p}).
\end{aligned} \tag{23}$$

where σ is a coordinate along these characteristic curves (rays). The phase function can be obtained by solving the following ordinary differential equation along the rays.

$$\frac{\mathrm{d}S}{\mathrm{d}\sigma} = p\mathcal{F}_p + q\mathcal{F}_q = 2(p^2+q^2) \tag{24}$$

The amplitude function a_0 follows from (19) by eliminating $\partial a_0/\partial z$. If transform the resuling equation to the characteristic coordinates and use x as the independent variable along the ray we obtain

$$\frac{\mathrm{d}a_0}{\mathrm{d}x} = f(x)a_0 \tag{25}$$

with

$$f(x) = \frac{-\frac{1}{2}(MS)^{-2}\Delta S + M^2 S - \nabla\zeta_r\cdot\nabla S - \frac{\Phi_{rzz}}{U}MS}{2MS\Phi_{rx} - (MS)^{-2}S_x}$$

As soon as the characteristics are computed and the initial conditions are known the amplitude follows by means of a simple integration along these characteristics. The initial condition follow from the asymptotic evaluation as indicated in the section on the near-field. Special care is needed for the numerical evaluation of the second derivatives of the phase function S. It is also possible to write differential equations for the second derivatives of S along the characteristics and to integrate them numerically.
The ray patterns for a lens-shaped body with bow and stern angle equal $\beta_b = \pi/8$ are shown in figure (2).
[2ex] The wave resistance is computed with the help of the far field expansion.

$$\mathcal{R}_w - \frac{1}{2}\rho\pi U^2 \int_{-\pi/2}^{+\pi/2} |A(\theta)|^2 \cos^3\theta \,\mathrm{d}\theta \tag{26}$$

where $A(\theta)$ is the free wave spectrum in the far field behind the ship. This formulation of the wave resistance follows from conservation of momentum

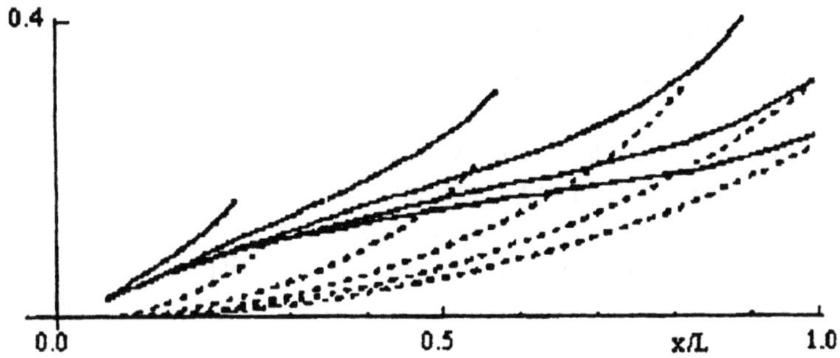

Figure 2. Some wave-fronts for the diverging bow-wave system $\beta_b = \pi/8$, ___ nonlinear results and - - - - linear results.

around the ship. Numerical results for the the wave resistance are shown in figure (3).

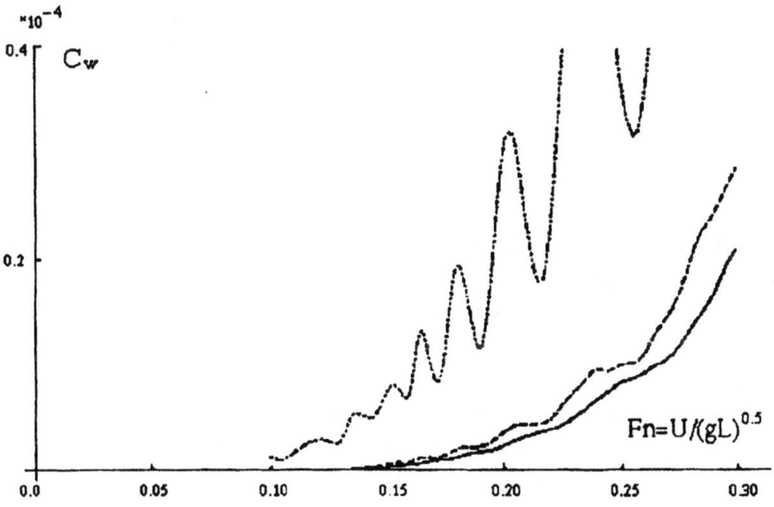

Figure 3. Comparison of $C_w = \frac{R_w}{\frac{1}{2}\rho U^2 L^2}$ for $\beta_b = 0.1974$, ······ Michell, - - - - Baba, ___ present theory.

3. A very large floating flexible platform in short waves.

At the eleventh workshop on water waves and floating bodies *Ohkusu* et.al. [8] described recent developments of the design of floating airports. These floating airports consist of a thin mat configuration of very large horizontal size. One must think of dimensions of about several kilometers by several hundred meters, while the thickness of the mat is several meters. For these configurations the natural bending rigidity is relatively small and the elastic deflection due to wave action will be dominant compared with the rigid body motions. This paper treats in principle the same problem as Ohkusu did but with a different mathematical method, a similar approach can be found in *Stoker* [12] for the motion of a floating elastic beam in shallow water. If one looks at the operational conditions of the airport it is expected that in general the deflections are generated by short waves, this means waves with a wavelength short with respect to the horizontal dimensions if the platform, but such that the thickness of the structure is small with respect to this wavelength. This motivates us to treat the mat as an infinite thin plate at the free-surface and to neglect its thickness. On the other hand if one wishes to study survival conditions, like an incident tsunamis wave, the analysis presented here brakes down because the wave length is larger or of the same size as the horizontal dimensions. The analyses here may serve as an analytic tool to predict the major effects at normal seastate conditions.

The problem has great resemblance with the deflection of a floating ice plate. There is a lot of literature about this topic, hence, some more information can be distracted from these sources. The papers of *Schulkes* et. al. [10],[11] and *Meylan* et.al. [6] are mentioned for further reference.

3.1. MATHEMATICAL FORMULATION

We consider the situation that the platform is positioned in an area where a tidal current is present and the incident waves are long crested. Here present the general frame work and we work out the formulation where the effect of a tidal currents is neglected. We assume that the paltform is oriented in the direction of the current, it is expected that this is the optimal situation with respect to minimal steady forces. The waves will be incident with a arbitrary angle of incidence. To keep the formulae simple we treat the case of infinite water depth, it will be clear from the analysis that the case of finite water depth is a straight forward extension. Viscous effects are neglected as well. In the fluid domain we introduce the velocity potential $\vec{V}(\vec{x},t) = \nabla\bar{\Phi}(\vec{x},t)$. We assume that the potential function can be split up in a steady and an unsteady part, where the steady part is approximated by Ux, hence we write

$$\bar{\Phi}(\vec{x},t) = Ux + \Phi(\vec{x},t)$$

The incident wave will be written as,

$$\Phi_{inc}(\vec{x},t) = e^{ik_0(x\cos\alpha + y\sin\alpha) + k_0 z - i\omega t}, \tag{27}$$

where $k_0 = \omega_0^2/g$ is the dispersion relation for deep water, α the angle of incidence and $\omega = \omega_0 + k_0 U \cos\alpha$ is the relative frequency, where the second term takes care of the Doppler shift due to the current. The waves are short with respect to the length L and the beam B of the platform, i.e. $k_0 B \gg 1$ and $L/B \gg 1$. The quantity $k = \omega^2/g$ will be the proper large parameter or *wave number* suited for the asymptotic expansions. Usually in the application of the *ray method* the expansions are made with respect to k, in the final results it becomes evident what dimensionless parameter plays a dominant role. In our case it is expected that in beam seas the parameter kB is essential, while for pure head seas we have to consider kL. So we don't bother about the two large parameters.

In the fluid domain we have the potential equation

$$\Delta\Phi = 0 \tag{28}$$

together with the linearized free-surface condition for (x, y) outside the platform

$$\left\{\frac{\partial^2}{\partial t^2} + 2U\frac{\partial^2}{\partial x\partial t} + U^2\frac{\partial^2}{\partial x^2}\right\}\Phi + g\frac{\partial\Phi}{\partial z} = 0 \text{ at } z = 0 \tag{29}$$

The platform is assumed to be a thin layer at the free-surface $z = 0$, this seems to be a good model for a shallow draft platform. The platform is modelled as an elastic plate with zero thickness. To describe the defelction w we apply the thin plate theory, this leads to an equation for w of the form

$$m\frac{\partial^2 w}{\partial t^2} = -EI\left(\frac{\partial^2}{\partial x^2} + \frac{\partial^2}{\partial y^2}\right)^2 w + p|_{z=0} \tag{30}$$

together with the kinematic condition $\Phi_z = w_t + Uw_x$ and the dynamic condition $p/\rho = -\Phi_t - U\Phi_x - gw$ at $z = 0$, where m is the mass of unit area of the platform while EI is its equivalent flexural rigidity. We apply the operator $\frac{\partial}{\partial t} + U\frac{\partial}{\partial x}$ to (30) and use the kinematic and dynamic condition to arrive at the following equation for Φ at $z = 0$ in the platform area.

$$\left\{\frac{EI}{\rho g}\left(\frac{\partial^2}{\partial x^2} + \frac{\partial^2}{\partial y^2}\right)^2 + \frac{m}{\rho g}\frac{\partial^2}{\partial t^2} + 1\right\}\frac{\partial\Phi}{\partial z} +$$

$$\frac{1}{g}\left\{\frac{\partial^2}{\partial t^2} + 2U\frac{\partial^2}{\partial x\partial t} + U^2\frac{\partial^2}{\partial x^2}\right\}\Phi = 0 \tag{31}$$

We now introduce harmonic waves in the form of the ray expansion

$$\Phi(\vec{x}, t; k) = a(\vec{x}, k)e^{ikS(\vec{x}) - i\omega t} \tag{32}$$

where $S(\vec{x})$ is the phase function and $a(\vec{x},k)$ the amplitude function. The latter is written as a regular series expansion with respect to inverse powers of ik,

$$a(\vec{x},k) = \sum_{j=0}^{N} \frac{a_j(\vec{x})}{(ik)^j} + o((ik)^{-N}) \tag{33}$$

To simplify the analysis we restrict ourselves to the case where the tidal current $U = 0$, the following analysis can be extended easily to the general case. Insertion of (32) into the Laplace equation (28) gives

$$-k^2\nabla_3 S \cdot \nabla_3 S a + ik(2\nabla_3 a \cdot \nabla_3 S + a\Delta_3 S) + O(1) = 0 \tag{34}$$

The subscript 3 is used to indicate the three-dimensional ∇ and Δ operator. If no subscript is used the operators are two-dimensional in the horizontal plane. Next we insert (33) and compare orders of magnitude in (35). This leads to a set of equations for S and a_0 to be satisfied in the fluid region:

$$O(k^2) : \nabla_3 S \cdot \nabla_3 S = 0, \tag{35}$$

$$O(k^1) : 2\nabla_3 a_0 \cdot \nabla_3 S + a_0 \Delta_3 S = 0. \tag{36}$$

Next we insert (32) into the free-surface condition (29) outside the platform and apply (33) to get the following :

$$O(k^1) : iS_z = 1 \text{ and } O(k^{-j}) : a_{jz} = 0 \text{ for } j = 0 \cdots N \text{ at } z = 0 \tag{37}$$

The next step is to insert (32) and (33) into the condition at the platform. At this stage we have to make some estimates of the order of magnitude of the parameters of the platform. We introduce

$$\frac{EI}{\rho g} = \frac{\mathcal{E}}{k^4} \text{ and } \frac{m}{\rho} = \frac{\mathcal{M}}{k}.$$

In this case the elastic properties and the mass of the platform play a role in the diffraction of the waves. The parameters $\mathcal{E}$ and $\mathcal{M}$ are of order one in k, this means they stay finite if k tends to infinity. This sounds like a contradiction, but it makes sense, it will be shown the values of these parameters may be large or small. The first two terms in the asymptotic evaluation of (31) become

$$O(k^1) : \left\{\mathcal{E}(S_x^2 + S_y^2)^2 - \mathcal{M} + 1\right\} iS_z = 1 \text{ at } z = 0 \tag{38}$$

and for the $O(k^0)$ term:

$$a_0 \left[\mathcal{E}\frac{\partial}{\partial z}(S_x^2 + S_y^2)^2 + 2S_z \left\{\frac{\partial}{\partial x}(S_x(S_x^2 + S_y^2)) + \frac{\partial}{\partial y}(S_y(S_x^2 + S_y^2))\right\}\right] + \tag{39}$$

$$a_{0z}\left\{\mathcal{E}(S_x^2+S_y^2)^2-\mathcal{M}+1\right\}+4a_{0x}S_xS_z(S_x^2+S_y^2)+4a_{0y}S_yS_z(S_x^2+S_y^2)=0$$

At first we combine the equations for the phase function at the free-surface with those in the water domain, we obtain:

$$S_z=-i \text{ or } S_x^2+S_y^2=1 \text{ outside the platform} \tag{40}$$

$$\left\{\mathcal{E}S_z^4-\mathcal{M}+1\right\}S_z=-i \text{ for } 0<x<L \text{ and } 0<y<B \tag{41}$$

Equation (41) has five solutions for S_z : $\{r_1, \pm r_2+ir_3, \pm r_4+ir_5\}$, we give some examples of numerical values in table (1). Only the values of

$\mathcal{E}$	$\mathcal{M}$	r_1	$\pm r_2+ir_3$	$\pm r_4+ir_5$ or r_4, r_5
0	0	$-i$	0	0
1	0.5	$-0.889i$	$\pm0.899-0.398i$	$\pm0.674+0.842i$
1	1.5	$-1.909i$	$\pm1.015-0.236i$	$\pm0.483+0.782i$
10	5	$-0.848i$	$\pm0.806-0.059i$	$0.253i, 0.714i$
100	10	$-0.573i$	$\pm0.551-0.027i$	$0.111i, 0.515i$
100	50	$-0.842i$	$\pm0.837-0.005i$	$0.020i, 0.831i$
1000	100	$-0.563i$	$\pm0.561-0.003i$	$0.010i, 0.558i$

TABLE 1. Examples of values of S_z

S_z with negative imaginary part, i.e. the third and the fourth column of table (1), have to be taken into account, because the potential tends to zero for large negative values of z. At $z=0$ we have to solve the eikonal equation $S_x^2+S_y^2=-S_{jz}^2:=n_j^2$, with $-\{n_j^2\}=\{r_1^2,(r_2+ir_3)^2,(-r_2+ir_3)^2\}$ for $j=1,2,3$.

3.2. INFINITELY LONG PLATFORM

We consider plane waves incident at $y=0$. For convenience we assume the platform infinitely long, hence the waves are diffracted by a half-plane or a strip of width B. The wave field for $y<0$ consists of an incident and reflected wave:

$$\Phi(\vec{x},t)=e^{ik(x\cos\alpha+y\sin\alpha)+kz-i\omega t}+Re^{ik(x\cos\alpha-y\sin\alpha)+kz-i\omega t} \tag{42}$$

where R is the reflection coefficient.

Let us first solve the problem for the half-plane in other words with $B=\infty$. We then have (42) for $y<0$ and for $y>0$:

$$\Phi(\vec{x},t)=\sum_{j=1}^{3}a_je^{ik\left(x\cos\alpha+y\sqrt{n_j^2-\cos^2\alpha}\right)+kir_jz-i\omega t} \tag{43}$$

where the square root is such it is positive or that its imaginary part is positive. This is to guarantee that the waves are either outgoing or evanescent. In this expression different types of solutions are combined. For all the components the amplitude function is a constant in the whole domain, in the case of a plane incident wave. For instance, for the real $n_1 < 1$ the solution consists, for angles of incidence with $|\cos\alpha| < n_1$ of a transmitted wave. This wave gives rise to reflections at $y = B$, if B is finite. If $|\cos\alpha| > n_1$ we have total reflection at $y = 0$, however the formulation (43) still gives a proper description of the evenescent mode. The angle for wich $|\cos\alpha_c| = n_1$ is called the critical angle α_c. The other two contributions are always of evenescent type, because the arguments of the sin and cos terms are always complex.

The four unknown coefficients $\{R, a_j\}$ for $j = 1, 2, 3$ are completely determined by the boundary conditions at the edge of the platform. We require continuity of the wave elevation and its inclination, hence, we assume platform very flexible. We now obtain

$$\Phi|_{0-} = \Phi|_{0+} \text{ and } \Phi_y|_{0-} - = \Phi_y|_{0+} \text{ at } y = 0 \text{ and } z = 0, \tag{44}$$

furthermore at the edge of the platform we have the condition of zero moment and zero shear force

$$\frac{\partial^3}{\partial y^3}\left(\frac{\partial\Phi}{\partial z}\right) = \frac{\partial^2}{\partial y^2}\left(\frac{\partial\Phi}{\partial z}\right) = 0 \text{ at } y = 0+ \text{ and } z = 0 \tag{45}$$

We now have to solve four linear equations for the four unknown coefficients. As an example we give the values for the reflection coefficient and the transmission coefficients a_j in table (2). We have chosen the value of $\alpha = \pi/2$, hence we have no total reflection. In this case $\sqrt{n_1^2 - \cos^2\alpha} = \beta_1$ is real and $\sqrt{n_j^2 - \cos^2\alpha} = \pm\beta_2 + i\beta_3$ for $j = 2, 3$ respectively. Because the incident wave is in the $y-$direction the values $\{\beta_1, \beta_2, \beta_3\} = \{-r_1, r_2, -r_3\}$ are given in table (2).
In the figures [4,5] the elevation is shown for several values of the angle of incidence for $\mathcal{E} = 1$ and $\mathcal{M} = 0.5$. For the angle of incidence $\alpha = \pi/3$ we notice a plane wave propagating along the platform and two evanescent modes, while for small values $\alpha = \pi/7$ total reflection occurs and the deflection of the platform consists of three evanescent modes.
Next we consider an infinitely long strip of width B. For $y > B$ the wave field consider of an outgoing transmitted wave:

$$\Phi(\vec{x}, t) = Te^{ik(x\cos\alpha + y\sin\alpha) + kz - i\omega t} \tag{46}$$

$\mathcal{E}$	$\mathcal{M}$	R	a_1	a_2	a_3
1	0.5	$0.207 - 0.140i$	$0.612 - 0.343i$	$0.481 + 0.217i$	$0.114 - 0.014i$
1	1.5	$0.280 - 0.016i$	$0.033 - 0.103i$	$0.957 + 0.003i$	$0.290 + 0.084i$
10	5	$0.144 - 0.015i$	$0.456 - 0.495i$	$0.643 + 0.485i$	$0.035 - 0.004i$
100	10	$0.312 - 0.006i$	$0.560 - 0.593i$	$0.723 + 0.589i$	$0.029 - 0.002i$
100	50	$0.092 + 0.000i$	$0.533 - 0.539i$	$0.555 + 0.539i$	$0.003 + 0.000i$
1000	100	$0.284 - 0.001i$	$0.632 - 0.635i$	$0.648 + 0.635i$	$0.003 - 0.001i$

TABLE 2. The reflection coefficient R and transmission a_j for $\alpha = \pi/2$

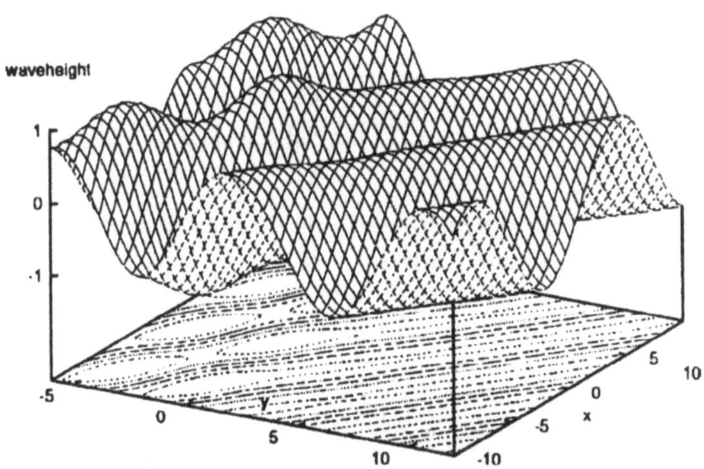

Figure 4. Waveheight for $\mathcal{E} = 1$, $\mathcal{M} = 0.5$ and angle of incidence $\alpha = \pi/3$.

At the plate the field exist six components:

$$\Phi(\vec{x}, t) = e^{ikx\cos\alpha - i\omega t} \sum_{j=1}^{3} e^{ikS_{jz}z} \left\{ a_j \cos\left(ky\sqrt{n_j^2 - \cos^2\alpha}\right) + \right.$$

$$\left. + \, b_j \sin\left(ky\sqrt{n_j^2 - \cos^2\alpha}\right) \right\} \qquad (47)$$

The eight unknown coefficients $\{R, T, a_j, b_j\}$ for $j = 1, 2, 3$ are completely determined by the boundary conditions at the boundaries of the platform. We impose the continuity relation (44) and the conditions of zero moment and zero shear force at $y = B$ also. We now have to solve eight linear equations for the eight unknown coefficients.

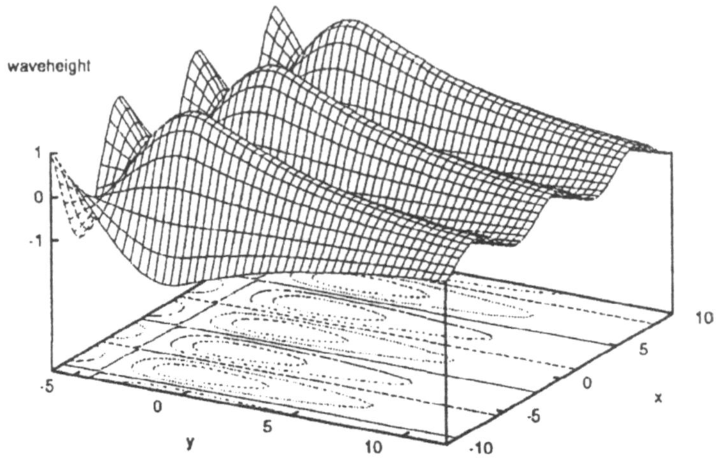

Figure 5. Waveheight for $\mathcal{E} = 1$, $\mathcal{M} = 0.5$ and angle of incidence $\alpha = \pi/7$.

Some results for the transmission and reflection coefficients are presented in figure [6] for angles of incidence $\alpha = \pi/2, \pi/4$ and $\pi/6$ and several values of kB. We see the effect of the the finite width of the strip in the form of the oscillatory behaviour. In figure [7] the same results are shown for smaller angles of incidence, $\alpha = \pi/6, \pi/7$ and $\pi/8$. It is clearly seen that for the two smaller angles total reflection occurs.

One must keep in mind that the results shown in the figure [6,7] are valid for an infinitely long platform, the wave pattern at the platform is a combination of waves, generated by incident wave, and bouncing partly between the edges of the platform. If one takes the front and the end of the platform into account one has to trace the rays generated at the edge. One gets a solution that is discontinuous along the *shadow* boundaries emanating at the corner points. To obtain a continuous solution one has to solve the *canonical* problem of the corner diffraction. This problem is not addressed in this paper. The method to construct this *canonical* problem can be found in classical text books, see for instance *Morse and Feshbach* [7], on wave diffraction. One also may take the asymptotic result for corner diffraction of the two-dimensional Helmholtz equation as a first approximation, where n_1 is taken as the refraction coefficient in the wedge region.

3.3. MOVING PRESSURE POINT

In this section we consider the respons of the platform to a moving point source. We consider the effect of a accelerating aircraft before take off or

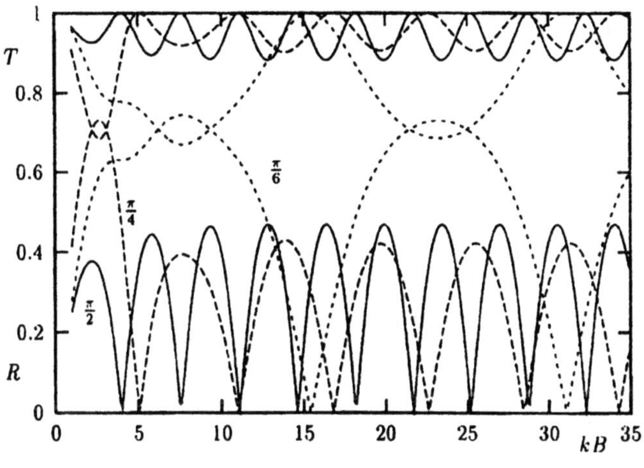

Figure 6. Reflection and transmission coefficients for $\mathcal{E} = 1$, $\mathcal{M} = 0.5$ (k =constant) with $\alpha = \pi/2$, $\pi/4$ and $\pi/6$.

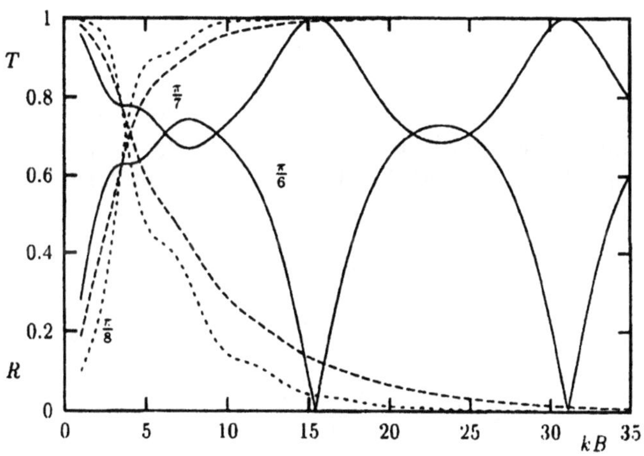

Figure 7. Reflection and transmission coefficients for $\mathcal{E} = 1$, $\mathcal{M} = 0.5$ (k =constant) with $\alpha = \pi/6$, $\pi/7$ and $\pi/8$.

a decelerating one after landing. The process we describe is instationary. The position of the pressure disturbance is along the x-axis at the position

$x = a(t)$. The condition at the platform surface becomes for $t > 0$

$$\left\{\frac{EI}{\rho g}\left(\frac{\partial^2}{\partial x^2}+\frac{\partial^2}{\partial y^2}\right)^2+\frac{m}{\rho g}\frac{\partial^2}{\partial t^2}+1\right\}\frac{\partial \Phi}{\partial z}+\frac{1}{g}\frac{\partial^2 \Phi}{\partial t^2}=-\frac{1}{\rho g}\frac{\partial}{\partial t}\delta(x-a(t))\delta(y) \tag{48}$$

at $z = 0$ together with the Laplace equation. We apply a Fourier transform with respect to x and a cosine transform with respect to y:

$$\bar{\Phi}(\alpha,\beta,z,t)=\int_{\infty}^{\infty}\int_{0}^{\infty} e^{i\alpha x}\cos\beta y\ \Phi(x,y,z,t)\ \mathrm{d}x\ \mathrm{d}y \tag{49}$$

This leads to a surface condition of the form

$$\left\{\frac{EI}{\rho g}(\alpha^2+\beta^2)^2+\frac{m}{\rho g}\frac{\partial^2}{\partial t^2}+1\right\}\frac{\partial \bar{\Phi}}{\partial z}+\frac{1}{g}\frac{\partial^2 \bar{\Phi}}{\partial t^2}=-\frac{1}{\rho g}\frac{\partial}{\partial t}e^{i\alpha a(t)}:=-\frac{1}{\rho g}\frac{\partial}{\partial t}f(t) \tag{50}$$

at $z = 0$ together with

$$-(\alpha^2+\beta^2)\bar{\Phi}+\bar{\Phi}_{zz}=0 \ \text{ for } z<0 \tag{51}$$

We assume that for $t \leq 0$ the whole system is in rest, and we apply the Laplace transform with respect to t.

$$\phi(\alpha,\beta,z,s)=\int_{0}^{\infty}\bar{\Phi}(\alpha,\beta,z,t)e^{-st}\ \mathrm{d}t \tag{52}$$

We solve the resulting ordinary differential equation for $\phi(\alpha,\beta,z,s)$ and make use of the kinematic condition at $z = 0$ reads $\Phi_z = w_t$ at $z = 0$ to obtain the transform $W(\alpha,\beta,s)$ of the elevation of the platform,

$$W(\alpha,\beta,s)=\frac{k\mathcal{F}(s)}{(s^2+K^2(k))(mk+\rho)} \tag{53}$$

where $\mathcal{F}(s)$ is the Laplace transform of $f(t)$, $k^2 := \alpha^2+\beta^2$ and

$$K^2(k):=\frac{EIk^5+k\rho g}{mk+\rho}$$

We write the inverse Laplace transform as a convolution integral, introduce $\alpha = k\cos\theta$, $\beta = k\sin\theta$ and carry out the integration with respect to θ to obtain:

$$w(x,y,t)=-\frac{1}{2\pi}\int_{k=0}^{\infty}\int_{\tau=0}^{t} k\mathcal{H}(k)\sin\left(K(k)(t-\tau)\right) J_0\left(k\sqrt{(x-a(\tau))^2+y^2}\right)\ \mathrm{d}k\ \mathrm{d}\tau \tag{54}$$

where

$$\mathcal{H}(k) := \frac{k}{\{(EIk^5 + \rho k)(mk + \rho)\}^{1/2}}$$

In the case $EI = 0$ and $m = 0$ we have obtained the classical free surface result as can be found in *Wehausen* ([13]). To compute the integral (54) the path of integration in the complex k-plane has to be choosen such that the integrand behaves like a function with a negative exponential as much as possible, where we write for the zero's order Bessel function J_0 its asymptotic expression for large values of its argument. Some analytic contour deformations can be carried out for the case that we neglect the effect of the density of the water on the fase of the exponential function, $K(k) = \sqrt{EI/m}\ k^2 := \mu k^2$. The integral can be computed directly by an accurate Gauss integration, this is straight forward but rather time consuming. Fast computer codes can be designed if necessary.

As an example some computations are carried out for a decelerating unit point source, starting at $t = 0$ in the origin and moving along the x-axis with $a(t) = t - 0.1t^2$. We show in figure (8) the wave elevation at $t = 2$. The parameters choosen are such that a clear wave phenomena can be shown. If realistic values for a specific construction are available the results can be computed. In the figures we see a combination of effects. In front of the point source we see the effect of an impulsive start combined with the moving pattern. It seems that the velocity of the radial wave component is larger than the speed of the point-source.
The effect of the impulsive landing becomes apparent if one looks at the effect of a starting pressure point. In (9) we show the surface elevation, at $t = 2$, of an accelerating point source starting at zero speed in the origin. The position of the point source is given by $a(t) = 0.1t^2$. The field is a superposition of a steady part and an unsteady part described by an initial value problem. The steady part of a unit point source in the origin has the form

$$w(x,y) = -\frac{1}{2\pi}\int_{k=0}^{\infty} \frac{k}{EIk^4 + \rho g} J_0(kR)\, \mathrm{d}k \qquad (55)$$

It becomes obvious that the circular wave pattern is not present in this case. It is noticed that the gravity does not give a substantial contribution in the form of a Kelvin-like pattern. The pattern is dominated by the elastic effects.

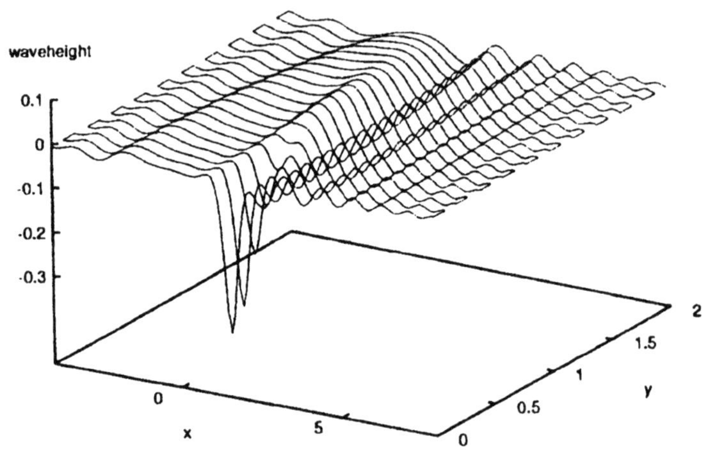

Figure 8. Waveheight for a decelerating unit point source at $t = 2$, starting at $t = 0$, with $EI = 0.01$ and $m = 0.09$.

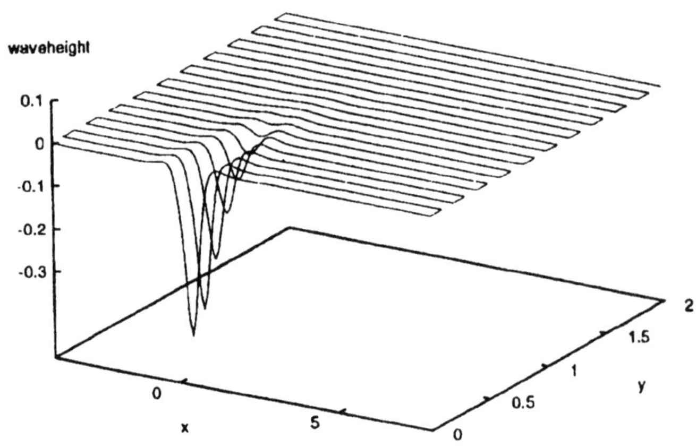

Figure 9. Waveheight for a decelerating unit point source at $t = 5$, starting at $t = 0$, with $EI = 0.01$ and $m = 0.09$.

References

1. Baba, E., Wave resistance of ships at low speed. *Mitsubishi Technical bulletin*, Vol 109, 1976.
2. Baba, E. and Hara, M., Numerical evaluation of a wave resistance theory for slow ships. *Mitsubishi Technical bulletin*, Vol 125, 1978.
3. Eggers, K., Non-Kelvin dispersive waves around non-slender ships. *Schiffstechnik* 28

(1981) 223-251.
4. Hermans, A.J. and Brandsma, F.J., Nonlinear ship waves at low Froude number. *J. of Ship Research*, Vol. 33, pp. 176-193, 1989.
5. Hermans, A.J. and Van Gemert, P.H., A linearized surface condition in low speed hydrodynamics. *Schiffstechnik* 36 (1989) 181-196.
6. Meylan, M. and Squire, V.A., The response of ice floes to ocean waves. *J. Geophysical Research*, Vol. 99 No. C1
7. Morse, P.M. and Feshbach, H., *Methods of Theoretical Physics, Part. 1,2*, McGraw-Hill, New York, 1953.
8. Ohkusu, M. and Nanba, M., Hydroelastic behavior of a very large floating platform in waves. *Proceedings of the eleventh Workshop on Water Waves and Floating Bodies*, Hamburg, 1996.
9. Sakamoto, T. and Baba, E., Minimization of resistance of slowly moving full hull forms in short waves. *Proc. 16th Symposium on Naval Hydrodynamics*, Berkeley, USA (1986) 598-612.
10. Schulkes, R.M.S.M., Hosking, R.J. and Sneyd, A.D., Waves due to a steady moving source on a floating ice plate. Part 2. *J.Fluid Mech.*, Vol. 180, pp. 297-318, 1987.
11. Schulkes and Sneyd, A.D., Time-dependent response of floating ice to a steadily moving load. *J.Fluid Mech.*, Vol. 186, pp. 25-46, 1988.
12. Stoker, J.J., *Water Waves*, Interscience Publ., New York, 1957.
13. Wehausen, J.V. and Laitone E.V., Surface Waves, *Handbuch der Physik*, Band IX, Strömungsmechanik III, Springer-Verlag, Berlin, 1960.

NON CLASSICAL SOLITARY WAVES

JEAN CLAUDE SAUT
Analyse Numérique et EDP
CNRS et Université Paris Sud, Bat 425
91405 Orsay, France

Abstract. This lecture will survey some recent results, mainly due to Anne de Bouard and the author, on solitary waves solutions to nonlinear dispersive equations with weak transverse effects. Typical examples are the generalized Kadomtsev Petviashvili equation or the equation for Langmuir waves in a weakly magnetized plasma. Those solitary waves are not "classical" in the sense that (contrarily to the solitary waves of the Korteweg de Vries or the usual nonlinear Schrödinger equations), they are not radial and do not decay rapidly at infinity. This is due to the breaking of radial symmetry which leads to the anisotropy of the underlying partial differential equation.

1. Introduction

It is appropriate to start a lecture devoted to "non classical" solitary waves by explaining what do we mean by "classical" ones. Let us consider two examples.

The first one is the generalized Korteweg de Vries (GKdV) equations

$$u_t + u^p u_x + u_{xxx} = 0, \qquad u = u(x.t), x \in \mathbb{R}, t \geq 0. \tag{1}$$

Equation (1) when $p = 1$ has been derived first by Boussinesq [11, pg. 360](this not too well known fact has been pointed to me by Bob Pego(see [32]) and rediscovered later by Korteweg and de Vries [24] to model the propagation of one dimensional, small amplitude, long surface waves on a channel. Later on , together with its extensions (different nonlinearities and/or dispersion), it has become a universal model to describe weakly nonlinear waves in many physical situations.

It is well known that (1) posseses, for any positive integer p and any velocity $c > 0$, a solitary wave solution $u(x,t) = \varphi_c(x - ct)$, where

$$\varphi_c(x) = \left(\frac{c(p+1)(p+2)}{2}\right)^{1/p} \operatorname{sech}^{2/p} \frac{\sqrt{c}}{2} px \tag{2}$$

This solution is unique (up to space translations), smooth, positive, radial (even) and decays exponentially to 0 as $|x| \to +\infty$.

Moreover, the solitary wave is orbitally stable (nonlinear stability modulo space translations) if and only if $p < 4$ (see Benjamin [3], Bona [5], Albert, Bona and Henry [2], Bona, Souganidis and Strauss [6]). Similar results are also valid for models where the dispersive term u_{xxx} in (1) is replaced by $-u_{xxt}$ (the generalized Benjamin-Bona-Mahony equation) or by Lu_x where L is a (generally) non local operator. Familiar examples are the Benjamin-Ono or the intermediate long wave equation. In particular Bona and Li [7] have recently derived sharp conditions on L insuring the smoothness and exponential decay of the solitary wave.

The second example is the familiar focusing nonlinear Schrödinger equation (NLS)

$$iu_t + \Delta u + |u|^p u = 0, \qquad u = u(x,t), x \in \mathbb{R}^n, t \geq 0, \tag{3}$$

where u is complex valued.

A solitary wave solution of (3) is a solution of the form $u(x,t) = e^{iwt}\varphi(x)$, $w > 0$, $\varphi \in H^1(\mathbb{R}^n)$, the usual Sobolev space.

The following facts are classical (see the book by Cazenave [9] and the references therein).

For any p such that $0 < p < \frac{4}{n-2}$ (no upper bound if $n = 1, 2$), there exists a unique (up to translations) positive and radial solitary wave, the ground state. It is smooth and decays exponentially to 0 as $|x| \to +\infty$. The ground state minimizes the action

$$S(u) = \frac{1}{2}\int_{\mathbb{R}^n} |\nabla u|^2\,dx - \frac{1}{p+2}\int_{\mathbb{R}^n} |u|^{p+2}\,dx + \frac{w}{2}\int_{\mathbb{R}^n} |u|^2\,dx$$

among all non trivial H^1 solutions of the elliptic equation

$$-wu + \Delta u = |u|^p u. \tag{4}$$

Furthermore, the ground state is orbitally stable if and only if $0 < p < \frac{4}{n}$. Of course there exists other radial, non positive solitary waves, the bound states, whose stability properties are unknown so far.

These two problems are clearly invariant by rotation. Our aim here is to investigate the effect of breaking this spherical invariance, specially on the solitary waves.

We will consider two examples which could be seen as multidimensional generalizations of (1) or (3).

(i) Generalized Kadomtsev-Petviashvili(KP) equations
We distinguish the 2 and 3 dimensional case.

$$\begin{cases} u_t + u^p u_x + u_{xxx} + \varepsilon v_y = 0 & (x,y) \in \mathbb{R}^2, t \geq 0, \varepsilon = \pm 1 \\ u_y = v_x \end{cases} \tag{5}$$

$$\begin{cases} u_t + u^p u_x + u_{xxx} + a v_y + b w_z, & & & (x,y,z) \in \mathbb{R}^3, \\ u_y = v_x, & u_z = w_x, & t \geq 0, & a, b = \pm 1. \end{cases} \tag{6}$$

In (5) (6), $p = m/n$, $n \geq 1$, m and n relatively prime, n odd.
Historically (4) has been first derived (in the case $p = 1$, $\varepsilon = \pm 1$) by Kadomtsev and Petviashvili [23] who were looking for a 2-dimensional equation to study the stability of the "line soliton"of the KdV equation (that is (2) when $p = 1$, viewed as a two dimensional function, constant in y) to weak transverse perturbations. Since then, (5) or (6) have appeared as "universal" models for dispersive, weakly nonlinear waves, which are essentially one dimensional, when weak transverse effects are taking into account (see for instance [33]).
Note that the case $\varepsilon = -1$ in (5) appears for $p = 1$ for the two dimensional long surface waves in gravity capillary waves where surface tension is strong enough (that is in a sufficiently thin film of water) or to describe the propagation of magnetosonic waves in plasma [36].

(ii) "KP-Schrödinger" type equations for Langmuir waves in a weakly magnetic field.
We refer to [25] for a precise description of the model and for the relevant references. The equation for the complex envelope ψ of high frequency waves in a weakly magnetized plasma writes in dimensionless variables

$$\Delta(i\psi_t + \Delta\psi) - \sigma\Delta_\perp\psi + \nabla \cdot (|\nabla\psi|^2 \nabla\psi) = 0, \tag{7}$$

where $\sigma > 0$, Δ is the Laplacian $\partial_x^2 + \partial_y^2 + \partial_z^2$ and $\Delta_\perp$ is the transverse Laplacian $\partial_y^2 + \partial_z^2$ (x is the direction of the magnetic field). If the wave condensate has characteristic longitudinal scales much smaller than the transverse ones (i.e. $|k_\perp| \ll |k|$, where k and $k_\perp$ are respectively the total and transverse wave number), 7 simplifies into

$$\frac{\partial^2}{\partial x^2}\left[i\psi_t + \frac{\partial^2}{\partial x^2}\psi\right] - \Delta_\perp\psi + \frac{\partial}{\partial x}\left[\left|\frac{\partial\psi}{\partial x}\right|^2 \frac{\partial\psi}{\partial x}\right] = 0. \tag{8}$$

Actually we will consider a slightly more general equation, namely

$$\begin{cases} \frac{\partial}{\partial x}(i\varphi_t + \varphi_{xx}) - \Delta_\perp \psi + \frac{\partial}{\partial x}[|\varphi|^\alpha \varphi] = 0 \\ \varphi = \frac{\partial \psi}{\partial x} \end{cases} \tag{9}$$

where now α is any positive number.

We will call solitary wave any "localized in all directions" solution of (5) or (6) (resp. (9)) of the form

$$u = u(x - ct, y), \qquad c > 0 \ \text{ or } \ u(x - ct, y, z)$$

(resp. $\varphi = e^{iwt}.\Phi(x, y, z), w > 0$). A precise definition will be given in the second section. It is worth noticing that when plugged into (5) or (6) (resp. (9)) those formulas lead to the very similar equations

$$-cu_{xx} + \varepsilon u_{yy} + \partial_x^4 u + \frac{1}{p+1}(u^{p+1})_{xx} = 0 \tag{10}$$

$$-cu_{xx} + au_{yy} + bu_{zz} + \partial_x^4 u + \frac{1}{p+1}(u^{p+1})_{xx} = 0, \tag{11}$$

$$-w\Phi_{xx} - \Delta_\perp \Phi + \partial_x^4 \Phi + (|\Phi|^\alpha \Phi)_{xx} = 0 \tag{12}$$

By a simple scaling argument c and w can be scaled to 1, a value which we assume to hold from now on.

Let us now describe the content of the paper. Section 2 is devoted to non existence and existence results for solitary waves. In Section 3, we give qualitative properties of the solitary waves: smoothness, rate of decay, cylindrical symmetry. In Section 4 we investigate the orbital stability of "ground states" solitary waves. The last paragraph deals with some remarks and open problems. Most of the results presented here have been obtained in collaboration with Anne de Bouard.

2. Non existence and existence

To give a rigorous definition of solitary wave, we need first to introduce a basic functional space which plays the role of the energy space for KP type equations.

We shall denote for $d = 2, 3$, Y the closure of $\partial_x\left(C_0^\infty(\mathbb{R}^d)\right)$ for the norm

$$\|\partial_x \psi\|_Y = \left(\|\nabla \psi\|_{L^2}^2 + \|\partial_x^2 \psi\|_{L^2}^2\right)^{1/2}$$

where $\partial_x\left(C_0^\infty(\mathbb{R}^d)\right)$ denotes the space of functions of the form $\partial_x \psi$ with $\psi \in C_0^\infty(\mathbb{R}^d)$ (i.e. the space of functions φ in $C_0^\infty(\mathbb{R}^d)$ such that

$\int_{-\infty}^{\infty} \varphi(x, x')\, dx = 0$ for every $x' \in \mathbb{R}^{d-1}$).

By classical imbedding theorems, if $\varphi \in Y$ and $d = 3$, then $\varphi = \partial_x \psi$ where $\psi \in L^6(\mathbb{R}^3)$; if $\alpha = 2$ and $\varphi \in Y$, then $\varphi = \partial_x \psi$ where $\psi \in L^q_{loc}(\mathbb{R}^2)$, $\forall q\ 2 \le q < +\infty$. Note that for $d = 2$, the choice of $\psi \in L^q_{loc}$ such that $\varphi = \partial_x \psi$ is not unique but two such ψ will differ by a function independent of x. Hence, only one of them (up to a constant), satisfies $v = \partial_x \psi \in L^2(\mathbb{R}^2)$. We assume in all what follows that when $\varphi \in Y$ and when we take $\psi \in L^q_{loc}$ with $\partial_x \psi = \varphi$, we also have $v = \partial_y \psi \in L^2$. We then denote $v = \partial_y \psi$ by $D_x^{-1} \varphi_y$.

Definition 2.1 *A solitary wave of (5) or (6) (resp. (9)) is a solution of the type $u(x - ct, x')$, $c > 0$, $u \in Y$ (resp. $e^{iwt}\Phi(x, x')$, Φ real valued, $\Phi \in Y$ and $w > 0$).*

Remark 2.1 *The fact that a solitary wave has to belong to Y is a weak formulation that it decays in all directions.*

Concerning the nonexistence of solitary waves, we have the following.

Theorem 2.1 *[12, 14, 11]*

(i) Assume that $d = 2$. The equation (5) does not admit any nontrivial solitary wave satisfying $u = \partial_x \varphi \in Y$, $u \in H^1(\mathbb{R}^2) \cap L^\infty_{loc}(\mathbb{R}^2)$, $\partial_x^2 u$ and $\partial_y^2 \varphi \in L^2_{loc}(\mathbb{R}^2)$ if

$$\varepsilon = -1 \ \textit{and} \ p \ge 4$$

or

$$\varepsilon = 1 \textit{ and } p \textit{ is arbitrary.}$$

A similar result holds true for equation (9) with $d = 2$ if $\alpha \ge 4$

(ii) Assume that $d = 3$. The equation (6) does not admit any nontrivial solitary wave satisfying $u = \partial_x \varphi \in Y$, $u \in H^1(\mathbb{R}^3) \cap L^{2(p+1)}(\mathbb{R}^3) \cap L^\infty_{loc}(\mathbb{R}^3)$, $\partial_x^2 u$, $\partial_y^2 v$ and $\partial_z^2 u \in L^2_{lox}(\mathbb{R}^3)$ if

$$ab = -1 \ \textit{(resp. } a = b = 1\textit{) and } p \textit{ is arbitrary}$$

or

$$a = b = -1, \quad p \ge 4/3$$

The proof of Theorem 2.1 is given in [14]. It is based on Pohojaev identities. The regularity assumptions of Theorem 2.1 are needed to justify them rigorously.

Before stating the existence results, let us write, for the remaining values of the parameters, the elliptic equation satisfied by the solitary waves. They are respectively

$$-\Delta u + \partial_x^4 u + \frac{1}{p+1}(u^{p+1})_{xx} = 0, \tag{13}$$

for the generalized KP case, and

$$-\Delta\Phi + \partial_x^4\Phi + (|\Phi|^\alpha\Phi)_{xx} = 0 \tag{14}$$

for the case of the KP Schrödinger equation.

Theorem 2.2 *[12, 14, 11, 15]*

(i) For $d = 2$ (resp. $d = 3$), the systems (5) (resp. (6)) and (9) admit a solitary wave if $0 < p < 4$ or $0 < \alpha < 4$ (resp. $0 < p, \alpha < 4/3$).

(ii) Any solitary wave is continuous and tends to zero at infinity. If p is an integer, or if $\alpha = 2$, it belongs to $H^\infty(\mathbb{R}^d) = \cap_{k\geq 0} H^k(\mathbb{R}^d)$

The proof of (i) consists in solving the minimization problem, for $\lambda > 0$

$$I_\lambda = \inf\{\|u\|_Y^2, u \in Y, \text{ with } \int_{\mathbb{R}^d} u^{p+2}\,dx\,dx' = \lambda\} \tag{15}$$

in the generalized KP case, or

$$I_\lambda = \inf\{\|\Phi\|_Y^2, \Phi \in Y, \text{ with } \int_{\mathbb{R}^d} |\Phi|^{\alpha+2}\,dx\,dx' = \lambda\} \tag{16}$$

in the KP Schrödinger case. The two problems are essentially identical and are solved by using the concentration compactness principle of P. L. Lions (see [40, 41] for alternative approaches).

The smoothness property in (ii) is proven by a bootstrapping argument which involves imbedding results for anisotropic Sobolev spaces [4] and the L^q regularity of elliptic equations with anisotropic symbols which follows from a Fourier multiplier result due to Lizorkin [28]. In fact, (ii) is valid for any Y solution of the elliptic equation (13) or (14), as the decay property that we will describe in the next section.

3. Qualitative properties of the solitary waves

To start with, we focus on decay properties of the solitary waves.

Theorem 3.1 *[15, 11]*

(i) Any non trivial solitary wave f of (5) or (9) in $\mathbb{R}^2$ *satisfies*

$$r^2 f \in L^\infty(\mathbb{R}^2), \qquad r^2 = x^2 + y^2. \tag{17}$$

(ii) Any nontrivial solitary wave f of (6) or (9) in $\mathbb{R}^3$ *satisfies*

$$r^\delta f \in L^2(\mathbb{R}^3), \qquad \forall \delta, 0 \leq \delta < 3/2, r = (x^2 + y^2 + z^2)^{1/2} \tag{18}$$

The proof of Theorem 3.1 is based on a careful analysis of the convolution equation equivalent to the elliptic equation (13) or (14). Let us explain why it is sharp. First, one can think of the explicit lump solution of KPI (see [1]), namely.

$$u(x - ct, y) = \frac{\gamma c(1 - \frac{c}{3}(x - ct)^2 + \frac{c^2}{3}y^2)}{(1 + \frac{c}{3}(x - ct)^2 + \frac{c^2}{3}y^2)^2}, \tag{19}$$

which decays exactly as r^{-2}.

On the other hand, any Y solution of (13) or (14) can not decay fast at infinity. Indeed, let take the Fourier transform of (13) or (14). If ξ is the dual variable of $\vec{x} = (x, x')$, it writes

$$\hat{f} = -\frac{\xi_1^2}{|\xi|^2 + \xi_1^4}\hat{g} \tag{20}$$

where $f = u$ (resp. Φ) and $g = \frac{1}{p+1}u^{p+1}$ (resp. $g = |\Phi|^\alpha \Phi$).

If f were exponentially decaying at infinity, $\hat{f}$ and $\hat{g}$ would be analytic functions by the Paley Wiener theorem. But, this is absurd (since the function $\xi \mapsto \frac{\xi_1^2}{|\xi|^2 + \xi_1^2}$ is not analytic at the origin), except if some moments of g vanish. Moreover, it is unlikely that f is integrable because if it were the case $\hat{f}$ would be continuous, which is absurd in view of (20) (one can easily show that $g \in L^1$ for $f \in Y$) except if $\int_{\mathbb{R}^d} g = 0$. This later condition is never satisfied in the KP case, $g = \frac{1}{p+1}u^{p+1}$, when $p = \frac{m}{n}$, m odd. So, in this case, if the solitary wave were to decay with an algebraic rate $r^{-\delta}$, then necessarily $\delta \leq d$.

It is worth noticing that the very same decay estimates are valid for "general" dispersion in x, namely when replacing the u_{xxx} term in (5) or (6) by Lu_x where L is a linear operator defined in Fourier variables by $\hat{Lu}(\xi) = p(\xi_1)\hat{u}(\xi)$ where p is an even polynomial (when $p(\xi) = -\xi^2$, i.e. $L = \partial_x^2$ one gets the classical KP type equations). They are essentially imposed by the singularity of the function $\xi \mapsto \frac{\xi_1^2}{|\xi|^2 + \xi_1 p(\xi_1)}$ at the origin, namely $\frac{\xi_1^2}{|\xi|^2}$.

We turn now to symmetry properties of solitary waves. Contrarily to the results above, we have to restrict to a subclass of solitary waves, the ground states that we will define now.

A ground state is a solitary wave which minimizes the action

$$S(u) = E(u) + \frac{c}{2}\int_{\mathbf{R}^d} |u|^2 \qquad d = 2,3$$

(resp.

$$S(\Phi) = E(\Phi) + \frac{w}{2}\int_{\mathbf{R}^d} |\Phi|^2 \qquad d = 2,3)$$

among all the non zero solutions of (13) (resp. (14)).

Here, the energy E is defined in Y by

$$E(u) = \frac{1}{2}\int_{\mathbf{R}^2} \left(u_x^2 + (D_x^{-1}u_y)^2\right) dx\, dy - \frac{1}{(p+1)(p+2)}\int_{\mathbf{R}^2} u^{p+2}\, dx\, dy$$

or

$$\begin{aligned} E(u) &= \frac{1}{2}\int_{\mathbf{R}^3} \left(u_x^2 + (D_x^{-1}u_y)^2 + (D_x^{-1}u_z)^2\right) dx\, dy\, dz \\ &\quad - \frac{1}{(p+1)(p+2)}\int_{\mathbf{R}^3} u^{p+2}\, dx\, dy\, dz \end{aligned}$$

(resp. $E(\Phi) = \frac{1}{2}\int_{\mathbf{R}^2} (\Phi_x^2 + (D_x^{-1}\Phi_y)^2)\, dx\, dy - \frac{1}{(\alpha+2)}\int_{\mathbf{R}^2} |\Phi|^{\alpha+2}\, dx\, dy$ or $E(\Phi) = \frac{1}{2}\int_{\mathbf{R}^3} (\Phi_x^2 + (D_x^{-1}\Phi_y)^2 + (D_x^{-1}\Phi_z)^2)\, dx\, dy\, dz - \frac{1}{(\alpha+2)}\int_{\mathbf{R}^3} |\Phi|^{\alpha+2}\, dx\, dy\, dz$ One can actually show that any solution constructed in Theorem 2.2 is a ground state and furthermore

Theorem 3.2 *[15, 11] Let $x' = y \in \mathbb{R}$ if $d = 2$ and $x' = (y,z) \in \mathbb{R}^2$ if $d = 3$. Then, up to a translation of the origin of coordinates in x', any ground state of (5) or (6) (resp. (9)) is radial in x', that is it depends only on x and $|x'|$.*

The proof consists, after giving alternate variational characterizations of the ground states, in using a "moving plane" method due to O. Lopes [29] in the context of the nonlinear Schrödinger equation, together with a global unique continuation result for linear equations with principal part $-u_{yy} + \partial_x^4 u$ (resp. $-u_{yy} - u_{zz} + \partial_x^4 u$. Note that the local unique continuation result of Isakov [21] is useless here since we need continuation across a characteristic hyperplane. Instead, one reduces to a backward uniqueness result for a parabolic type equation where one of the transverse variables plays the role of time.

4. Stability issues

In this section we will restrict by the sake of simplicity to the solitary waves of (5).

In order to give a theorem stating the main result related to the initial value problem for (5), we define for $s \geq 0$ $X_s = \{f \in H^s(\mathbb{R}^2), \mathcal{F}^{-1}(\frac{\hat{f}}{\xi_1}) \in H^s(\mathbb{R}^2)\}$, where $H^s(\mathbb{R}^2)$ is the usual Sobolev space of order s and $\mathcal{F}^{-1}(f)$, $\hat{f}$ denote respectively the inverse and direct Fourier transform of f.

The following theorem is a consequence of results in [39, 34, 22, 37].

Theorem 4.1 *Let $u_0 \in X_s$, $s \geq 3$. There exists $T > 0$ such that (5) with initial data u_0 has a unique solution*

$$u \in C([0,T]; H^s(\mathbb{R}^2)) \cap C^1([0,T]; H^{s-3}(\mathbb{R}^2))$$
$$v \in C([0,T]; H^{s-1}(\mathbb{R}^2))$$

In addition, $\int_{\mathbf{R}^2}(u(x,y,t))^2\,dx\,dy$ and

$$E(u(t)) = \int_{\mathbf{R}^2} \left[\frac{1}{2}(u_x^2 + u^2) - \frac{1}{(p+1)(p+2)}u^{p+2}\right] dx\,dy$$

are independent of $t \in [0,T]$, and if $0 < p < \frac{4}{3}$, then $\sup_{t\in[0,T]} \|u(.,t)\|_Y \leq c(\|u_0\|_{L^2}, E(u_0))$.

Remark 4.1 *In [34] the conservation of the energy $E(u(t))$ was established under the extra assumption that $\partial_y^2 u_0 = \partial_x^2 \psi$ for some $\psi \in L^2(\mathbb{R}^2)$. This assumption is not necessary as pointed out by L. Molinet [30].*

Due to the invariance by translations, the good notion of stability for solitary waves is that of orbital stability in the following sense

Definition 4.1 *Let φ_c be a solitary wave of (5). Then φ_c orbitally stable if for all $\varepsilon > 0$, there is a $\delta > 0$ such that for every $u_0 \in X_s$, $s \geq 3$ with $\|u_0 - \varphi_c\|_Y < \delta$, the corresponding solution $u(t)$ of (5) with $u(0) = u_0$ satisfies*

$$\sup_{t\geq 0} \inf_{(r,q)\in\mathbf{R}^2} \|u(t) - \varphi_c(. - r, . - q)\|_Y < \varepsilon.$$

We will show that in the case $p < 4/3$, The set $\mathcal{G}_c$ of ground states of (5) is stable. However, we have not been able to prove that the ground states themselves are orbitally stable, because we do not know whether or not they are isolated point in Y, up to translations (i.e. whether they are unique up to translations in a sufficiently small neighborhood of their orbit).

Theorem 4.2 *[13] Let $p < 4/3$, $c > 0$ and $\varphi_c \in \mathcal{G}_c$; then for all positive ε, there is positive δ such that if $u_0 \in X_s$, $s \geq 3$, and $\|u_0 - \varphi_c\|_Y \leq \delta$, then the solution $u(t)$ of (5) with $u(0) = u_0$ satisfies*

$$\sup_{t \geq 0} \inf_{\psi \in \mathcal{G}_c} \|u(t) - \psi\|_Y < \varepsilon$$

Remark 4.2 *A similar result have been obtained in [27] by a different method.*

Note that if $\varphi \in \mathcal{G}_c$, then all the translates $\varphi_c(.-c,.-y)$ with $(x,y) \in \mathbb{R}^2$ are also elements of $\mathcal{G}_c$. On the other hand, if for some cylindrical symmetric function $\varphi_c \in Y$, $\mathcal{G}_c$ is reduced to the set of the spatial translates of φ_c, then Theorem 4.1 says that the solution $\varphi_c(x - ct, y)$ of equation (5) is orbitally stable. However, as pointed out above, we do not know whether the ground sate solution of (5) (or equivalently the solution of the minimization problem I_λ) is unique up to translations or not.

The key point in the proof of Theorem 4.1 is the fact that for $p < 4/3$, the ground state solutions are characterized as minimizers of energy constrained by constant L^2 norm. This characterization was first used for the ground states of the subcritical nonlinear Schrödinger equation by Cazenave and Lions [10] (see also [9]). It is stated in the following Lemma.

Lemma 4.1 *[13] There is a positive number μ^* such that*

(i) The minimization problem

$$I = \inf\{E(u), u \in Y, \int_{\mathbf{R}^2} |u|^2\, dx\, dy = \mu^*\}$$

has at least one solution.

(ii) Any minimizing sequence u_n for I is relatively compact in Y up to translations, i.e. there are subsequences u_{n_k} and $(x_{n_k}, y_{n_k}) \in \mathbb{R}^2$ such that $u_{n_k}(\cdot - x_{n_k}, \cdot - y_{n_k})$ converge strongly in Y to some u which is a minimum of I.

(iii) The set of minima of I is exactly equal to $\mathcal{G}_{\rfloor}$

We refer to [13] for a proof of Lemma 4.1 (which is essentially based on the concentration-compactness principle) and of Theorem 4.1.

Concerning the instability of solitary waves we have the following result.

Theorem 4.3 *[13] Let $4/3 < p < 4$ and φ_c be a cylindrical symmetric solitary wave of (5). Then φ_c is orbitally unstable.*

Note that the stability results holds for any cylindrical solitary wave (not only ground states). The proof uses in a crucial way the decay properties proven in Section 3.

Remark 4.3 *1. It is very likely that Theorem 4.3 is valid for the solitary waves of equation (9) in* $\mathbb{R}^2$.

2. It is conjectured that the instability of the solitary waves for $p > 4/3$ *is associated to blow up in finite time of the solutions to the corresponding Cauchy problem. This is supported by some numerical simulations [40, 17]. A blow up result has been established (formally in [38] and rigorously in [34]) when* $p \geq 4$. *This blow up is irrelevant to the conjecture above: it depends only on the transverse dispersion (the very same result holds for other dispersive terms in* x *[35]).*

5. Final remarks and open problems

The results presented in this lecture can be extended in several ways. For instance those of Sections 2 and 3 are valid when adding a higher order dispersive term to (5) (see [14, 15]). In [41], Willem has extended the existence part of Theorem 2.2 to "general" nonlinearities (not necessary homogeneous) by using a variant of the mountain pass lemma.

On the other hand, all the results of Section 2 and 3 can be immediately translated for the regularized (BBM type) version of the KP equaitons, that is

$$(u_t + u_x + u^p u_x - u_{xxt})_x - u_{yy} = 0. \tag{21}$$

In fact, it is readily checked that if $u_c(x - ct, y)$, $c > 1$, is a solitary wave of (21), then

$$U_{\tilde{c}}(X - \tilde{c}t, Y) = \left(\frac{1}{c}\right)^{1/p} u(x - ct, c^{1/2}y)$$

is a solitary wave of (5) with $\tilde{c} = \frac{c-1}{c}$, $X = x$, $Y = c^{1/2}y$. Conversely if $u_c(x - ct, y)$ is a solitary wave of (5), $c > 0$, then

$$U_{\tilde{c}}(X - \tilde{c}t, Y) = \left(\frac{\tilde{c}-1}{c}\right) u\left(\left(\frac{c\tilde{c}}{\tilde{c}-1}\right)^{1/2} x, \frac{c\sqrt{\tilde{c}}}{\tilde{c}-1} y\right)$$

$X = \left(\frac{c\tilde{c}}{\tilde{c}-1}\right)^{1/2} x$, $Y = \frac{c\sqrt{\tilde{c}}}{\tilde{c}-1} y$ is a solitary wave of (21) with velocity $\tilde{c} > 1$.

The stability of these ground states for (21) will be studied elsewhere.

As mentioned previously, the question of uniqueness (modulo translations) of the ground states solutions is open (for the classical KPI equation, $p = 1$, $\varepsilon = -1$ in (5), one even does not know if the lump solution (19) is a ground state!) The problem for uniqueness is the following. In the case of the nonlinear Schrödinger equation (3), the question is reduced to the uniqueness of the radial positive solution of the nonlinear elliptic equation

$$-wu + \Delta u + |u|^p u = 0,$$

and one can therefore use (tricky) ODE techniques. In the KP situation, such a reduction is impossible (since the solitary waves are not radial) and one has to face the difficult problem of proving the uniqueness of the ground state solutions of the PDE (13) or (14). In [26] a very crude argument is given in favor of the stability of the lump solution (boundedness from below of the Hamiltonian for fixed charge). This argument certainly does not prove the orbital stability of the lump solution.

We have so far considered solitary waves which are localized in $\mathbb{R}^2$ or $\mathbb{R}^3$. There are however other interesting geometries. For instance, one can look for solitary waves which are localized in x and periodic in y. In the case of KPI equation, such solutions have been obtained by superposition of rational lumps (see Zaitsev [42] or [31] for a study of their interaction). These solitary waves are exponentially decaying in x. It would be interesting to study the existence (and the properties) of such solitary waves for the generalized KP equations. Some results in this direction have been obtained in [42, 19]. Also it would be quite interesting to justify rigorously the emergence of "lump like" solitary waves from a perturbation of the line soliton (the KdV solitary wave viewed as a 2D object) evolving according to KPI type equations [20, 17, 16].

References

1. M. J. Ablowitz and J. Satsuma. Solitons and rational solutions of nonlinear evolution equations. *J. Math. Phys.*, 19(10):2180–2186, 1978.
2. J. Albert, J. L. Bona, and D. Henry. Sufficient conditions for stability of solitary wave solutions of model equations of long waves. *Physica D*, 24:343–366, 1987.
3. T. B. Benjamin. The stability of solitary waves. *Proc. Roy. Soc.*, A328:153–183, 1972.
4. O. V. Besov, V. P. Il'in, and S. M. Nikolskii. *Integral representation of functions and imbeddings theorems*, volume I. J. Wiley, 1978.
5. J. L. Bona. On the stability of solitary waves. *Proc. Roy. Soc.*, A344:363–374, 1975.
6. J. L. Bona, Souganidis P. E., and Strauss W. A. Stability and instability of solitary waves. *Proc. Roy. Soc.*, A411:395–417, 1987.
7. J. L. Bona and A. Li Yi. Decay and analyticity of solitary waves. Preprint, 1995.
8. M. J. Boussinesq. Essai sur la théorie des eaux courantes. In *Mémoires présentés par divers savants á l'Académie des Sciences. Inst. France*, volume 23 of *2*, pages 1–680, 1877.
9. T. Cazenave. An introduction to nonlinear Schrödinger equations. *I.M.U.F.R.J.*, 1992.
10. T. Cazenave and P. L. Lions. Orbital stability of standing waves for some nonlinear Schrödinger equations. *Comm. Math. Phys.*, 85:549–561, 1982.
11. A. de Bouard and J. C. Saut. Langmuir solitary waves in a weakly magnetized plasma. In *Proceedings of the Fourth MSJ International Research Institute*, Sapporo, 1995. To appear.
12. A. de Bouard and J. C. Saut. Sur les ondes solitaires des équations de Kadomtsev-Petviashvili. *C. R. Acad. Sci. Paris*, 320:315–320, 1995.
13. A. de Bouard and J. C. Saut. Remarks on the stability of generalized KP solitary waves. In *Mathematical Problems in the Theory of Water Waves*, pages 75–84. Contemporary Mathematics 200 AMS, 1996.
14. A. de Bouard and J. C. Saut. Solitary waves of the generalized Kadomtsev-

Petviashvili equations. *Annales IHP, Analyse Non lineaire*, 1997.
15. A. de Bouard and J. C. Saut. Symmetries and decay of the generalized KP solitary waves. *SIAM J. Math. Analysis*, 1997.
16. K. A. Gorshkov and D. E. Pelinovsky. Asymptotic theory of plane soliton self focusing in two dimensional wave media. *Physica D*, 85:468–484, 1995.
17. F. Hamidouche. *These de Doctorat.* PhD thesis, Université Paris Sud, Orsay, 1997.
18. M. Haragus and K. Kirchgassner. Breaking of the dimension of a steady wave : some examples. Technical report, 1995. Preprint.
19. M. Haragus and R. L. Pego. In preparation.
20. E. Infeld and A. A. Senatorski. Decay of Kadomtsev-Petviashvili solitons. *Phys. Rev. Letters*, 72(9):1345–1347, 1994.
21. V. Isakov. Carleman type estimates in an anisotropic case and applications. *J. Diff. Eq.*, 105:217–238, 1993.
22. P. Isaza, J. Mejia, and V. Stallbohm. Local solution for the Kadomtsev-Petviashvili equation in $\mathbb{R}^2$. *J. Math. Anal. and Appl.*, 196:566–587, 1995.
23. B. B. Kadomtsev and V. I. Petviashvili. On the stability of solitary waves in weakly dispersive media. *Sov. Phys. Dok.*, 15:539–541, 1970.
24. D. J. Korteweg and G. de Vries. On the change of form of long waves advancing in a rectangular channel and on a new type of solitary waves. *Phil. Mag.*, 39:422–443, 1895.
25. E. A. Kuznezov and M. M. Skoric. Hierarchy of collapse regimes for upper hybrid and lower hybrid waves. *Phys. Rev. A.*, 38(3):1422–1426, 1988.
26. E. A. Kuznezov and S. K. Turitsyn. Two and Three dimensional solitons in weakly dispersive media. *Sov. Phys. JETP*, 55(5):844–847, 1982.
27. Y. Liu and X. P. Wang. Nonlinear stability of solitary waves of a generalized Kadomtsev-Petviashvili equation. Preprint, 1996.
28. P.I. Lizorkin. Multipliers of Fourier integrals. *Proc. Steklov Inst. Math.*, 89:269–290, 1967.
29. O. Lopes. A constrained minimization problem with integrals on the entire space. *J. Diff. Eq.*, 124:378–388, 1996.
30. L. Molinet. *Thése de Doctorat.* PhD thesis, Université Paris Sud, Orsay, 1996.
31. Y. Murakami and M. Tajiri. Interaction between two y-periodic solitons solutions the Kadomtsev Petviashvili with positive dispersion. *Wave Motion*, 14:169–185, 1991.
32. R. L. Pego and M. I. Weinstein. Strong spectral stability of solitary waves for Boussinesq equations. Preprint, 1995.
33. V. I. Petviashvili and V. V. Yanikov. Solitons and turbulence. *Review of Plasma Physics XIV*, pages 1–61, 1989.
34. J. C. Saut. Remarks on the generalized Kadomtsev Petviashvili equations. *Indiana Math. J. 42*, 3:1011–1026, 1993.
35. J. C. Saut. Recent results on the generalized Kadomtsev-Petviashvili. *Acta Applicandae Mathematicae*, 39:477–487, 1995.
36. T. Taniuti and A. Hasegawa. Wave Motion 13. 1991.
37. M. Tom. On a generalized Kadomtsev-Petviashvili equation. In *Mathematical Problems in the Theory of Water Waves*, pages 193–210. Contemporary Mathematics 200 AMS, 1996.
38. S. K. Turitsyn and Fal'Kovich G. E. Stability of magnetoelastic solitons and self focusing of sound in antiferromagnets. *Sov. Phys. JETP 62*, 1:146–152, 1985.
39. S. Ukai. Local solutions of the Kadomtsev-Petviashvili equation. *J. Fac. Sci. Univ. Tokyo Sect. IA Math*, 36:199–209, 1989.
40. X. P. Wang, M. J. Ablowitz, and M. Segur. Wave collapse and unstability of solitary waves for a generalized Kadomtsev-Petviashvili. *Physica D*, 78:97–113, 1994.
41. M. Willem. *Minimax Theorems.* Birkhauser, 1996.
42. A. A. Zaitsev. Formation of stationary nonlinear waves by superposition of solitons. *Sov. Phys. Dokl.*, 28(9):720–722, 1983.

MATHEMATICAL MODELLING OF TIDAL MOTION IN REGIONAL WATERS OF SINGAPORE

N. JOTHI SHANKAR, H. F. CHEONG AND C.T. CHAN*
Department of Civil Engineering
The National University of Singapore
Kent Ridge, Singapore

Abstract. This paper describes briefly the development of a two-dimensional depth integrated numerical model 2DTIDFLO based on the numerical scheme of Stelling (1984) for the solution of shallow water wave equations. The application of the numerical model to regional waters of Singapore is demonstrated for the case of Singapore Straits by comparing the results of numerical simulation with the field measurements of tidal water levels and currents.

As a further enhancement of the regional model a finer grid nested model is established for the central region of Singapore waters for a more detailed study of the nearfield hydrodynamics. Some results of simulation using the nested modelling approach are presented.

INTRODUCTION

Numerical models of shallow water wave equations play an important role in coastal engineering applications. With these models, the impact on the coastal marine system as a result of large scale coastal development projects such as coastal reclamation, port and harbour construction, shore protection works, navigation channel and dredging activities can be evaluated. The rapid advances in computer technology and numerical techniques have encouraged the development of computer-based mathematical models for the analysis of complex interactive hydrodynamic processes involved in the coastal zone. The formulation and the solution procedures for mathematical models have been greatly improved and applied to finer scale problems.

*Presently Doctoral Student, Imperial College of Science, Technology & Medicine, U.K.

Reliable results are now obtainable from numerical simulations with an accuracy that is required for practical applications.

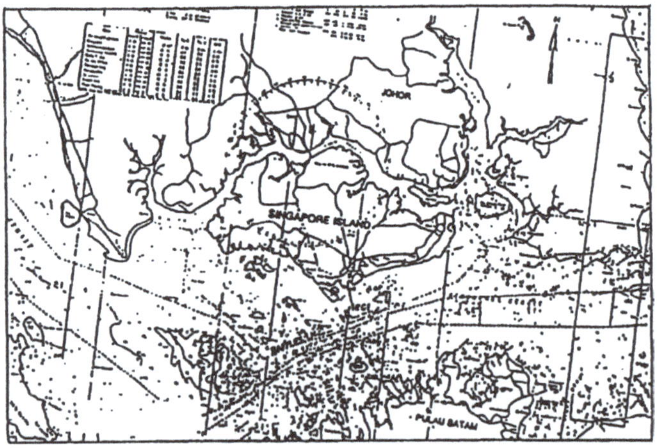

Figure 1. Study Area showing the Straits of Singapore

Singapore (Fig. 1) lies in a low wave energy environment but a strong current environment due to tides. It is protected by the Malay Peninsula in the north, Sumatra in the west and Rhio Archipelago in the south. The tidal variations in the Singapore Straits are significant with a tidal range in the order of 2.5 m - 3.0 m during spring tides and 0.7 m - 1.2 m during neap tides. These tidal occurences are semi-diurnal with one strong westerly stream and a strong easterly stream within a day. The velocities of the associated tidal streams vary from 0.5 - 1.0 m/s in the open waters of the Straits of Singapore to as much as 1.5 - 2.0 m/s in the constricted channels between the islands.

In view of the large magnitude in tidal level variations and the associated currents in the regional waters of Singapore careful evaluation of tidal hydrodynamic characteristics has become essential for the successful implementation of coastal engineering developments in the region. In order to obtain an accurate and reliable knowledge of tidal hydrodynamics involving tidal waves and currents and the extent to which they are altered as a result of large scale coastal engineering projects, mathematical/numerical models have become necessary tools for application to coastal waters.

This paper presents briefly the development of a two-dimensional depth integrated numerical model, 2DTIDFLO based on the numerical scheme of Stelling (1984) for solution of shallow water wave equations. The application of the numerical model is demonstrated for the case of Singapore Straits by comparing the results of numerical simulation with the data obtained from field measurements on tidal water levels and currents. As a further enhancement of the regional hydrodynamic model, a nested model of finer grid resolution was established for the central region of Singapore coastal waters for a more detailed characterization of the near-field hydrodynamics. The setting up and implementation of nested modelling approach for study of near-field problems are also briefly described.

TWO-DIMENSIONAL NUMERICAL MODELS OF SHALLOW WATER EQUATIONS

Developments and applications of 2-D numerical models to simulate tidal circulation in coastal seas and estuaries have been carried out rigorously over the last three decades (Dronkers, 1964; Leendertse, 1967; Reid and Bodine, 1968; Butler, 1978; Lean and Weare, 1979; Abbot, 1979); Stelling, 1984; Falconer and Owens, 1987.

The main advantage of using the finite difference scheme is its simplicity and ease of programming. A number of techniques have been devised to improve the stability and resolution of finite difference schemes. These include the Alternating Direction Implicit (ADI) scheme developed by Leendertse (1967), a fully implicit scheme by Butler (1978) and an economical fully implicit scheme, the Douple Sweep method by Abbot (1979) and an improved finite difference approximation scheme (ADI) proposed by Stelling (1984). All the above numerical finite difference models have been applied to prototype situations with varied degrees of success. With the successful application 2-D numerical models for shallow water flow problems, 2-D finite difference numerical models have established themselves as standard tools for simulation of 2-D hydrodynamic flow behaviour in coastal seas and estuaries.

A Finite Difference Method for Nonlinear Shallow Water Wave Equations

The numerical model developed in the present work is based on the numerical scheme introduced by Stelling (1984). The model utilizes the vertically integrated equations of motion and continuity for incompressible fluid. The long wave approximation is used with the neglect of vertical acceleration and velocities. This results in a set of nonlinear equations to solve shallow water wave problems.

The governing shallow water equations referred to a rectangular coordinate system with x-axis directed eastward and y-axis directed northward are given by:

$$\frac{\partial u}{\partial t}+u\frac{\partial u}{\partial x}+v\frac{\partial u}{\partial y}-fv+g\frac{\partial \varsigma}{\partial x}+gu\frac{(u^2+v^2)^{\frac{1}{2}}}{C^2(d+\varsigma)}-\nu\nabla^2 u=F^{(x)} \quad (1)$$

$$\frac{\partial v}{\partial t}+u\frac{\partial v}{\partial x}+v\frac{\partial v}{\partial y}+fu+g\frac{\partial \varsigma}{\partial x}+gu\frac{(u^2+v^2)^{\frac{1}{2}}}{C^2(d+\varsigma)}-\nu\nabla^2 v=F^{(y)} \quad (2)$$

$$\frac{\partial \varsigma}{\partial t}+\frac{\partial (Hu)}{\partial x}+\frac{\partial (Hv)}{\partial y}=0 \quad (3)$$

where:

u, v : depth averaged velocity components in $x-$ and $y-$ direction

ς : water elevation above some plane of reference

d : water depth below some plane of reference (bottom profile)

$H = d + \varsigma$: total water depth

f : Coriolis parameter

ν : eddy viscosity

g : acceleration due to gravity

C : Chezy coefficient for bed resistance

$F^{(x)}, F^{(y)}$: external forces (eg. wind stress) in $x-$ and $y-$ direction

The above equations are based on the assumption that the variations in water density and temperature are insignificant and the pressure variation is hydrostatic.

The solution of Eqs. (1) to (3) is achieved numerically using an alternating direction implicit (ADI) method. While maintaining the similar structure of the Leendertze method the improved finite difference approximation scheme proposed by Stelling (1984) satisfies the following important requirements:

(i) The scheme is linearly unconditionally stable with the numerical solution at least of second order accuracy and not greatly subjected to spurious and rounding off errors.

(ii) The scheme is able to handle wide range of practical 2-D flow problems in Civil Engineering such as tidal flow simulation in coastal seas and steady state problems in free surface flows.

(iii) The numerical treatment of boundary conditions are such that overall accuracy and efficiency of the scheme are not greatly decreased.

(iv) The model is numerically robust and the solution method is computationally efficient.

The finite-difference approximations are based on a fully staggered grid scheme shown in Fig. 2 for the spatial discretization. The indices m and n in the finite difference formulation run from 1 to the respective limits of the computational domain in the x and y directions. All space derivatives are approximated by central differences except for the cross-advective terms $\partial u/\partial y$ and $\partial v/\partial x$ which are approximated by a weighted central difference (explicit) and a second order upwind difference (implicit). The time derivatives are approximated using forward difference scheme with half time steps in two stages.

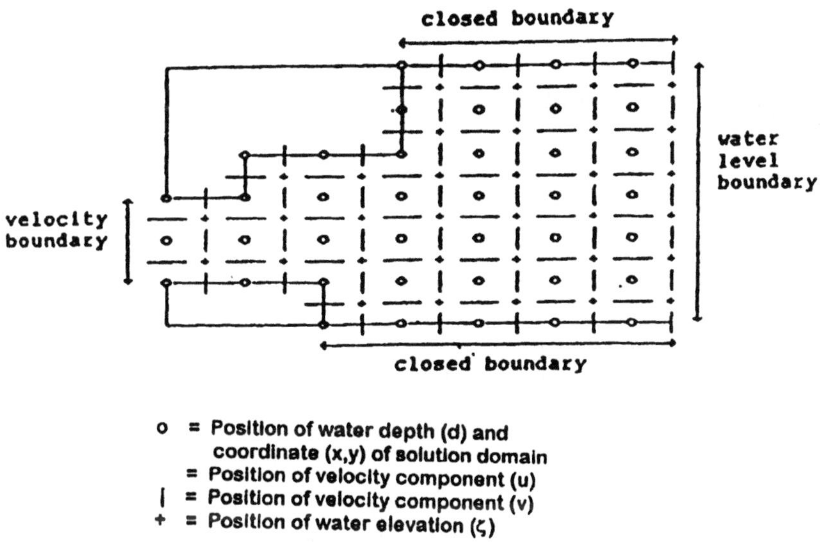

Figure 2. Space Staggered Grid with Variable Definitions

The complete forms of the finite difference method for the approximation of nonlinear shallow water equations are furnished below.

In the computational scheme, for a variable A at node (m,n) and time step k, the following nomenclature is used.

$$A_{0x} = [A^k_{m+\frac{1}{2},n} - A^k_{m-\frac{1}{2},n}] \;/\; \Delta X \quad \text{(simple difference along } x\text{)} \qquad (4)$$

$$A_{0y} = [A^k_{m,n+\frac{1}{2}} - A^k_{m,n-\frac{1}{2}}] \;/\; \Delta Y \quad \text{(simple difference along } y\text{)} \qquad (5)$$

$$\overline{A}^x = [A^k_{m+\frac{1}{2},n} + A^k_{m-\frac{1}{2},n}] \,/\, 2 \qquad \text{(simple average along } x\text{)} \tag{6}$$

$$\overline{A}^y = [A^k_{m,n+\frac{1}{2}} + A^k_{m-\frac{1}{2},n-\frac{1}{2}}] \,/\, 2 \qquad \text{(simple average along } y\text{)} \tag{7}$$

$$\overline{\overline{A}} = [A^k_{m+\frac{1}{2},n+\frac{1}{2}} + A^k_{m-\frac{1}{2},n+\frac{1}{2}} + A^k_{m+\frac{1}{2},n-\frac{1}{2}} + A^k_{m-\frac{1}{2},n-\frac{1}{2}}]/4 \tag{8}$$

Stage 1:

$$u^{[o]} = u^k \quad v^{[o]} = v^k \quad \varsigma^{[o]} = \varsigma^k$$

for p=1,2 and q=1,2, ...Q:

$$\left(u^{[q]} - u^k \right) / \tfrac{1}{2}\tau + u^{[q]}\overline{u^k_{ox}}^y + S_{oy}\left(\overline{\overline{v}}^{k+\frac{1}{2}}, u^k \right) - f\overline{\overline{v}}^{k+\frac{1}{2}}$$

$$+g\zeta^{[q]}_{0x} + gu^{[q]}\left[(\overline{\overline{v}}^{k+\frac{1}{2}})^2 + (u^k)^2 \right]^{\frac{1}{2}} /(C^2H^k) - \nu(u^k_{oxx} + u^k_{oyy}) = 0,$$

at $m+\frac{1}{2}, n$

(9)

$$(v^{[p]} - v^k)/\tfrac{1}{2}\tau + v^k\,\overline{v^{[p]}_{oy}}^x + S_{+x}\left[\overline{\overline{u}}^k, v^{[p]}, s(p+p') \right] +$$

$$f\overline{\overline{u}}^k + g\zeta^k_{oy} + gv^{[p]}\left[(\overline{\overline{u}}^k)^2 + (v^k)^2 \right]^{\frac{1}{2}} /(C^2H^k) - \nu(u^{[*]}_{oxx} + u^{[p]}_{oyy}) = 0,$$

at $m, n+\frac{1}{2}$

(10)

$$(\zeta^{[q]} - \zeta^k)/\tfrac{1}{2}\tau + (\overline{h}^y u^{[q]})_{0x} + \zeta^{[q-1]}u^{[q]}_{0x} + \overline{u^{[q-1]}\zeta^{[q]}_{0x}}^x + (H^k v^k)_{oy} = 0,$$

at m, n

(11)

$$u^{k+\frac{1}{2}} = u^{[Q]}, v^{k+\frac{1}{2}} = v^{[2]}, \zeta^{k+\frac{1}{2}} = \zeta^{[Q]}$$

where

$$S_{oy}\left(\overline{\overline{v}}^{k+\frac{1}{2}}, u^k \right) \text{ at } m+\tfrac{1}{2}, n = \overline{\overline{v}}^{k+\frac{1}{2}}_{m+\frac{1}{2},n}\left(u^k_{m+\frac{1}{2},n+2} + \right.$$

$$\left. 4u^k_{m+\frac{1}{2},n+1} - 4u^k_{m+\frac{1}{2},n-1} - u^k_{m+\frac{1}{2},n-2} \right)/12\Delta y$$

$$S_{+x}\left[\ \overline{\overline{u}}^k, v^{[p]}, s\ \right] \text{ at } m, n+\tfrac{1}{2} = \begin{cases} \overline{\overline{u}}^k_{m,n+\frac{1}{2}}(3v^{[p-s]}_{m,n+\frac{1}{2}} - 4v^{[p-s]}_{m-1,n+\frac{1}{2}} + \\ v^{[p-s]}_{m-2,n+\frac{1}{2}}/2\Delta x \quad \text{if } \overline{\overline{u}}^k_{m,n+\frac{1}{2}} > 0 \\ \overline{\overline{u}}^k_{m,n+\frac{1}{2}}(-3v^{[p-1+s]}_{m,n+\frac{1}{2}} + 4v^{[p-1+s]}_{m+1,n+\frac{1}{2}} - \\ v^{[p-1+s]}_{m+2,n+\frac{1}{2}}/2\Delta x \quad \text{if } \overline{\overline{u}}^k_{m,n+\frac{1}{2}} \leq 0 \end{cases}$$

$$s(p+p') = \tfrac{1}{2}[1+(-1)^{p+p'}]$$

$$p' = \begin{cases} 0, \text{ if } \sum\limits_{m,n} u^k > 0 \\ 1, \text{ if } \sum\limits_{m,n} u^k \leq 0 \end{cases}$$

$\left(\sum\limits_{m,n} u \text{ denotes the sum of } u \text{ over all grid points} \right)$

and

$$v^{[*]}_{0xx} \text{ at } m, n+\tfrac{1}{2} = \left(v^{[p-1+s(p+p')]}_{m+1,n+\frac{1}{2}} - 2v^{[p]}_{m,n+\frac{1}{2}} + v^{[p-s(p+p')]}_{m-1,n+\frac{1}{2}} \right) / \Delta x^2$$

$$u^{[0]} = u^{k+\frac{1}{2}}, v^{[0]} = v^{k+\frac{1}{2}}, \zeta^{[0]} = \zeta^{k+\frac{1}{2}},$$

for $p = 1, 2$ and $q = 1, 2, \ldots., Q$:

$$\begin{aligned} &\left(u^{[p]} - u^{k+\frac{1}{2}} \right) / \tfrac{1}{2}\tau + u^{k+\frac{1}{2}}\overline{u^p}_{ox} + S_{+y}\left[\ \overline{\overline{v}}^{k+\frac{1}{2}}, u^{[p]} \right. \\ &s(p+p') - f v^{=k+\frac{1}{2}} + g\zeta_{ox}^{k+\frac{1}{2}} + g u^{[p]}\left[\ (\overline{\overline{v}}^{k+\frac{1}{2}})^2 + (u^{k+\frac{1}{2}})^2\ \right]^{\frac{1}{2}} \\ &/(C^2 H^{k+\frac{1}{2}}) - \nu(u^{[p]}_{oxx} + u^*_{oyy}) = 0, \qquad \text{at } m+\tfrac{1}{2}, n \end{aligned} \tag{12}$$

$$\begin{aligned} &(v^{[q]} - v^k)/\tfrac{1}{2}\tau + \quad \overline{v^{[q]} v_{0y}^{k+\frac{1}{2}}}^{x} + S_{0x}\left(\overline{\overline{u}}^{k+1}, v^{k+\frac{1}{2}} \right) \\ &+ f\overline{\overline{u}}^{k+1} + g\zeta^{[q]}_{oy} + g v^{[q]}\left[\ (v^{k+\frac{1}{2}})^2 + (\overline{\overline{u}}^{k+1})^2\ \right] / (C^2 H^{k+\frac{1}{2}}) \\ &-\nu\left(v_{oxx}^{k+\frac{1}{2}} + v_{oyy}^{k+\frac{1}{2}} \right) = 0, \qquad \text{at } m, n+\tfrac{1}{2} \end{aligned} \tag{13}$$

$$u^{k+1} = u^{[k]}, v^{k+1} = v^{[Q]}, \zeta^{k+1} = \zeta^{[Q]}$$

where

$$S_{+y}\left[\,\overline{\overline{v}}^{k+\frac{1}{2}}, v^{[p]}, s\,\right] \text{ at } m, n+\frac{1}{2} = \begin{cases} \overline{\overline{v}}^{k+\frac{1}{2}}_{m+\frac{1}{2},n}(3v^{[p-s]}_{m+\frac{1}{2},n} - 4v^{[p-s]}_{m+\frac{1}{2},n-2} + \\ u^{[p-s]}_{m+\frac{1}{2},n-2})/2\Delta y \qquad \text{if } \overline{\overline{v}}^{k+\frac{1}{2}}_{m+\frac{1}{2},n} > 0 \\ \overline{\overline{v}}^{k+\frac{1}{2}}_{m+\frac{1}{2},n}(-3u^{[p-1+s]}_{m+\frac{1}{2},n} + 4u^{[p-1+s]}_{m+\frac{1}{2},n+1} - \\ u^{[p-1+s]}_{m+\frac{1}{2},n+2})/2\Delta y \qquad \text{if } \overline{\overline{v}}^{k+\frac{1}{2}}_{m+\frac{1}{2},n} \leq 0 \end{cases}$$

$$s(p+p') = \tfrac{1}{2}[1+(-1)^{p+p'}]$$

$$p' = \begin{cases} 0, \text{ if } \sum\limits_{m,n} v^{k+\frac{1}{2}} > 0 \\ 1, \text{ if } \sum\limits_{m,n} v^{k+\frac{1}{2}} \leq 0 \end{cases}$$

($\sum\limits_{m,n} v$ denotes the sum of all grid points)

$$S_{ox}\left(\,\overline{\overline{u}}^{k+1},\ v^{k+\frac{1}{2}}\,\right) \text{ at } m, n+\tfrac{1}{2} =$$

$$\overline{\overline{u}}^{k+1}_{m+\frac{1}{2}}\left(\,v^{k+\frac{1}{2}}_{m+\frac{1}{2},n+\frac{1}{2}} + 4v^{k+\frac{1}{2}}_{m+1,n+\frac{1}{2}} - \right.$$

$$\left. 4v^{k+\frac{1}{2}}_{m-1,n+\frac{1}{2}} - v^{k+\frac{1}{2}}_{m-2,n+\frac{1}{2}}\,\right)/12\Delta x$$

and

$$u^{[*]}_{oyy} \text{ at } m+\frac{1}{2}, n = \left(\,u^{[p-1+\delta(p+p')]}_{m+\frac{1}{2},n+1} - 2u^{[p]}_{m+\frac{1}{2},n} + u^{[p-\delta(p+p')]}_{m+\frac{1}{2},n-1}\,\right)/\Delta y^2$$

At stage 1, Eq. (10) is an implicit equation. Because of the definition of p', Eq. (10) is solved column by column in the dominant flow direction of u. If the sign of u is constant then Eq. (10) is solved in one sweep, otherwise second iteration is necessary with sweep in the opposite direction. After Eq. (10) is solved, Eqs. (9) and (11) are solved. Since these two equations are coupled implicitly, they are solved simultaneously. By substitution of Eq. (9) at $(m-\frac{1}{2}, n)$ and $(m+\frac{1}{2}, n)$ into Eq. (11) at (m,n) the implicit equations are tri-diagonal and of the same size as the tridiagonal equation of (10).

At stage 2, Eq. (12) is implicit and is solved similarly as Eq. (10). The coupled implicit equations (13) and (14) are solved according to Eqs. (9) and (11). By solving Eqs. (9) to (14) in the sequence (10), [(9) and (11)], (12), [(13) and (14)] the computer implementation needs only one array per dependent variable (u, v, ζ) and one working array of the size of the member of "ζ - points" of the grid. Therefore, this finite difference method is efficient with respect to storage.

Boundary Conditions

To complete the shallow water equation given in Eqs. (1) to (3), boundary conditions need to be imposed. In general, two different type of boundaries would be encountered in a computational domain, closed boundaries and open boundaries. Closed boundaries are referred to physical or existing land-water boundaries whereas open boundaries are mathematical boundaries introduced to restrict the size of the domain of the problems.

Closed boundaries

At closed boundaries, the following boundary conditions are prescribed:

$$u_{\perp} = 0 \tag{14}$$

$$(1 - \alpha)u_{\shortparallel} + \Delta\alpha\frac{\partial}{\partial n}(u_{\shortparallel}) = 0 \tag{15}$$

where $u_{\perp}$ denotes the velocity normal to the boundary
$u_{\shortparallel}$ denotes the velocity parallel to the boundary
$\frac{\partial}{\partial n}$ denotes the derivative normal to the boundary
Δ denotes spatial discretization ΔX or ΔY

Equation (15) shows that when $\alpha = 1$, a 'perfect slip' boundary condition is presented while $\alpha = 0$, a 'non-slip' boundary condition will be prescribed. In the present programme development, only perfect slip condition is adopted. These boundary conditions can be readily implemented on the space staggered grid domain (Fig. 2).

Open boundaries

Open boundaries are artificial water-water boundaries that are arbitrarily drawn across a wider flow field to restrict the domain of the problem. The boundary conditions at open boundaries are prescribed as:

$$u_{\perp} = f^{u}(t) \text{ for velocity boundary} \tag{16}$$
$$\zeta = f^{\zeta}(t) \text{ for water level boundary} \tag{17}$$
$$u_{//} = 0 \tag{18}$$
$$\partial(u_{//})/\partial n = 0 \tag{19}$$

where $f^{u}(t)$ denotes the velocity boundary values and $f^{\zeta}(t)$ denotes the water level boundary values. The boundary conditions Eqs. (18 - 19) are prescribed at inflow situations as extrapolations procedures are sensitive to instabilities in practical problems. The well-posedness of the boundary conditions Eqs. (16) to (19) is treated by Verboom et al. (1981). Because of the use of staggered grid, special boundary treatment for the prescribed boundary conditions is only necessary for the advection terms. The open boundary conditions can be readily implemented on the space staggered grid domain.

Structure of the implicit equations

It is known that the stability and accuracy of a numerical model will depend on the method of discretization and also the sensitivity of the solving algorithm to rounding errors (Wilkinson [1965]). In this context the structure of the present FDM matrix and the restriction imposed on its solution method are briefly discussed.

The structure of the finite difference procedures described previously can be seen by looking at the coefficients of the implicit part of Eqs. (9) and (11) given as follows:

$$A'_{m-\frac{1}{2}}\zeta^{[p+1]}_{m-1,n} + u^{[p+1]}_{m-\frac{1}{2},n} + B'_{m-\frac{1}{2}}\zeta^{[p+1]}_{m,n} = C'_{m-\frac{1}{2}}$$

$$A_{m-1}\zeta^{[p+1]}_{m-1,n} + B_{m-\frac{1}{2}}u^{[p+1]}_{m-\frac{1}{2},n} + C_m\zeta^{[p+1]}_{m,n} + D_{m+\frac{1}{2}}u^{[p+1]}_{m+\frac{1}{2},n} + E_{m+1}\zeta^{[p+1]}_{m+1,n} = F_{m,n}$$

$$A'_{m+\frac{1}{2}}\zeta^{[p+1]}_{m,n} + u_{m+\frac{1}{2},n} + B'_{m+\frac{1}{2}}\zeta^{[p+1]}_{m+1,n} = C'_{m+\frac{1}{2}} \tag{20}$$

where :

$$A'_{m+\frac{1}{2}} = (-g/\Delta s)/\left\{2/\Delta t + (u^k_{m+3/2,n} - u^k_{m-\frac{1}{2},n})/(2\Delta S) + g\left[(u^k_{m+\frac{1}{2},n})^2 + (\overline{v^{k+\frac{1}{2}}_{m+\frac{1}{2},n}}^{m,n})^2\right]^{\frac{1}{2}} C^2(\overline{d}^n + \overline{\zeta^k}^m)m + \tfrac{1}{2}, n\right\}$$

$$B'_{m+\frac{1}{2}} = A'_{m+\frac{1}{2}}$$

$$C'_{m+\frac{1}{2}} = \left\{ \overline{v^{k+\frac{1}{2}}_{m+\frac{1}{2},n}}^{m,n} (u^k_{m+\frac{1}{2},n+\frac{1}{2}} + 4u^k_{m+\frac{1}{2},n-1} - u^k_{m+\frac{1}{2},n-2})/ (12\Delta s) + f\overline{v^{k+\frac{1}{2}}_{m+\frac{1}{2}}}^{m,n} + u^k_{m+\frac{1}{2},n}(2/\Delta t) + \nu[(u^k_{m+\frac{3}{2},n} - 2u^k_{m+\frac{1}{2},n} + u^k_{m-\frac{1}{2},n})/(\Delta s^2) - (u^k_{m+\frac{1}{2},n+1} - 2u^k_{m+\frac{1}{2},n} + u^k_{m+\frac{1}{2},n-1})/\Delta s^2] + R^{(x)}_{m+\frac{1}{2},n} \right\} / \left\{ 2/\Delta t + (u^k_{m+\frac{3}{2},n} - u^k_{m-\frac{1}{2},n})/(2\Delta s) + g\left[(u^k_{m+\frac{1}{2},n})^2 + \left(\overline{v^{k+\frac{1}{2}}_{m+\frac{1}{2},n}}^{m,n}\right)^2 \right]^{\frac{1}{2}} / C^2(\overline{d}^n + \overline{\zeta^k}^m)_{m+\frac{1}{2},n} \right\}$$

$$A_{m-1} = -(\tfrac{1}{2}\Delta s)u^k_{m-\frac{1}{2},n}$$

$$B_{m-\frac{1}{2}} = \left(\overline{d}^n_{m-\frac{1}{2},n} + \zeta^{[p]}_{m,n} \right)/(2\Delta s)$$

$$C_m = (2\Delta t) + \left(u^k_{m-\frac{1}{2},n} - u^k_{m+1,n} \right)/(2\Delta s)$$

$$D_{m+\frac{1}{2}} = \left(\overline{d}^n_{m+\frac{1}{2},n} + \zeta^{[p]}_{m,n} \right)/(2\Delta s)$$

$$E_{m+1} = -(\tfrac{1}{2}\Delta s)u^k_{m+\frac{1}{2},n}$$

Equation (20) is penta-diagonal, but by substitution of u from the tridiagonal parts into the penta-diagonal part Eq. (20) can be reduced to

$$\zeta^{[p+1]}_{m-1,n}\left(A_{m-1}+B_{m-\frac{1}{2}}A'_{m-\frac{1}{2}}\right)+\zeta^{[p+1]}_{m,n}\left(C_{m-1}+B_{m-\frac{1}{2}}B'_{m-\frac{1}{2}}+\right.$$

$$\left.D_{m+\frac{1}{2}}A'_{m+\frac{1}{2}}\right)+\zeta^{[p+1]}_{m+1,n}\left(E_{m+1}+D_{m+\frac{1}{2}}B'_{m+\frac{1}{2}}\right)=F_{m-n}-$$

$$B_{m-\frac{1}{2}}C'_{m-\frac{1}{2}}-D_{m+\frac{1}{2}}C'_{m+\frac{1}{2}} \tag{21}$$

Equation (21) represents a system of tridiagonal equations and therefore can be solved by a simple recursive algorithm. However, to prevent amplification of rounding off errors the following relations must be satisfied when the recursive double sweep method is to be employed (Stelling 1984):

$$\text{Max}\left(\ |u_{m+\frac{1}{2},n}|(\zeta/\Delta s), |v_{m,n+\frac{1}{2}}|(\zeta/\Delta s)\ \right)\leq 4 \tag{22}$$

In order to avoid the restriction of allowable timestep by Eq, (22), Gauss-elimination with partial pivoting is adopted in the present program.

NUMERICAL TIDAL MODEL - VALIDATION AND PRACTICAL APPLICATIONS REGIONAL HYDRODYNAMIC MODEL

A regional spatially two-dimensional numerical tidal hydrodynamic model, 2DTIDFLO covering the whole of Singapore with the southern part of Johore forming the northern boundary, the Indonesian Rhio islands forming the southern boundary with the Malacca Straits and the South China Sea forming the West and East boundaries respectively. The regional model has four open boundaries, one each at the eastern and western boundaries and two at the southern boundary. The regional model is oriented at 7.5 degrees counter clockwise relative to geographical north and covers an area of 107 km x 72 km with a grid size of 1000m x 1000 m. Figure 3 shows the discretization of the model. The seabed topography is extremely complex as depicted in Fig. 4.

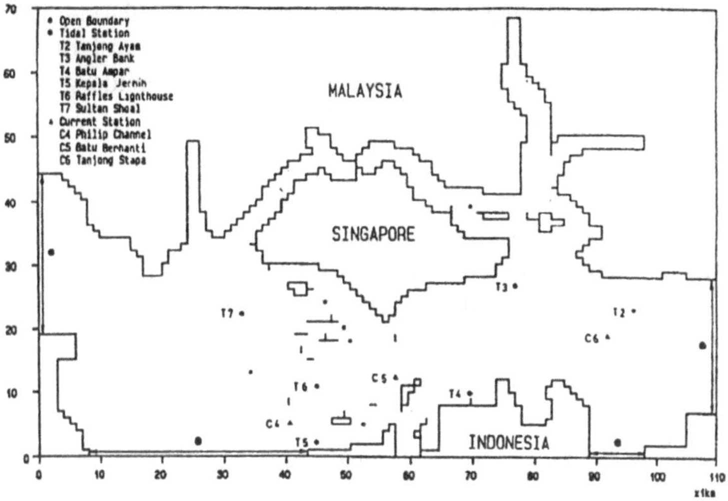

Figure 3. Computational Domain of Regional Model Showing Locations of Tidal Current Stations.

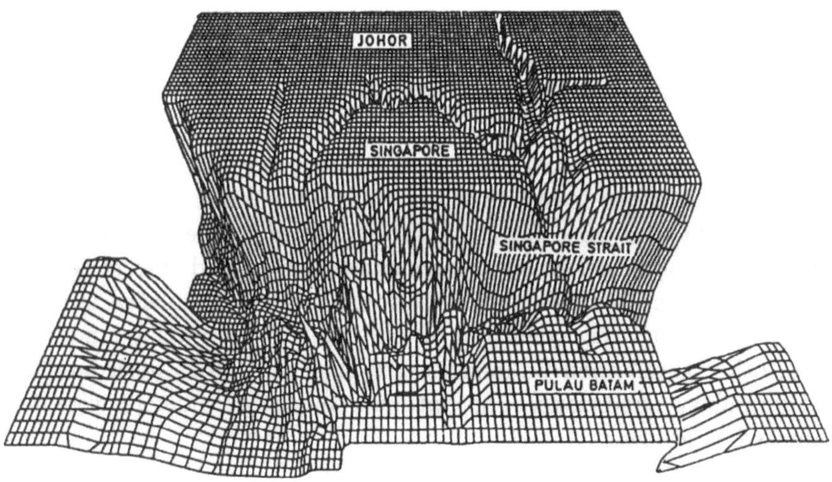

Figure 4. Seabed Topography of the Study Area

MODEL TESTING-CALIBRATION AND VERIFICATION OF NUMERICAL REGIONAL MODEL

The testing of 2DTIDFLO commenced with the establishment of a regional hydrodynamic model on IBM 3081 mainframe. This large area of the regional model is established to allow for a far field simulation of regional tidal hydrodynamics. Four open boundaries - one at the western boundary, one at the eastern boundary, and two at the southern boundary constituted the limits of the regional numerical model (Fig. 3).

The computer runs for the model calibration and verification process were carried out based on a 14 days spring-neap tidal cycle involving prototype measurements carried out in August 1978. Two separate periods were selected for the calibration and verification of the model.

- Calibration period: Spring tide condition

  1200 hrs 5 August 1978 to 0000 hrs 8 August 1978

- Verification period: Neap tide condition

  1000 hrs 12 August 1978 to 2200 hrs 14 August 1978

A series of test runs was conducted during the calibration process involving the adjustment of bottom roughness parameter (Chezy resistance coefficient, C), eddy viscosity coefficient and the numerical computational time step. The model was operated with water level boundary conditions prescribed at the four open boundaries. These boundary data and measurements necessary for the calibration were supplied by the Port of Singapore Authority (PSA) based on simultaneous water level and current recordings made during the Joint Tidal and Current Straits of Malacca and Singapore (1978). With the exception of the western open sea boundary condition, other three open boundaries are prescribed with horizontal water level boundary conditions. The optimum model parameters were finally calibrated as follows:

- Chezy coefficient $c = 65m^{\frac{1}{2}}/s$

- eddy viscosity $\nu = 10m^2/s$

Six tidal stations and three current stations with continuous field data were used to examine the numerical results. The location of the nine stations corresponding to the model coordinates are presented in Table 1 below (also see Fig. 3).

STATION	NAME	MODEL COORDINATES
TID 2	TANJONG AYAM	96, 26
TID 3	ANGLER BANK	77, 30
TID 4	BATU AMPAR	70, 13
TID 5	KEPALA JERNIH	46, 4
TID 6	RAFFLES LIGHTHOUSE	46, 14
TID 7	SULTAN SHOAL	34, 25
CUR 4	PHILIP CHANNEL	41, 8
CUR 5	BATU BERHANTI	48, 15
CUR 6	TANJONG STAPA	92, 22

Table 1. Prototype Stations for Calibration

The results of the computed water surface elevations are depicted in Figs. 5a - 5c together with the field measurement of the tidal amplitudes. It can be observed from these figures that the simulated and measured water surface elevations at all the tidal stations show very good agreement for the calibration runs. The magnitude as well as the phase of the water surface elevation were well simulated for the spring tide conditions. Minor deviations of tidal variation can be observed from the comparisons of the results at the station and Batu Ampar (Fig. 5a). These discrepancies could be attributed to the coarse grid approximation error near the island group of Indonesia. The regional hydrodynamic model was further verified by comparison of computed water surface elevations against measured water levels for neap tide conditions.

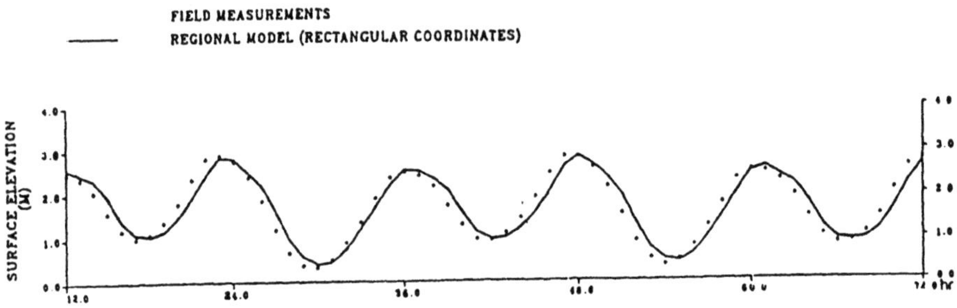

Figure. 5a. Measured and Computed Tidal Water Levels at Batu Ampar

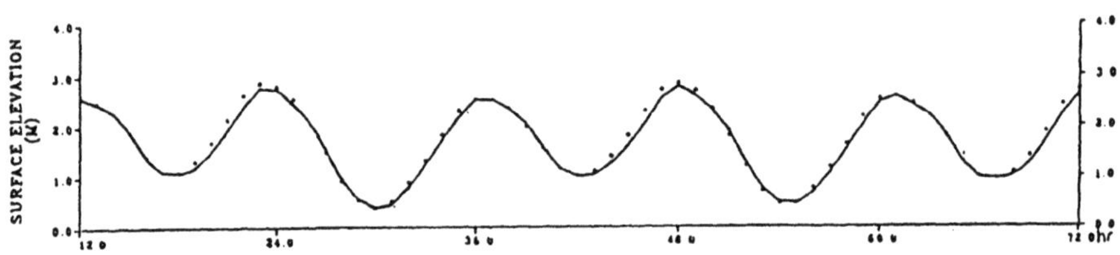

Figure. 5b. Measured and Computed Tidal Water Levels at Angler Bank

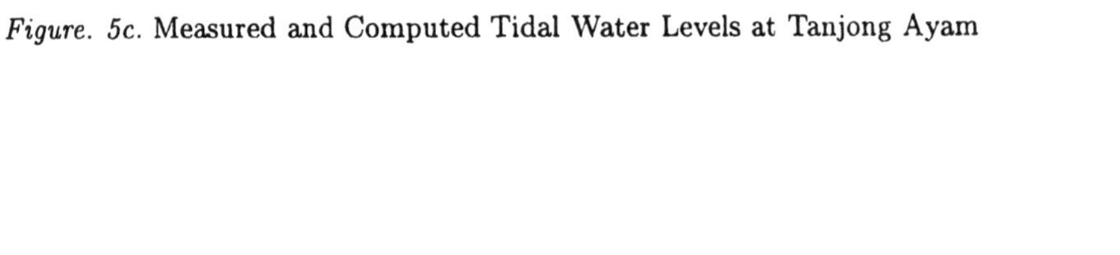

Figure. 5c. Measured and Computed Tidal Water Levels at Tanjong Ayam

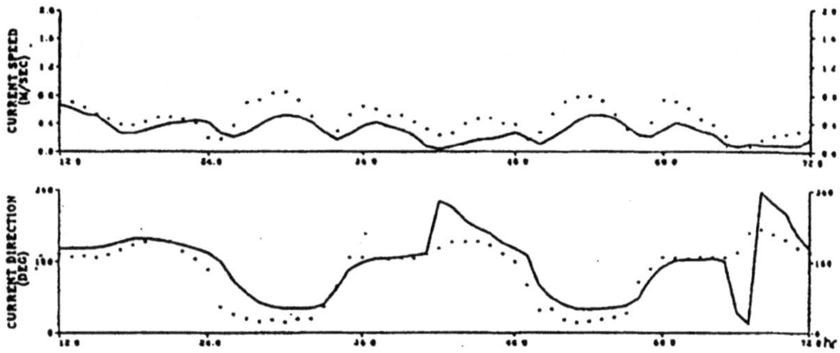

Figure 6a. Measured and Computed Currents at Phillip Channel

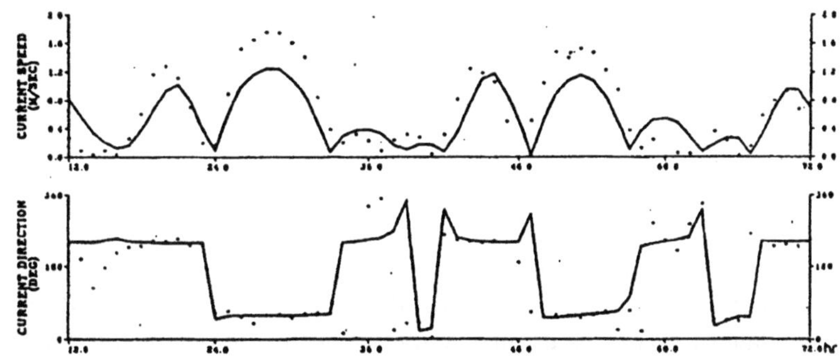

Figure 6b. Measured and Computed Currents at Batu Berhanti

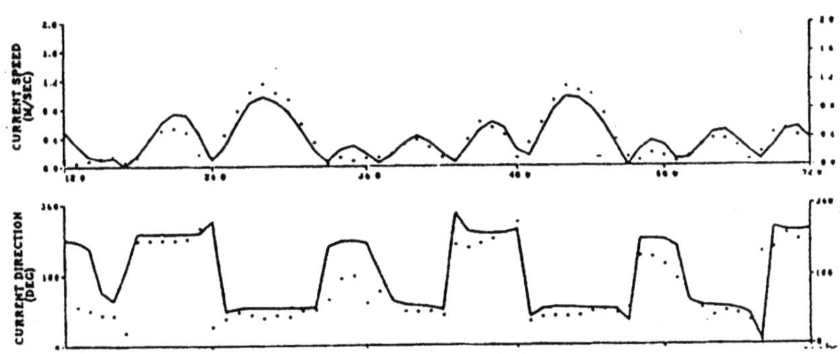

Figure 6c. Measured and Computed Currents at Tanjong Stapa

The difficult part in the model calibration and verification processes is that of achieving accurate simulation of tide induced currents with regard to both magnitude and direction at the selected stations. As can be observed from Figs. 6a - 6c, that although good correlation is obtained between the computed and measured currents at all the three current stations, the agreement is inferior to that observed in the comparison of water surface elevation. While the current directions are simulated satisfactorily the computed magnitudes are relatively lower than the measured values. Possible reasons for this deviation could be the association of stronger current flow with mid-depth measurements as compared to the depth averaged computed results and the coarse grid resolution.

It was found in the calibration runs that current magnitudes were relatively more sensitive to the Chezy C parameter than the viscosity coefficient. An increase in Chezy value would lead to measurable increase in current speeds, and vice-versa. The calibrated Chezy coefficient of 65 presented previously is indeed physically meaningful. This value is consistent with the Manning sea bed roughness in the clayey-silt range, along the nearshore region South-East of Singapore Coastal Waters. Chezy C parameter was derived from the Manning coefficient given by

$$C = \frac{1}{n} H^{\frac{1}{6}} \tag{23}$$

where n is the Manning roughness coefficient and H is the water depth. It is obvious that the successful calibration of the Regional Model with a constant value of Chezy C implies a spatial variation of bottom roughness in the coastal region in view of the varying water depths.

During the calibration of the eddy coefficient (ν), no standard formulation was adopted. As the eddy coefficient in the present formulation of shallow water equations is treated as a semi-empirical factor describing the effects of non-uniform vertical velocity distribution, it was kept low to fulfill the requirement of nearly horizontal flow assumption. Because of its insignificant influence upon the computed water level and current magnitudes, a value of 10 has been adopted. This value has posed no numerical difficulty for the present coarse-grid-large-timestep Regional Model.

In order to have a spatial examination of simulated tidal currents within the model domain, instantaneous current vectors are output at every three hour interval. Figs. 7a and 7b show the current circulation patterns in the regional model corresponding to flood (west going) and ebb (east going currents). This simulation period corresponds to the input conditions of a typical spring tide covering 5-6 August 1978. A detailed examination of the current patterns shows that the tide induced current is favourably correlated with the input conditions (excitation tides at the model boundaries).

The results depict a stable trend of current flow from high water to low water. While the measured currents at the three current stations provided partial verification of the circulation pattern, complete verification of the current pattern is an expensive task as this would call for a large network of current stations. However based on PSAs experiences in Singapore coastal waters and schematic drawings from pilots the overall current patterns have been found to conform to actual flow patterns.

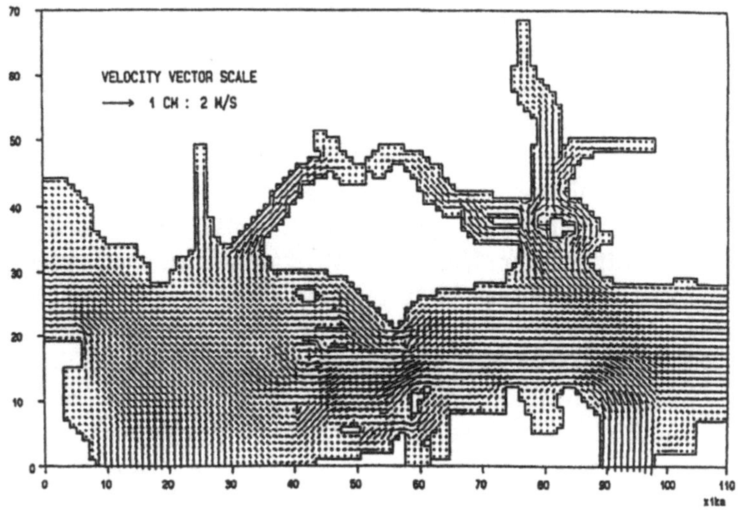

Figure 7a. Circulation Pattern at 2100 hrs, 5 August 1987 - Regional Model

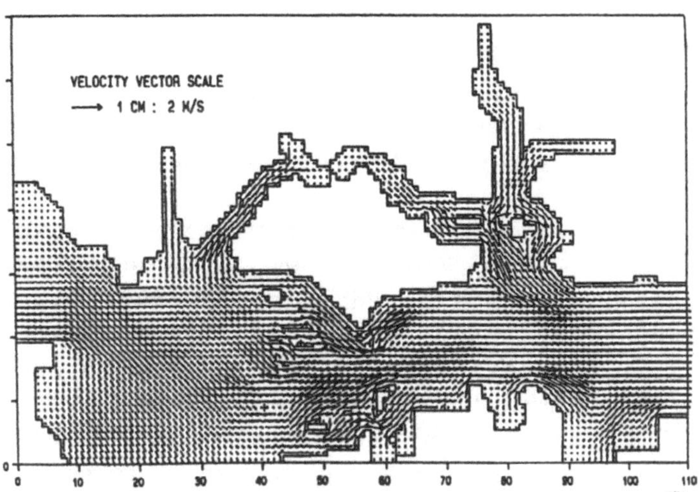

Figure 7b. Circulation Pattern at 0300hrs, 6 August 1987 - Regional Model

Based on the abovementioned calibration runs and verification of the regional model, the performance of the shallow water numerical model, 2DTIDFLO can be considered to be very satisfactory.

NESTED MODEL - NUMERICAL SIMULATIONS

The application of mathematical models to small scale problems requires a small grid increment in the numerical simulation procedure. Two common methods are the local grid refinement technique and the approach of nested modelling. Nested model covers only part of the problem area and get its boundary data from a model that covers a larger area. Nested model approach is adopted in the present studies as one can use the existing code with minor adjustment to facilitate the transfer of boundary data and to utilize the full available storage capacity for the detail model.

As the problem domain in a detail model covers only a small area and the inflow and outflow boundaries are not far apart, dissipation by bottom friction is relatively inactive. As a result disturbances introduced at the boundaries may persist for a long time in such a model. Further there are no rigorous guidelines for the prescription of boundary conditions with particular reference to the combination of transfer water level and velocity boundary conditions at the various model open boundaries. In this paper a brief discussion on the experience of the boundary conditions setup and their effect on the flow behaviour in the south-central region of Singapore coastal waters are presented. The data for the nested model (Central Model) is obtained from a calibrated regional model that covers the far-field region of Singapore Strait.

Model Setup and Boundary Conditions for Nested Models

Nested models are required when detailed information is needed in a specific region without having to consider the surrounding area. However, boundary conditions for such a local detail model usually cannot be obtained from field measurements.

When boundary conditions for the detail model are extracted from the results of a corresponding coarse grid model, the accuracy of detail model simulation should be such as to maintain the same order of accuracy as the coarse grid model and at the same time gain the advantage of improved spatial discretization. As the transfer of boundary data in a nesting process is internally consistent, viz, the surface elevation and discharges satisfy the equations of continuity, only minor adjustments are usually needed to calibrate the detail model. However, certain general guidelines should be borne in mind in order to reduce any boundary conditions problems that may arise. From our past experiences of shallow water flow simulations carried out in the region of Singapore coastal waters with system 2DTIDFLO at the National University of Singapore (Shankar et al [1990a, 1990b]), several features are found to be particularly important in establishing a well behaved nested model. In the order of significance these features are highlighted below as general guidelines:

(i) The boundaries of the detail model should be chosen so that the main flows are nearly perpendicular to the boundaries. If this is difficult to fulfill, then the boundaries should be located near regions of low velocity.

(ii) The schematization of local topography near open boundaries might differ appreciably from the coarse grid model, especially when island groups are present. A bottom adjustment zone for bathymetry is usually needed to reduce the discrepancies. A section of one or two grid intervals should be provided at these boundaries in which the bottom profile can be gradually adjusted so that a smooth transition region can be observed between the two grids.

(iii) Water level boundary conditions can be sensitive to small changes when they are prescribed at an almost parallel flow boundary. In this case velocity boundary conditions should be a better choice.

(iv) In a very small refined grid model, model bottom friction becomes rather inactive to dissipate unwanted wave components. Weakly reflective boundary conditions are usually used to get rid of the short wave components (Verboom et al [1981]).

Nested Model - A Case Study

In recent years there has been a phenomenal increase in the development activities in the foreshore regions of Singapore. This has led to a series of model studies concerning the influence of large scale coastal development projects such as nearshore land reclamation on the behaviour of streaming flow, sediment transport and the overall marine environment condition in the Singapore coastal region.

Figure 8 shows the overall coarse grid model - the Regional model covering the far-field of Singapore Strait and its surrounding waters together with the boundary layout of the nested model that covers the central coastal waters where extensive coastal developments have occured. The grid size of the regional model is 1 km x 1 km and for the central detail model (Central model) that covers an area of approximately 30 x 30 km^2, a grid size of 250 m x 250 m is adopted. Figure 9 shows the detail coastline layout of the nested Central model together with four selected stations where the time history of water level variation and tidal current flow will be examined. It can be seen that the nested model has three open boundaries and this obviously allows for the choice of several combinations of boundary conditions.

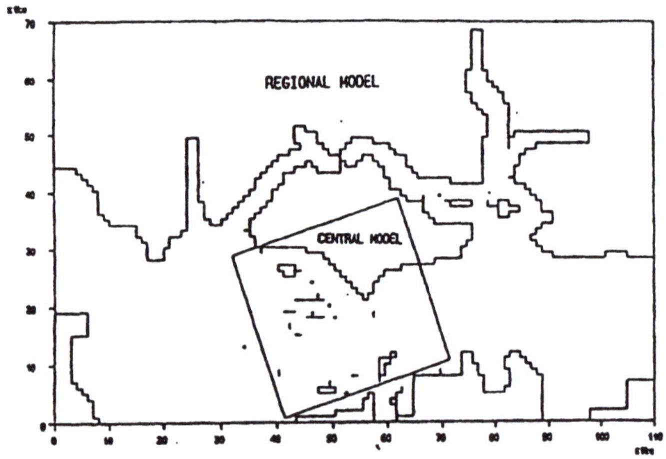

Figure 8. Regional and Nested Central Models

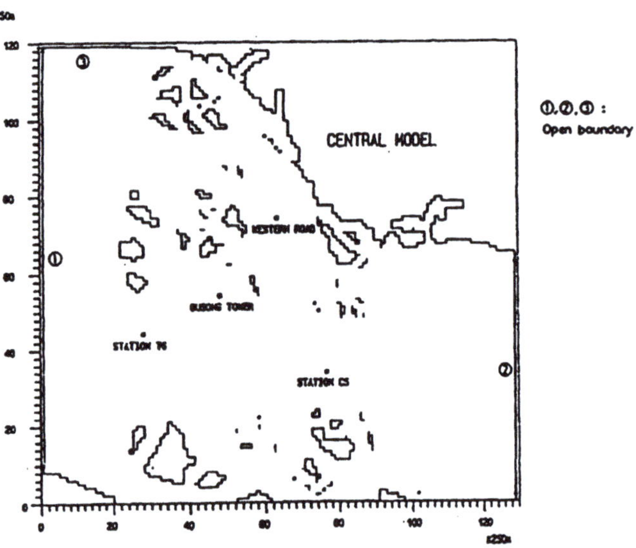

Figure 9. Central Model Domain Showing Model Open Boundaries and Field Measurement Stations

The Regional model was calibrated by adapting the boundary conditions (field measurements) and bottom roughness (Shankar et al [1990a and 1990b]). The boundary conditions for the nested model are taken from

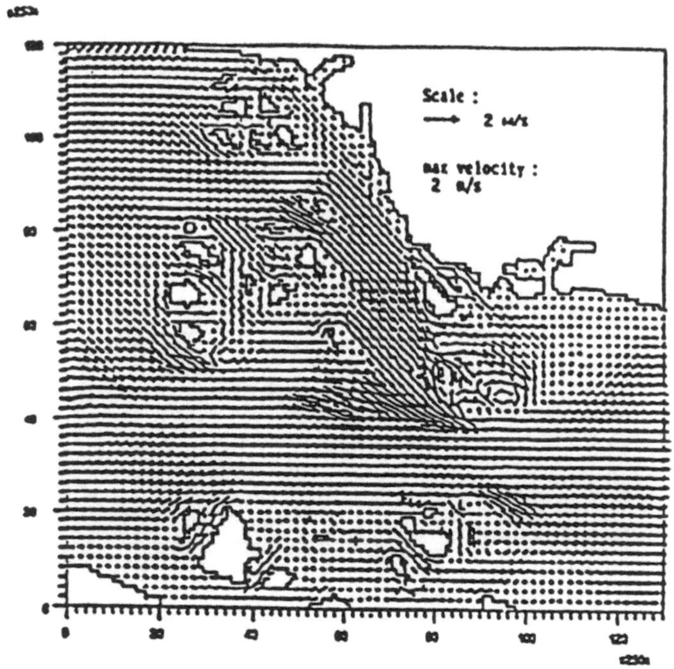

Figure 10 a

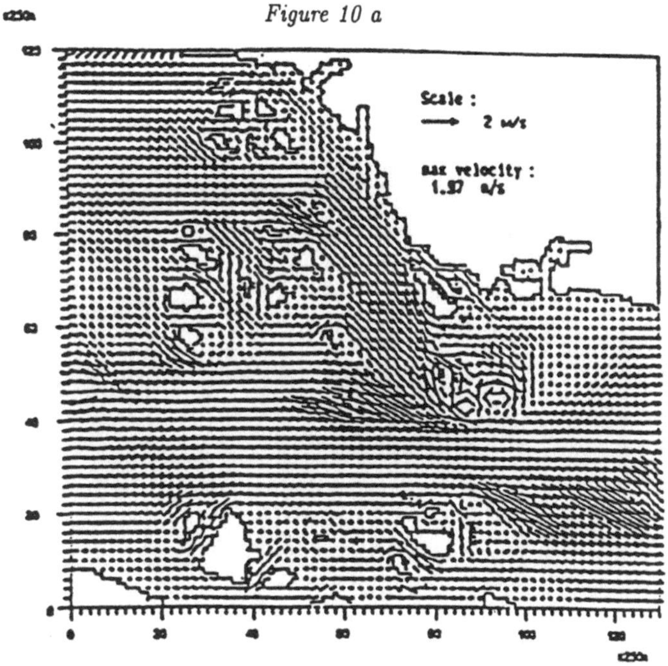

Figure 10 b

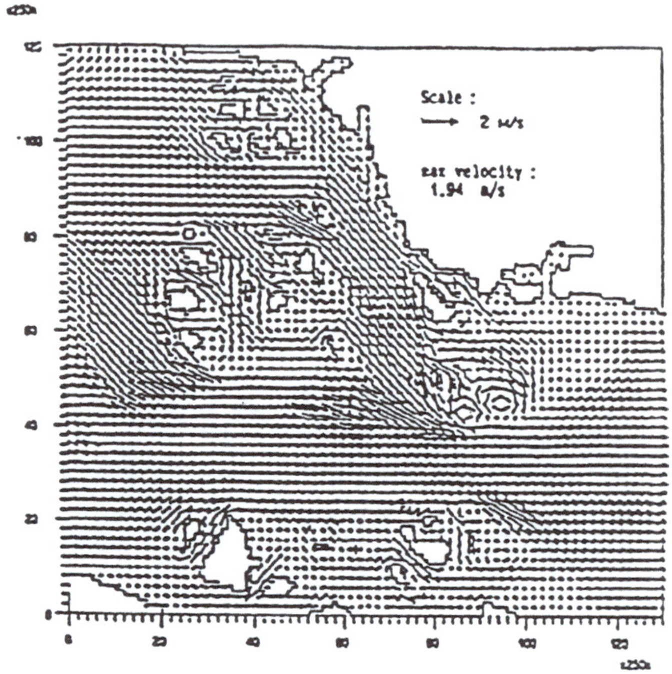

Figure 10 c

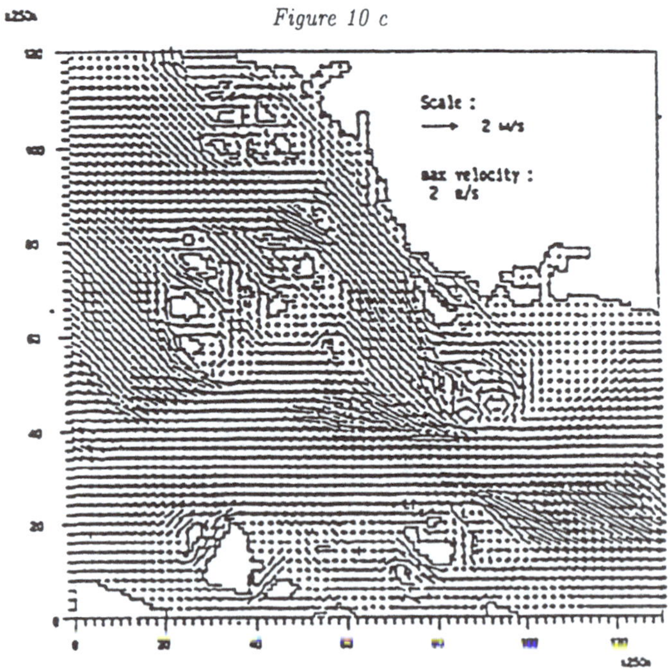

Figure 10 d

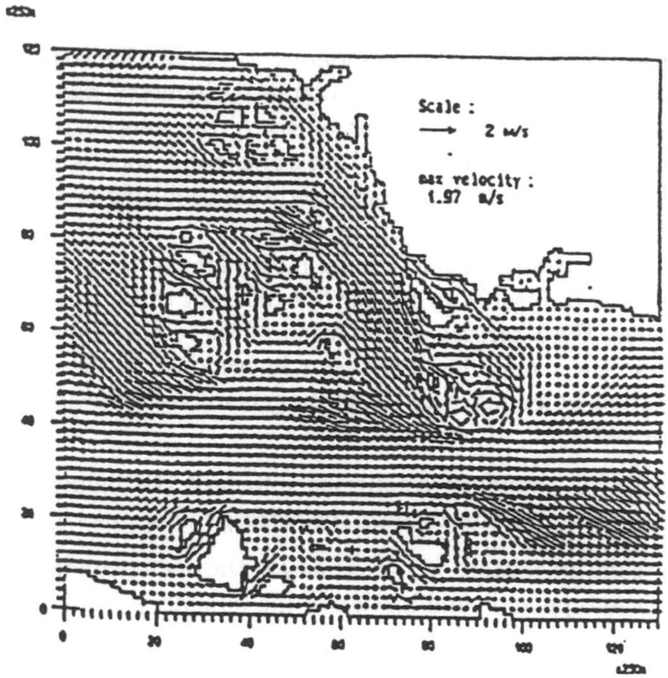

Figure 10 e

Boundary Condition / Figure	Open boundary ① Model ② (③ top)		
	①	②	③
10(a)	water level	water level	water level
10(b)	water level	velocity	water level
10(c)	velocity	water level	water level
10(d)	velocity	velocity	water level
10(e)	water level & velocity	velocity	water level

Figure 10. Flow Patterns Showing Eastward Currents for Different Combinations of Boundary Conditions

the coarse grid Regional model and linear interpolation is done among the finer grid points along the boundaries. Bottom schematization of the detail model is derived from the available bathymetry map so that the effect of bottom topography distortion by direct interpolation of coarse grid model can be avoided. No bottom adjustment zone is needed in the present nested model as no boundary condition problems were encountered (phenomena of local instability for example).

The current patterns in the Central model for typical maximum velocities are depicted in Fig.10a to 10e for five different combinations of boundary conditions. By a careful examination of these flow patterns, one can easily observe the model flow behaviour in response to the corresponding set of excitation boundary conditions. In general the following phenomena can be observed:

(i) The velocity boundary conditions influence the model flow with a higher velocity magnitude near the corresponding velocity boundaries.

(ii) The circulation pattern in the vicinity of the open boundaries are sensitive to the type and combination of the boundary conditions imposed.

(iii) When two different types of boundary conditions are prescribed at a boundary, the flow pattern near the boundary shows two distinct regions of flow behaviour pertaining to their respective type of boundary conditions with a transition region in between (see Fig. 10e) where the water level boundary and velocity boundary are combined at coordinates (0,90).

(iv) For regions with about twenty or more grid points away from the open boundaries, the influence of boundary conditions become minimal. In the present example,this implies that the model is set to preserve a 'boundary immune region of about 50 percent of the model size. The region is expected to be smaller for coarser grid and larger for finer grids.

Figure 11 depicts the time history of water level variations at tidal station T6 and Figs. 12 to 14 show the time series of current flow at current station C5, Gusong Tower and Western Road respectively. It can be seen that both tidal and current stations, besides a slight over estimation in the tidal variation at T6 for the simulation with model boundary combination VVW, there is no distinct merit among the computed values with the other combinations of the boundary conditions as compared to the prototype measurements. This is clearly consistent with the observation that these selected gauge stations are indeed located within the area of model 'boundary immune region.

LEGEND

W : Water level boundary condition
V : Velocity boundary condition
①②③ : Sequence of open boundary representation
+ : Field measurements
——— : Computed values with boundary conditions WWW
- - - : Computed values with boundary conditions WVW
····· : Computed values with boundary conditions VWW
□ : Computed values with boundary conditions VVW
Δ : Computed values with boundary conditions (W&V)VW

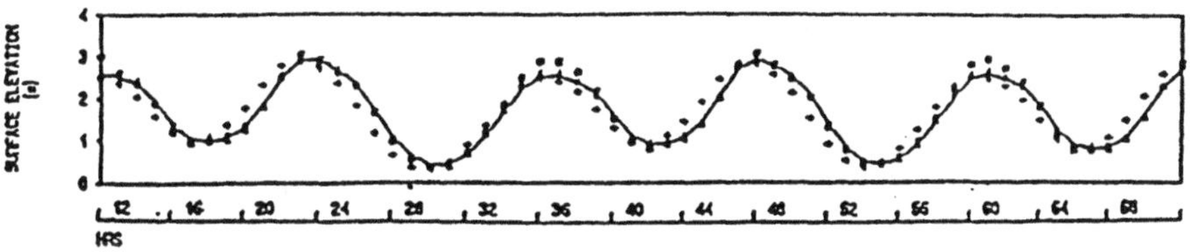

Figure 11. Measured and Computed Tidal Variations at Station T6

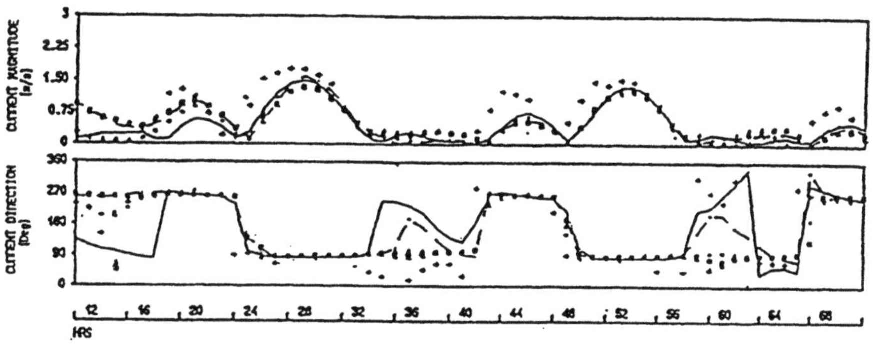

Figure 12. Measured and Computed Currents at Station C5

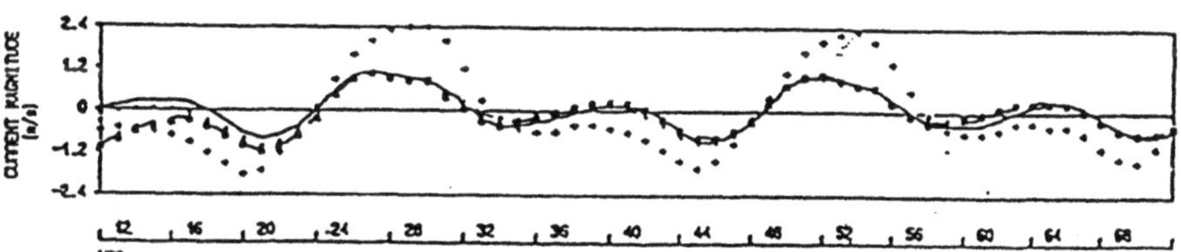

Figure 13. Measured and Computed Currents at Gusong Tower

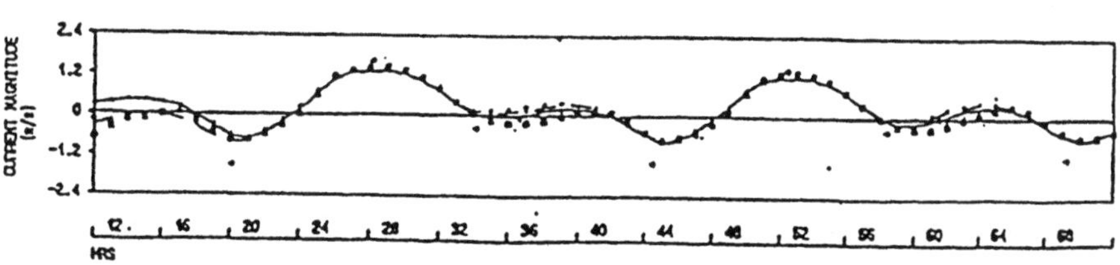

Figure 14. Measured and Computed Currents at Western Road

CONCLUSIONS

A numerical hydrodynamic model based on shallow water equations (SWE) has been developed and successfully applied to tthe regional waters of Singapore. The numerical model which is based on the finite difference formulation of Stelling (1984) has proven to be a useful tool for the study of tidal flow dynamics in coastal waters covering both far field and near-field phenomena. The latter is accomplished through a nested modelling approach. The nested modelling technique is successfully implemented on a finer grid model (central model) encompassing the central region of Singapore waters. Some guidelines are provided for efficient implementation of the nested modelling technique.

The results of the numerical simulations comprising time histories of tidal water levels and currents compare very well with the measured values of tidal water surface elevations and currents.

The successful implementation of the numerical hydrodynamic model has provided a reliable tool for hydraulic studies relating to coastal developments involving land reclamation, port and harbour development, causeway construction as well as waste disposal in coastal waters.

ACKNOWLEDGEMENT

The authors wish to extend their appreciation to the Port of Singapore Authority for their support relating to field measurements and hydrographic data. This study was supported by a research grant from the Science Council of Singapore (presently National Science & Technology Board) under the Research and Development Assistance Scheme.

References

1. Abbot, M.B. (1979) Computational Hydraulics: Elements of the Theory of Free-Surface Flows, Pitman, London.
2. Butler, H.L. (1978) *Numerical simulation of tidal hydrodynamics*, Great Egg Harbor and Corson Inlets, New Jersey, U.S. Army Engineer Waterways Experiment Station, Vicksburg, Miss., Tech. Rep H78-11.
3. Dronkers, J.J. (1964) *Tidal Computations in Rivers and Coastal waters* , North-Holland Publishing Co. Amsterdam.
4. Falconer, R.A, and Owens, P.H. (1984) *Mathematical Modelling of Tidal Currents in the Humber Estuary*. Journal of the Institution of Water Engineers and
5. Jothi Shankar, N., Cheong, H.F. and Chan, C.T. (1990a) *A Hybrid Model for Singapore Coastal Waters*. Proc. Eighth Conference of Asean Federation of Engineering Organization, Penang, Malaysia.
6. Jothi Shankar, N., Cheong, H.F and Chan, C.T. (1990b) *Coastal Modelling-Achievements and Future Challenges*. Proc. The 1990 Convention of Inst. Engrs. Singapore, May 1990.
7. Lean, G.H. and Weare, T.J. *Modelling Two-dimensional Circulating Flow*. Journal of Hydraulics Division, Vol. 105, HY1, Jan. 1979.

8. Leendertse, J.J. (1967) *Aspects of A Computational Model for Long-period Water-Wave Propagation*, RM-5294-PR, The Rand Corporation, Santa Monica.
9. Reid, R.O and Bodine,B.R. (1968) *Numerical Model for Storm Surges in Galveston Bay*. Journal of the Waterways and Harbors Division, ASCE, Vol. 94, No. WW1, February, 1968, pp33-57.
10. Report on Joint Tidal and Current Studies in *the Straits of Malacca and Singapore (1979)*, Indonesia, Japan, Malaysia and Singapore.
11. Stelling, G.S. (1984) *On the Construction of Computational Methods for Shallow Water Flow Probl;ems*, Rijkswaterstaat Communications No. 35/1984, The Hague, Netherlands.
12. Verboom, G.K., Stelling, G.K. and Officer, M.J. (1981) Non-reflective Boundary Conditions in Horizontal Flow Models. Proc. Int. Conf. Numerical Methods for Coupled Problems, Swansea.

PHASE PORTRAITS OF QUADRATIC SYSTEMS

J.W. REYN
Delft University of Technology,
Faculty of Technical Mathematics and Informatics,
P.O. Box 5031, 2600 GA Delft, The Netherlands

1. Introduction

As in the lecture given at the Conference, this paper intends to give an impression of what has become known so far about the phase portraits of quadratic systems in the plane. By a *quadratic system* is understood the system of two autonomous differential equations in the plane

$$\begin{aligned} \frac{dx}{dt} = \dot{x} &= a_{00} + a_{10}x + a_{01}y + a_{20}x^2 + a_{11}xy + a_{02}y^2 \equiv P(x,y), \\ \frac{dy}{dt} = \dot{y} &= b_{00} + b_{10}x + b_{01}y + b_{20}x^2 + b_{11}xy + b_{02}y^2 \equiv Q(x,y), \end{aligned} \tag{1}$$

, where $\dot{} = \frac{d}{dt}$, a_{ij}, b_{ij} $\in \mathbb{R}$ and $P(x,y)$ and $Q(x,y)$ are relative prime polynomials, which are not both linear. A *phase portrait* of (1) is the geometrical picture formed from the solution curves (orbits, trajectories or paths) of the equation

$$\frac{dy}{dx} = \frac{Q(x,y)}{P(x,y)} \quad \left(\text{or } \frac{dx}{dy} = \frac{P(x,y)}{Q(x,y)}\right), \tag{2}$$

in the x, y (or phase) plane.

The determination of the phase portraits of (1), although belonging to the Poincaré-Bendixson theory, which in a general sense, has assumed already an almost definite form, is still far from complete. Nevertheless, a large number of papers has appeared on the subject [14], the first ones, in fact, dating from the beginning of the century. Figure 1 shows the number of papers that appeared over the years. It may be seen that a large part of them have been published in the (former) Soviet Union and the People's Republic of China. Apart from the large theoretical interest that was taken in the subject later on, since the twenties, quadratic systems were also used as models in a variety of fields of applications. An example of an equation occuring frequently in applications is the (generalized) Volterra-Lotka equation

$$\begin{aligned} \dot{x} &= x(a_0 + a_1 x + a_2 y), \\ \dot{y} &= y(b_0 + b_1 x + b_2 y), \end{aligned} \tag{3}$$

where $a_i,\ b_i \in \mathbb{R}$. All phase portraits of (3) are known and given as function of the coefficients in [13]. Several applications are also given in [13].

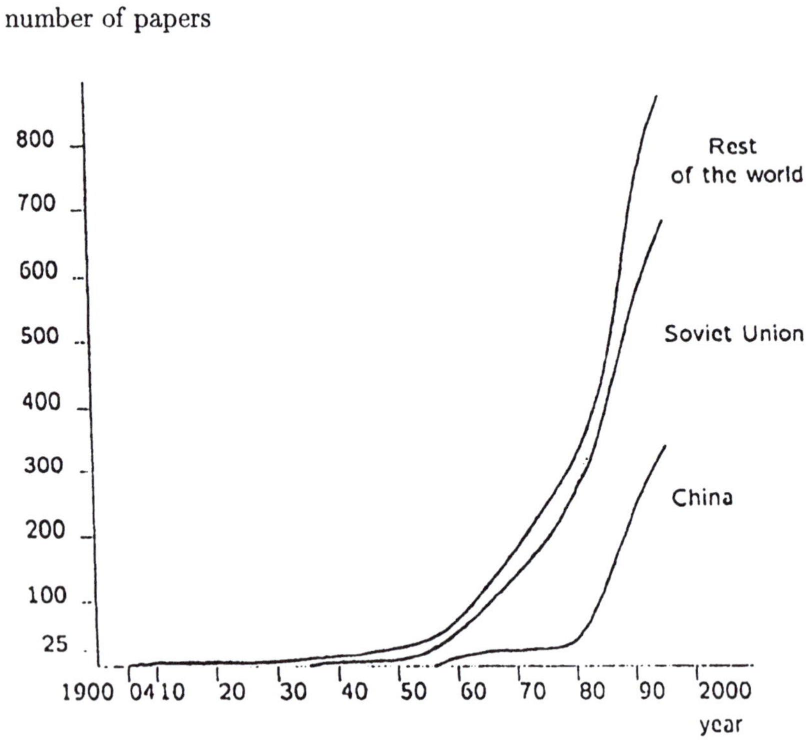

Figure 1. Number of paper on quadratic systems over the years

As an example of a phase portrait take in (3) $a_0 = 3,\ a_1 = -1,\ a_2 < -5,\ b_0 = -1,\ b_1 = 1,\ b_2 = 1$; then Figure 2 is obtained [1]. As usual the phase portrait is given on the Poincaré disk, where near the unit circle the behaviour at infinity is represented. The relation between the x, y or phase plane and the *Poincaré disk* is visualized in Fig. 3 [1]. Herein the (Poincaré) unit sphere touches the x, y plane in the origin (0,0). A point $M(x, y)$ in the x, y plane is connected with the center of the sphere through a straight line which intersects the sphere in the points M' and M''. Projection of

the point M' on the lower hemisphere unto the x, y plane then leads to the point $\tilde{M}$ on the Poincaré disk. In particular, if $M(x, y)$ is sent to infinity, $\tilde{M}$ approaches to the unit circle.
Through every ordinary point ($P^2 + Q^2 \neq 0$) there exists precisely one solution curve. In Figure 2 some of them are drawn; in the present example they all run from one *critical point* ($P^2 + Q^2 = 0$) to another, the arrows pointing into the direction of increasing t. It may be noticed that there are four finite critical points, which is the general situation. This is due to the fact that two conics, as are $P(x, y) = 0$ and $Q(x, y) = 0$, in general have four intersection points.

Figure 2.

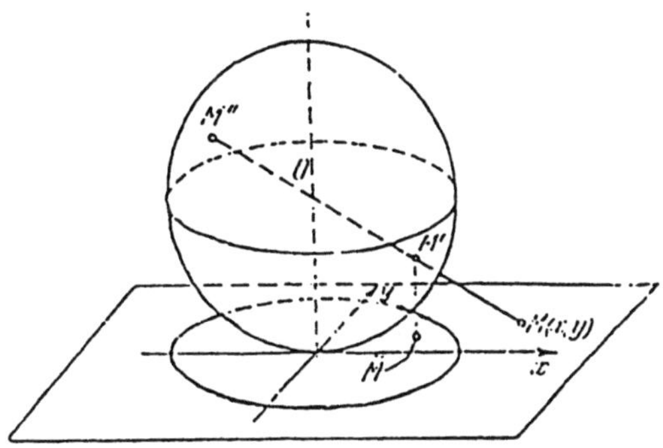

Figure 3.

Critical points at infinity can be characterized by the condition $\dot{\theta} = 0$, where θ is the polar angle of the polar coordinates r and θ. If (1) is written in polar coordinates there follows

$$\begin{aligned} \dot{r} &= A_1(\theta) + rB_1(\theta) + r^2C_1(\theta), \\ r\dot{\theta} &= A_2(\theta) + rB_2(\theta) + r^2C_2(\theta) \end{aligned} \tag{4}$$

where from $A_i(\theta)$, $B_i(\theta)$, $C_i(\theta)$, only $C_2(\theta)$ is explicitly needed, being the dominant term at large r. From

$$C_2(\theta) = b_{20}\cos^3\theta + (b_{11} - a_{20})\sin\theta\cos^2\theta + (b_{02} - a_{11})\sin^2\theta\cos\theta - a_{02}\sin^3\theta = 0 \tag{5}$$

follows that in general, there are six pairwise diametrically opposite points with $\dot{\theta} = 0$; they will be seen as three infinite critical points. The finite critical points on the axes are saddle points; to each one of them approach four separatrices, which serve as dividing lines in the phase portrait and are drawn in fat. The fourth finite critical point is a focus to which other orbits spiral in . The infinite critical points are nodal points.

Another example is given by

$$\begin{aligned} \dot{x} &= 2xy, \\ \dot{y} &= 1 + y - x^2 + y^2. \end{aligned} \tag{6}$$

Again there are four finite critical points: two real (in $x = \pm 1$, $y = 0$), and two complex. There are also three infinite critical points: one real in $\theta = \pm\frac{\pi}{2}$ and two complex. See Figure 4. [1]

Figure 4.

An example that shows that the number of finite critical points is not always equal to four and that of the infinite critical points not always equal to three is given by

$$\begin{aligned} \dot{x} &= x(y + \tfrac{3}{2}), \\ \dot{y} &= x + y - 2y^2. \end{aligned} \tag{7}$$

As shown in Figure 5 there is a nodal point in (0,0) and two saddle points: in $(0, \frac{1}{2})$ and in $(6, -\frac{3}{2})$. The fourth intersection point of the degenerate hyperbola $x(y + \frac{3}{2}) = 0$ and the parabola $x + y - 2y^2 = 0$ is "at infinity" in the direction of the common asymptotic direction $y \equiv 0$.

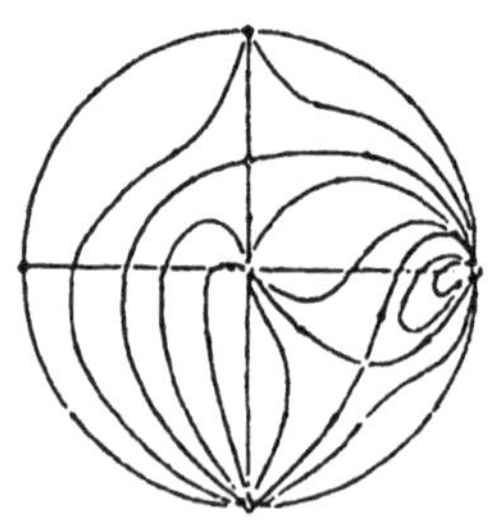

Figure 5.

The infinite critical point "at the ends of the x axis" is also a double root of (5) and can be viewed as the coalescence of three critical points with a more complicated structure [1].

2. Critical points in quadratic systems

Since the determination of the behaviour of orbits near a critical point is a local problem, i.e. the notion of a limit may be used, methods are available to investigate what kind of critical points are possible in quadratic systems and under what conditions on the coefficients in the system. In fact, application of these methods has led to a virtually definite body of knowledge on the critical points in quadratic systems. Below a sketchy survey will be given; further details may be found in the literature [6].
In a critical point (x_0, y_0), by definition there is $P(x_0, y_0) = Q(x_0, y_0) = 0$. For the determination of its properties use is made of the quantity

$$\Omega(x_0, y_0) = \frac{\partial P}{\partial x}(x_0, y_0)\frac{\partial Q}{\partial y}(x_0, y_0) - \frac{\partial P}{\partial y}(x_0, y_0)\frac{\partial Q}{\partial x}(x_0, y_0),$$

which, in fact, is the product of the eigenvalues of the locally linearized system, and

$$\Delta(x_0, y_0) = \frac{\partial P}{\partial x}(x_0, y_0) + \frac{\partial Q}{\partial y}(x_0, y_0),$$

which is equal to their sum or at the same time the divergence of the vector field in (x_0, y_0).
Finite critical points can be divided into the classes:

(i) Elementary critical points, then $\Omega(x_0, y_0) \neq 0$.
(ii) Semi-elementary critical points, then $\Omega(x_0, y_0) = 0$, $\Delta(x_0, y_0) \neq 0$.
(iii) Degenerate critical points, then $\Omega(x_0, y_0) = \Delta(x_0, y_0) = 0$.
They may be further divided into:
   (a) Nilpotent critical points, then at least one of the derivatives $\frac{\partial P}{\partial x}$, $\frac{\partial P}{\partial y}$, $\frac{\partial Q}{\partial x}$ or $\frac{\partial Q}{\partial y} \neq 0$ in (x_0, y_0).
   (b) Critical points in essentially homogeneous systems, then $\frac{\partial P}{\partial x} = \frac{\partial P}{\partial y} = \frac{\partial Q}{\partial x} = \frac{\partial Q}{\partial y} = 0$ in (x_0, y_0).

2.1. ELEMENTARY CRITICAL POINTS

Further details of these points can be found in many textbooks, i.e. [1], [6], see Figure 6. A critical point (x_0, y_0) is a saddle if $\Omega(x_0, y_0) < 0$ (Fig. 6a); its Poincaré index $i = -1$. If $\Omega(x_0, y_0) > 0$ then $i = 1$, reason why the point is sometimes called an anti saddle. Further distinction gives: if $\Omega(x_0, y_0) < \frac{1}{4}\Delta^2(x_0, y_0)$ the point is a nodal point with two tangents (Fig. 6b); if $\Omega(x_0, y_0) = \frac{1}{4}\Delta^2(x_0, y_0)$ the point is a nodal point with either one or an infinite number of tangents (Fig. 6c,d); if $\Omega(x_0, y_0) > \frac{1}{4}\Delta^2(x_0, y_0) \neq 0$ the point is a focus (Fig. 6e); and if $\Omega(x_0, y_0) > \frac{1}{4}\Delta^2(x_0, y_0) = 0$ the point is either a weak focus or a center (Fig. 6e,f).

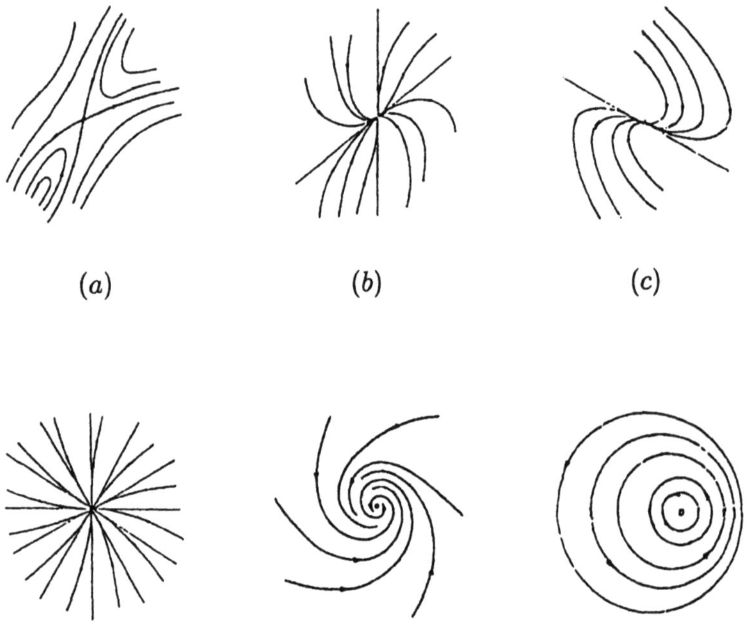

(a) (b) (c)

(d) (e) (f)

Figure 6. Elementary critical points in quadratics systems

A centerpoint is a critical point for which there exists a neighbourhood such that through every point therein there exists a closed orbit encircling the critical point. A closed orbit represents a periodic solution. Apart from the conditions stated above, further restrictions on the coefficients of (1) are needed to characterize the systems having (at least one) center. As a result of a series of papers since the beginning of the century the conditions on (1) to have a center are known, as well as the corresponding phase portraits, see e.g. [20]. For quadratic systems with a center point the first integral is known. As an example, the phase portrait in the region of closed orbits for the system

$$\begin{aligned} \dot{x} &= -1 + 2y + x^2 + y^2, \\ \dot{y} &= -2x \end{aligned} \tag{8}$$

is given in Figure 7, [1].
The condition $\Omega(x_0, y_0) \neq 0$ for an elementary critical point also implies that the ∞ isocline $P(x,y) = 0$ and the 0 isocline $Q(x,y) = 0$ intersect at an angle different from zero. Small changes in the coefficients of (1) do

not alter this fact. As a result there remains a single critical point and the point is said to be of first order or of multiplicity one. If $\Omega(x_0, y_0) = 0$ such is no longer than the case and the critical point is of higher order; its multiplicity is larger than one and depending on its value several critical points can bifurcate from it upon changing the coefficients of (1) slightly.

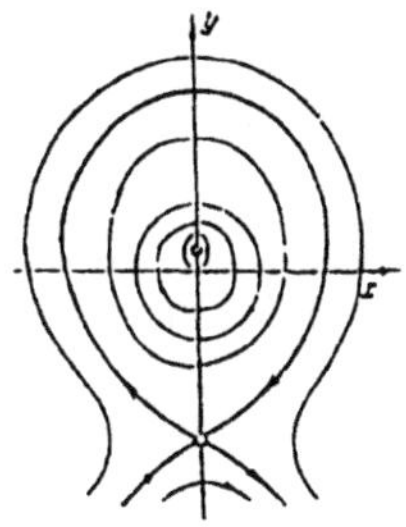

Figure 7.

2.2. SEMI-ELEMENTARY CRITICAL POINTS

A critical point (x_0, y_0) is semi-elementary if $\Omega(x_0, y_0) = 0$, $\Delta(x_0, y_0) \neq 0$, and by a linear transformation may be brought to the origin, using a normal form instead of (1):

$$\begin{aligned} \dot{x} &= a_{20}x^2 + a_{11}xy + a_{02}y^2, \\ \dot{y} &= y + b_{20}x^2 + b_{11}xy + b_{02}y^2. \end{aligned} \tag{9}$$

Using the theory in [1] there follows that the critical point in the origin is

(i) a second order saddle node iff $a_{20} \neq 0$, $(i = 0)$;see fig. 8
(ii) a third order saddle node iff $a_{20} = 0$, $a_{11}b_{20} > 0$, $(i = -1)$; see fig. 6a
(iii) a third order node iff $a_{20} = 0$, $a_{11}b_{20} < 0$, $(i = 1)$; see fig. 6b, and
(iv) a fourth order saddle node iff $a_{20} = a_{11} = 0$, $(i = 0)$; see fig. 8.

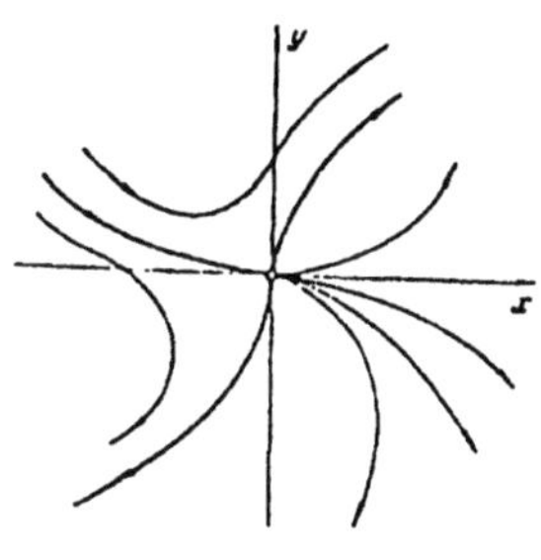

Figure 8.

The semi-elementary critical points of (9) in the origin have (a) center manifold(s) tangent to the x axis. The phase portraits for quadratic systems with a third or fourth order semi-elementary point are classified by [10].

2.3. NILPOTENT CRITICAL POINTS

A critical point (x_0, y_0) is nilpotent if $\Omega(x_0, y_0) = \Delta(x_0, y_0) = 0$ and at least one of the derivatives $\frac{\partial P}{\partial x}$, $\frac{\partial P}{\partial y}$, $\frac{\partial Q}{\partial x}$ or $\frac{\partial Q}{\partial y} \neq 0$ in (x_0, y_0).
A normal form having a nilpotent critical point in the origin can be written as

$$\begin{aligned} \dot{x} &= y + a_{20}x^2 + a_{11}xy + a_{02}y^2, \\ \dot{y} &= b_{20}x^2 + b_{11}xy + b_{02}y^2. \end{aligned} \tag{10}$$

Using the theory in [[1], p. 346] there follows, that the critical point in (0,0) is:

(i) a second order cusp iff $b_{20} \neq 0$, $(i = 0)$; see fig. 9a
(ii) a third order saddle iff $b_{20} = 0$, $a_{20}b_{11} < 0$, $(i = -1)$; see fig. 9b
(iii) a third order point, having an elliptic and a hyperbolic sector iff $b_{20} = 0$, $a_{20}b_{11} > 0$, $(i = 1)$, see fig. 9c, and
(iv) a fourth order saddle node iff $b_{20} = b_{11} = 0$, $(i = 0)$; see fig. 9d.

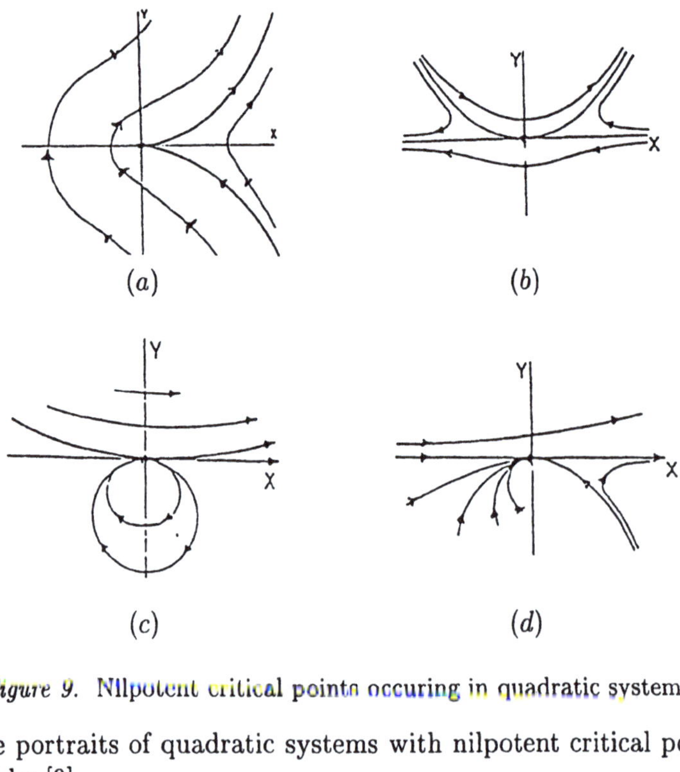

Figure 9. Nilpotent critical points occuring in quadratic systems

Phase portraits of quadratic systems with nilpotent critical points are classified by [9].

2.4. CRITICAL POINTS IN HOMOGENEOUS SYSTEMS

If in a critical point (x_0, y_0) there is $\Omega(x_0, y_0) = \Delta(x_0, y_0) = 0$, and moreover, $\frac{\partial P}{\partial x} = \frac{\partial P}{\partial y} = \frac{\partial Q}{\partial x} = \frac{\partial Q}{\partial y} = 0$ in this point, by a shift it can be transferred to the origin, and (1) becomes a homogeneous system. Figure 10 shows the various possible topological configurations of the orbits near such a critical point. A geometrical classification of homogeneous quadratic systems was given by [23]. A later classification was given by [5].

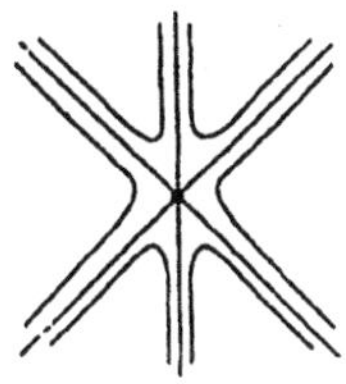

$A < 0,\ c_{45} < 0,\ c_{56} < 0,\ m = 4,\ i = -2$

$A < 0,\ c_{45} > 0,\ c_{56} > 0,\ m = 4,\ i = 2$

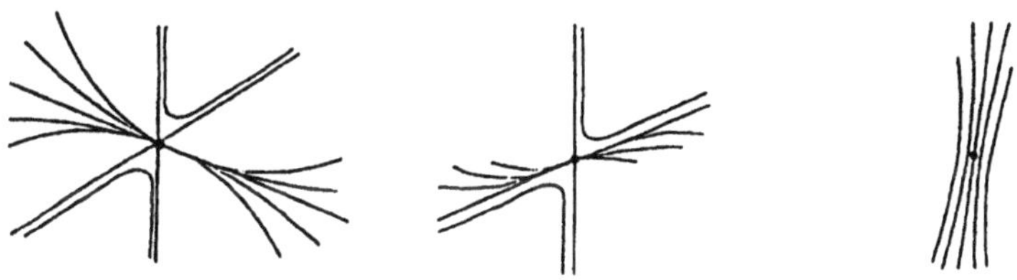

$A > 0,\ m = 4,\ i = 0$

Figure 10. Critical points in homogeneous quadratic systems.
$A = c_{46}^2 - c_{45}c_{56}$, $c_{45} = a_{20}b_{11} - a_{11}b_{20}$, $c_{46} = a_{20}b_{02} - a_{02}b_{20}$, $c_{56} = a_{11}b_{02} - a_{02}b_{11}$
m = multiplicity, i=Poincaré index

2.5. INFINITE CRITICAL POINTS

As stated before there are in general three infinite critical points, their location being given by (5), which also may be written as

$$-(a_{20}x^2 + a_{11}xy + a_{02}y^2)y + (b_{20}x^2 + b_{11}xy + b_{02}y^2)x = 0. \qquad (11)$$

They also may be seen to include the critical points obtained by the intersection of the ∞ isocline $P(x,y) = 0$ and the 0 isocline $Q(x,y) = 0$ in the case that they have one or two common asymptotic directions so that

$$\begin{aligned} a_{20}x^2 + a_{11}xy + a_{02}y^2 &= 0, \\ b_{20}x^2 + b_{11}xy + b_{02}y^2 &= 0, \end{aligned} \tag{12}$$

are simultaneously satisfied.
Most of the types of finite critical points discussed before can also occur at infinity. Moreover, several types of critical points of multiplicity 5, 6 and 7 of more complicated structure are possible. A description of infinite critical points and the conditions for which they occur is given by [15].

2.6. SOME GLOBAL PROPERTIES OF FINITE CRITICAL POINTS

Tung's lemma [22]
In a quadratic system an arbitrary straight line is either a trajectory or the sum of points on that line which are either critical or points of contacts is at most equal to two.
Proof
Without loss of generality we may take $y \equiv 0$. There

$$\begin{aligned} \dot{x} &= a_{00} + a_{10}x + a_{02}x^2, \\ \dot{y} &= b_{00} + b_{10}x + b_{20}x^2. \end{aligned} \tag{13}$$

So, either $b_{00}^2+b_{10}^2+b_{20}^2 = 0$, then $y \equiv 0$ is a solution curve or $b_{00}^2+b_{10}^2+b_{20}^2 \neq 0$ and there are 0,1 or 2 points with $\dot{y} = 0$. They are points of contacts if $\dot{x} \neq 0$ in them and critical points if $\dot{x} = 0$.

Corollary 1
A line through two critical points is either a solution curve or an isocline. On the segment connecting the two critical points the flow is opposite to the flow on the remaining segments of the line.

Proof This follows from writing (13) in the form

$$\begin{aligned} \dot{x} &= a_{20}(x - x_1)(x - x_2), \\ \dot{y} &= b_{20}(x - x_1)(x - x_2), \end{aligned} \tag{14}$$

where $(x_1, 0)$ and $(x_2, 0)$ are the coordinates of the critical points.

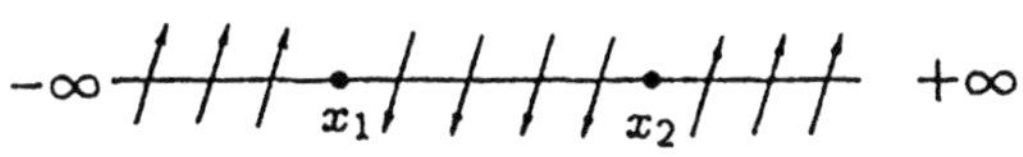

Figure 11.

Corollary 2
In a quadratic system there are at most two critical points which are either a focus or a center.

Proof
The situation sketched in Figure 12 including three critical points and their connecting straight lines (which are isoclines) shows a contradiction since for a given flow on P_1P_2 the flow on P_1P_3 and P_2P_3 is determined in such a way that there cannot be a center or focus around P_3.

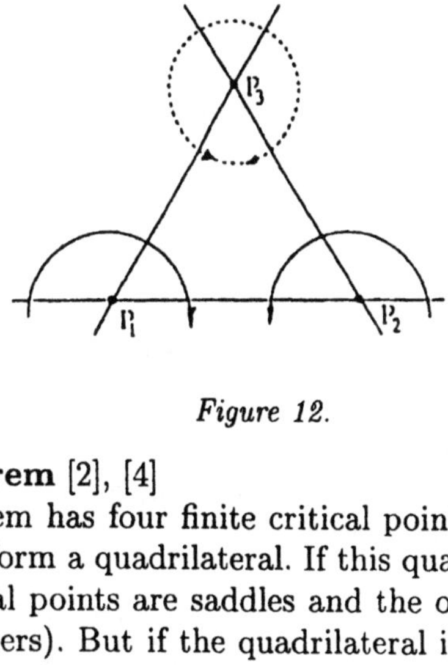

Figure 12.

Berlinskii's theorem [2], [4]
If a quadratic system has four finite critical points, then they are all elementary and they form a quadrilateral. If this quadrilateral is convex then two opposite critical points are saddles and the other two are antisaddles (nodes, foci or centers). But if the quadrilateral is not convex then either the three exterior vertices are saddles and the inferior vertex an antisaddle or the exterior vertices are antisaddles and the inferior vertex a saddle.

Proof
By a linear transformation the critical points may be placed at (0,0), (1,0), (0,1) and (α, β), where α, β, $\alpha + \beta - 1 \neq 0$. Then the equation may be written as

$$\begin{aligned} \dot{x} &= a_1 x(x-1) + b_1 y(y-1) + c_1 xy, \\ \dot{y} &= a x(x-1) + b y(y-1) + cxy, \end{aligned} \tag{15}$$

where $c = -\frac{\alpha-1}{\beta}a - \frac{\beta-1}{\alpha}b$, $c_1 = -\frac{\alpha-1}{\beta}a_1 - \frac{\beta-1}{\alpha}b_1$. Now, by calculation, follows

$$\Omega(0,0) = a_1 b - a b_1,\ \Omega(1,0) = -\frac{\alpha+\beta-1}{\alpha}\Omega(0,0)$$
$$\Omega(0,1) = -\frac{\alpha+\beta-1}{\alpha}\Omega(0,0),\ \Omega(\alpha,\beta) = (\alpha+\beta-1)\Omega(0,0).$$

Take $\alpha > 0,\ \beta > 0,\ \alpha+\beta-1 > 0$ then the quadrilateral formed by the critical points is convex. Then follows that $\Omega(\alpha,\beta)\Omega(0,0) > 0$ so (0,0) and (α,β) are both saddles or both antisaddles, whereas from $\Omega(1,0)\Omega(0,0) < 0$ and $\Omega(0,1)\Omega(0,0) < 0$ follows that (1,0) and (0,1) have the opposite index from (0,0) and (α,β). For $\alpha > 0,\ \beta > 0,\ \alpha+\beta-1 < 0$ the quadrilateral is not convex and the result follows from $\Omega(\alpha,\beta)\Omega(0,0) < 0,\ \Omega(1,0)\Omega(0,0) > 0,\ \Omega(0,1)\Omega(0,0) > 0$. The argument may be repeated to cover all the cases for α, β.

3. Periodic solutions in quadratic systems

Periodic solutions forming a nest of orbits in the neighbourhood of a centerpoint consisting only of periodic orbits were already met before and an example for that case was given in system (8) and illustrated in Figure 7. Apart from them, also *limit cycles* occur in quadratic systems, being periodic solutions, isolated in the sense that they have a neighbourhood such that no other periodic solution exists in that neighbourhood.
For example, the system

$$\begin{aligned} \dot{x} &= -1 + 2y + x^2 + xy + y^2, \\ \dot{y} &= -2x - x^2, \end{aligned} \tag{16}$$

has a limit cycle coinciding with the unit circle (Fig. 13) [1]. Two limit cycles involving two different critical points can be seen in Figure 14, which is the phase portrait of the system

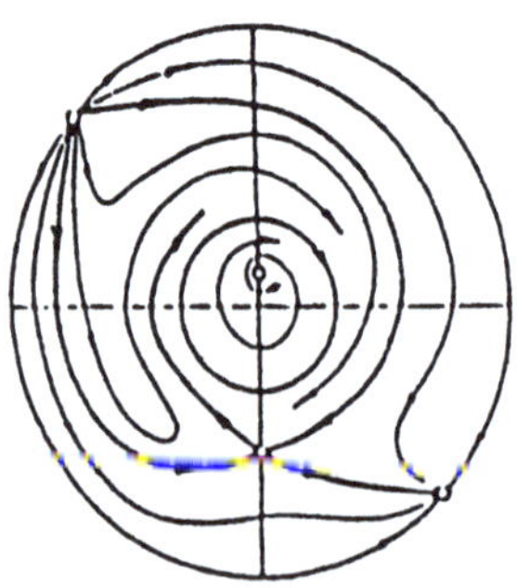

Figure 13.

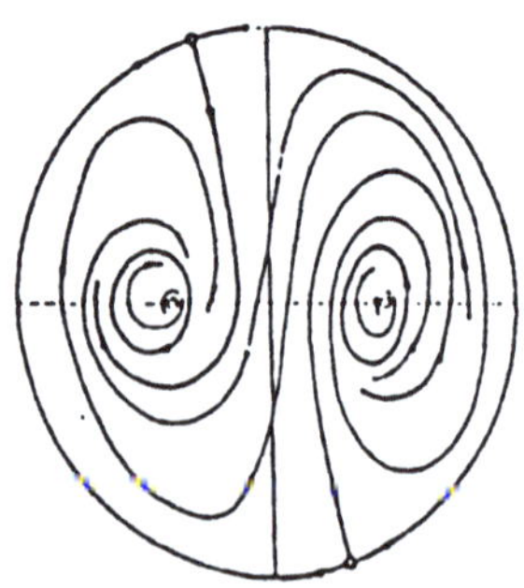

Figure 14.

$$\begin{aligned}\dot{x} &= xy - \varepsilon[-\tfrac{1}{3}(x-1)(x+2) + \tfrac{1}{2}y^2 + \tfrac{1}{3}xy + \tfrac{1}{3}y],\\ \dot{y} &= -\tfrac{1}{3}(x-1)(x+2) + \tfrac{1}{2}y^2 + \tfrac{1}{3}xy + \tfrac{1}{3}y + \varepsilon xy,\end{aligned} \tag{17}$$

where ε is a small parameter [1].
The study of limit cycles in quadratic systems has to take into account some basis properties common to all periodic solutions; they can be derived on the basis of Tung's lemma.

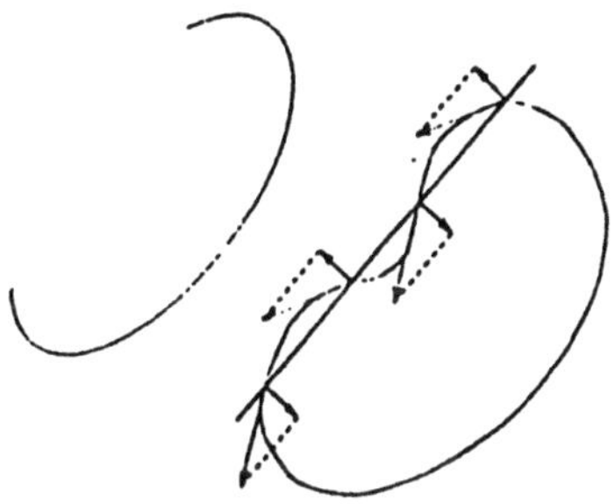

Figure15.

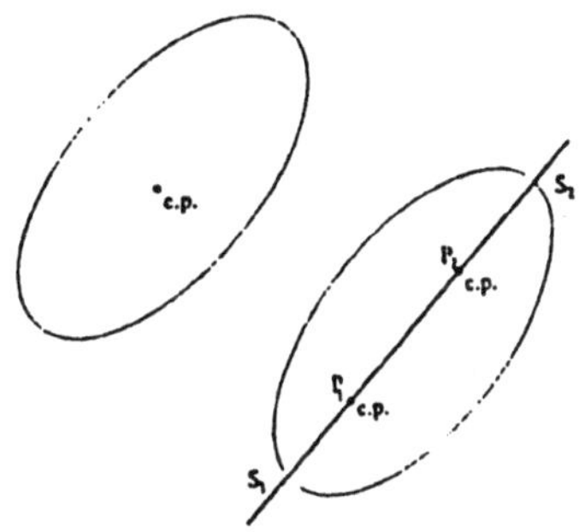

Figure 16.

Property 1
In a quadratic system a periodic orbit is convex and contains a unique critical point in its interior.

Proof
Suppose, as sketched in Figure 15, that the periodic orbit γ is not convex. Then there exists a line l such that there exists at least four points of intersection of l and γ Consider the normal component to the line l of the vector (P, Q) in (1). Since (P, Q) is continuous there are at least three zeros of this component, thus three points either critical or with contact with respect to l, in contradiction to Tung's lemma.
Now suppose there is more than one critical point in the interior of the periodic solution. Consider the line l through two critical points P_1 and P_2 in this interior as sketched in Figure 16. According to Tung's lemma l is an isocline and the flow on ∞P_1 should be in the same direction as on $P_2\infty$, which leads to a contradiction if the flow on the periodic orbit is considered.

Remark
It can, in fact, be shown that the critical point in the interior of the periodic orbit is either a centerpoint or a focus (the latter in case of a limit cycle) [4].

Property 2
Two periodic orbits are oppositely oriented if external to each other and similarly oriented if one is inside the other.

Proof
Consider two periodic orbits external to each other as in Figure 17. The direction

of flow on the segment P_1P_2 of the line l through the critical points P_1 and P_2 being an isocline, is such that the flow on the periodic orbits rotates in opposite direction to each other.
If a periodic orbit is situated inside another one and the flow on these orbits rotates in opposite direction, as in Figure 18, at least three points exist on a line l through the critical point P inside the orbits, which are either critical or points of contacts. This contradicts Tung's lemma.

Figure 17.

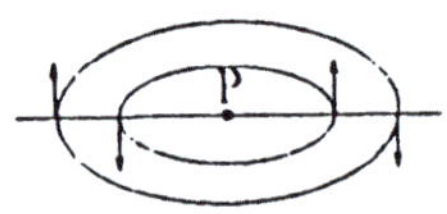

Figure 18.

Collorary
In a quadratic system there are at most two nests of periodic orbits.

What has been stated above for periodic orbits applies in particular to limit cycles. So far up till three limit cycles were found in a single nest but no proof is available that three is the maximum number of limit cycles in a nest. If two nests are present up till now precisely one limit cycle is found in one of the nests and again up till three limit cycles in the other nest, which seems to make the maximum number of limit cycles possible in a quadratic system equal to four. A proof of these statements has not been given yet and will probably remain a difficult problem for a long time to come.

4. Classification of phase portraits of quadratic systems

For a systematic classification of phase portraits of quadratic systems, it seems useful to use the notion of finite multiplicity m_f.

Definition
The *finite multiplicity* m_f of a quadratic system is the sum of multiplicities of the finite critical points. The *infinite multiplicity* m_i of a quadratic system is the sum of the multiplicities of the infinite critical points.

Property 3 [12]
$m_f + m_i = 7$

Property 4 [13]
The sum of the Poincaré indices of the (finite and infinite) critical points in a quadratic system is equal to one.

Along the line of increasing values of m_f classifications of phase portraits are made for which the following remarks can be made.

$m_f = 0$

A classification is given by [16]. Trivially, since there exist no finite critical points there exists no limit cycle.

$m_f = 1$

A classification is given by [17]. It is shown that in this class there exists at most one limit cycle.

$m_f = 2$

A classification is given by [18]. It is shown that in this class there exist at most two limit cycles, except that a proof has yet to be given that at most two limit cycles can encircle the origin for the equation

$$\begin{aligned} \dot{x} &= dx - y + lx^2 + mxy, \\ \dot{y} &= x + x^2, \end{aligned} \tag{18}$$

where $-2 < d < 0,\ l > 0,\ m \geq 2$. This equation was studied extensively, see e.g. [24].

$m_f = 3$

For systems with a degenerate infinite critical point the phase portraits were determined by [19] and contain at most one limit cycle. The remaining class consists of systems having a semi-elementary infinite critical point and includes bounded quadratic systems. The separatrix structure for this class was determined by [8]. Two nests of limit cycles are possible; then in at least one of the nests there exists precisely one limit cycle [7]. Up till three limit cycles can be constructed in case of a single nest. More information about the limit cycles is needed to complete the classification of phase portraits for $m_f = 3$.

$m_f = 4$

As mentioned before, systems with a fourth or third order finite critical point are classified as well as those with a (second order) cusp. At most one limit cycle can exist in these cases. For the other case with a second order finite critical point, i.e. with a second order saddle node no systematic classification exists. For the cases with four (elementary) finite critical points partial results exist. Among them are those having two real and two complex finite critical points. In this class examples are given with (at least) four limit cycles, see e.g. [3]. A lot of work still remains to be done for $m_f = 4$. On the basis of the classification so far the following conjecture is made.

Conjecture

The number of limit cycles in a quadratic system is at most equal to its finite multiplicity m_f.

References

1. Andronov, A.A., E.A. Leontovich, I.I. Gordon and A.G. Maier, (1973) Qualitative theory of second-order dynamic systems, *Translated from the Russian, Israel Program for Scientific Translations*, Jerusalem-London, Wiley New York.
2. Berlinskiĭ, A.N., (1960) On the behaviour of the integral curves of a differential equation, (Russian), *Izv. Vysš. Ucebn. Zaved. Matematika*, **no. 2, (15)**, p. 3-18
3. Chicone, C. and Tian Jinghuang, (1982) On general properties of quadratic systems, *The Amer. Math. Soc.*, **no. 3, (89)**, p. 167-178
4. Coppel, W.A., (1966) A survey of quadratic systems, *J. Diff. Eq.*, **(2)**, p. 293-304
5. Date, T., (1979) Classification and analysis of two-dimensional real homogeneous quadratic differential equation systems, *J. Diff. Eq.*, **(32)**, p. 311-334
6. Grimshaw, R., (1990) Nonlinear ordinary differential equations, *Appl. Math. and Eng. Science Texts*, Blackwell Scientific Publications, 328 pp.
7. Huang Xianhua and J.W. Reyn, (1995) On the limit cycle distribution over two nests in quadratic systems, *Bull. Austrl. Math. Soc.*, **(52)**, p. 461-474
8. Huang Xianhua, (1996) Qualitative analysis of certain nonlinear differential equations: Quadratic systems and Delay equations, Thesis, Delft University of Technology, 162 pp.
9. Jager, P. de, (1990) Phase portraits for quadratic systems with a higher order singularity with two zero eigenvalues, *J. Diff. Eq.*, **(87)**, p. 169-204
10. Jager, P. de, (1989)*Phase portraits of quadratic systems, higher order singularities and separatrix cycles,*, Thesis, Delft University of Technology, 139 pp
11. Kukles, I.S. and M. Khasanova, (1964) On the distribution of critical points of the first and second group, (Russian), *Izv. Vysš. Ucebn. Zaved. Matematika*, **no. 6 (43)**, p. 88-97
12. Reyn, J.W., (1991) Classes of quadratic systems of differential equations in the plane, Proc. of the Special Program at the Nakai Inst. of Mathematics, Tianjin, P.R. China (sept. 1990-june 1991), Nakai Series in Pure, Applied Mathematical and Theoretical Physics, **Vol. 4 Dyn. Syst.**, p. 146-180
13. Reyn, J.W., (1987) Phase portraits of a quadratic system of differential equations occuring frequently in applications, *Nieuw Archief voor Wiskunde (4) 5*, **no. 2**, p. 107-155
14. Reyn, J.W., (1994) A bibliography of the qualitative theory of quadratic systems of differential equations in the plane, Third edition, Reports of the Faculty of Technical Mathematics and Informatics, Delft University of Technology, **no. 94-02**
15. Reyn, J.W. and R.E. Kooij, (1995) Infinite singular points of quadratic systems in the plane, *J. of Nonlinear Analysis, Theory, Methods and Applications*, **Vol. 24, no. 6**, p. 895-927
16. Reyn, J.W., (1996) Phase portraits of quadratic systems without finite critical points, *J. of Nonlinear Analysis, Theory, Methods and Applications*, **no. 2 (27)**, p. 207-222
17. Reyn, J.W., Phase portraits of quadratic systems with finite multiplicity one, *J. of Nonlinear Analysis, Theory, Methods and Applications*, accepted for publication.
18. Reyn, J.W. and R.E. Kooij, (1995) Phase portraits of nondegenerate quadratic systems with finite multiplicity two, Proc. of the Symposium on Planar Nonlinear Dynamical Systems on the occasion of the 65th birthday of J.W. Reyn, 25 and 26 September 1995, Delft University of Technology, Faculty of Technical Mathematics and Informatics, The Netherlands, accepted for publication
19. Reyn, J.W. and Huang Xianhua, (1996) Phase portraits of quadratic systems with finite multiplicity three and a degenerate critical point at infinity, *Rocky Mountain J. of Mathematics*, accepted for publication
20. Schlomiuk, D.E., J. Guckenheimer and R. Rand, (1990) Integrability of plane quadratic vector fields, *Expositiones Mathematicae*, **8** , p. 3-25
21. Schlomiuk, D.E., (1993) Algebraic particular integrals, integrability and the problem

of the center, *Trans. Amer. Math. Soc.* **(338) no. 2**, p. 799-841
22. Tung Chinchu, (1959) Positions of limit cycles of the system $\frac{dx}{dt} = \sum_{0\le i+k\le 2} a_{ik}x^i y^k, \frac{dy}{dt} = \sum_{0\le i+k\le 2} b_{ik}x^i y^k$, *Sci. Sinica*, **8**, p. 151-171
23. Vulpe, N.I. and K.S. Sibirskiĭ, (1977) Geometric classification of quadratic systems, (Russian), *Diff. Uravn.*, **no. 5 (13)**, p. 803-814; Eng. transl. p. 548-556
24. Zegeling, A., (1994) Separatrix cycles and multiple limit cycles in a class of quadratic systems, *J. Diff. Eq.*, **Vol. 113**, p. 355-380

Part II: Contributed Papers

MICROWAVE HEATING: CRITICAL DEPENDENCE ON DATA AND PARAMETERS

ANDONOWATI
Department of Mathematics, Institut Teknologi Bandung
Jl. Ganesha 10, Bandung, Jawa Barat, Indonesia †

Abstract. A model for microwave heating is considered. The behaviour of the solution is analysed and a procedure to calculate this solution is presented. This procedure is based on an eigenfunction expansion. Combining both analytical and numerical results, it is shown that the qualitative behaviour of solution can be deduced from its fundamental mode. Using this fundamental mode, computation for the critical values of a parameter δ and initial data are carried for a sphere configuration with unit radius.

1. Introduction

Recently, a number of authors [2], [4], [6] have considered the problem of microwave heating. The governing equations consist of a reaction-diffusion type equation for temperature of the medium coupled with the Maxwell equation for electric field which causes the heating. After some simplification [4], the equation are decoupled and the relevant equation then has the form

$$\frac{\partial \theta}{\partial t} = \nabla \cdot (k(\theta)\nabla\theta) + \delta E(\mathbf{x}) f(\theta) . \tag{1}$$

Here, θ is the temperature; $k(\theta)$ is the diffusivity with the properties $k(\theta) > 0$, $k'(\theta) > 0$ and in this study is taken to be $k(\theta) = e^{\gamma\theta}$, for some positive parameter γ; δ is a positive parameter incorporating the geometry of the medium; $E(\mathbf{x})$ is the heating source due to the electric field; and $f(\theta)$ is the auto-catalytic chemical heating source with the properties $f(\theta) > 0$, $f'(\theta) > 0$. As observed by Smyth [6] it is realistic to take $f(\theta)$ to be Arrheniuss type,

†Address for correspondence: Department of Mathematics and Statistics, McGill University, 805 Sherbrooke Street West, Montréal, Québec, Canada H3A 2K6

that is $f(\theta) = e^{\frac{\alpha\theta}{\alpha+\theta}}$ for some $\alpha > 0$. For $k(\theta) = 1$ and $E(\mathbf{x}) = 1$ the equation (1) is a central problem in the combustion theory (see Frank-Kamenestskii [3]) and has been studied by many authors, among them Tam [7]. For $k(\theta) \neq 1$ the equation (1) has also been studied extensively, among them Lacey and Wake [5] and Tam [8].

In this note, we consider equation (1) in an open bounded domain D subject to the initial and the boundary condition

$$\theta(\mathbf{x}, 0) = h(\mathbf{x}) \geq 0 \quad , \quad \theta(\mathbf{x}, t) = 0 \ \ on \ \partial D . \tag{2}$$

The purpose of this study is to demonstrate that the qualitative behaviour of the solution can be approximated by a fundamental mode. We first study the behaviour of solution. We then develop an approximation procedure based on an eigenfunction decomposition. We observed that the qualitative bevaviour of solution is similar to that of its first eigen-mode and so we focus on this fundamental mode [1]. Let δ_{ucr} and δ^{ucr} be the critical values of δ where the steady state solution of (1) is $O(1)$, for $\delta < \delta_{ucr}$, and $O(e^{\alpha})$, for $\delta > \delta^{ucr}$. We show that the critical values δ_{cr} and δ^{cr} calculated from using the fundamental mode approximate the critical values δ_{ucr} and δ^{ucr}, respectively. Thus the critical phenomena of the solution can be predicted from the equation for this single mode. We also compute approximated values of upper and lower bounds of the initial data, say A_{ucr} and A^{ucr}, such that if the initial data is less than A_{ucr} then the steady state solution is of $O(1)$ and if the initial data is larger than A^{ucr} then the steady state solution is of $O(e^{\alpha})$.

In the next section, we present some preliminary results. We investigate the behaviour of the solution in Section 3. A procedure to calculate the solution as well as some justification of the fundamental mode approximation are described in Section 4. In Section 5 numerical results for the critical values δ_{cr} and δ^{cr} as well as the corresponding smallest steady states A_{cr} and A^{cr} are carried using the fundamental-mode approximation for a sphere configuration with unit radius. We present the concluding remark in the last section.

2. Preliminary Results

We introduce the transformation

$$v = \int_0^{\theta} k(s)ds . \tag{3}$$

With this transformation, Equation (1) becomes

$$\frac{\partial v}{\partial t} = K(v)\{\nabla^2 v + \delta E(\mathbf{x})F(v)\} , \tag{4}$$

where $K(v) = k(\theta(v))$ and $F(v) = f(\theta(v))$. Since $v(\theta)$ is monotonically increasing, we observe that both $K(v)$ and $F(v)$ have the same features as $k(\theta)$ and $f(\theta)$ respectively. The initial and the boundary condition (2) become

$$v(\mathbf{x},0) = H(\mathbf{x})\,, \tag{5}$$

and

$$v(\mathbf{x},t) = 0 \ on\ \partial D\,, \tag{6}$$

where $H(\mathbf{x}) = \int_0^{h(\mathbf{x})} k(s)ds$.

Let τ be such that $\frac{d\tau}{dt} = K(v(\mathbf{x},t))$ and $u(\mathbf{x},\tau) = v(\mathbf{x},t)$. Since $K(v(\mathbf{x},t))$ is positive then $\tau(t)$ is monotonically increasing and so $u(\mathbf{x},\tau)$ has the same feature as $v(\mathbf{x},t)$. Using this time scaling, then u satisties

$$\frac{\partial u}{\partial \tau} = \nabla^2 u + \delta E(\mathbf{x})F(u)\,. \tag{7}$$

For $E(x) \equiv 1$, using an eigenfunction expansion for u, Tam in [7] and [8] demonstrated that the feature of u can be obtained from its fundamental mode. Considering that $E(\mathbf{x})$ in this problem is positive and bounded, we shall show that the similar treatment would be adequate.

3. Properties of the Solution

We consider the following IBVP

$$\frac{\partial u}{\partial t} = \nabla^2 u + \delta E(\mathbf{x})F(u)\,. \tag{8}$$

$$u(\mathbf{x},0) = H(\mathbf{x}) \geq 0\,, \quad u(\mathbf{x},t) = 0 \text{ on } \partial D\,. \tag{9}$$

We wish to investigate the behaviour of solution $u(\mathbf{x},t)$. In doing so, we first consider the following IBVP

$$\frac{\partial U}{\partial t} = \nabla^2 U + \delta E(\mathbf{x})F(m)\,, \tag{10}$$

$$U(\mathbf{x},0) = H(\mathbf{x}) \geq 0\,, \quad U(\mathbf{x},t) = 0 \text{ on } \partial D\,, \tag{11}$$

for some $m \geq 0$.

Let $\bar{u}$ and $\bar{U}$ denote the steady state solution of (8) and (10), respectively. Let $m = max_{\mathbf{x}}\bar{u}(\mathbf{x})$. Using *the minimum principle* we can show that $\bar{U}(\mathbf{x},m) \geq m$.

From now on, we use the upper bar to denote the steady state solution. $\bar{u}$, for example, denotes the steady state solution of (8).

Let φ_n and λ_n be the normalized eigenfunctions and eigenvalues of the boundary-value problem

$$\nabla^2\varphi_n = -\lambda_n^2\varphi_n\,, \tag{12}$$
$$\varphi_n = 0 \ on \ \partial D\,, \tag{13}$$

Let $\lambda_1 < \lambda_2 \leq \lambda_3 \leq ....$

The steady state solution for (10) is

$$\bar{U}(\mathbf{x}, m) = \delta F(m) \sum_i \frac{B_i}{\lambda_i^2}\varphi_i(\mathbf{x})\,, \tag{14}$$

where $B_i = \int_D E(\mathbf{x})\varphi_i(\mathbf{x})d\mathbf{x}$. So $max_{\mathbf{x}}\bar{U}(\mathbf{x}, m) = \delta MF(m)$, where $M = max_{\mathbf{x}} \sum_i \frac{B_i}{\lambda_i^2}\varphi_i(\mathbf{x})$. In Figure (1) we show the graph of $max_{\mathbf{x}}\bar{U}(\mathbf{x}, m)$ vs m, for different δ.

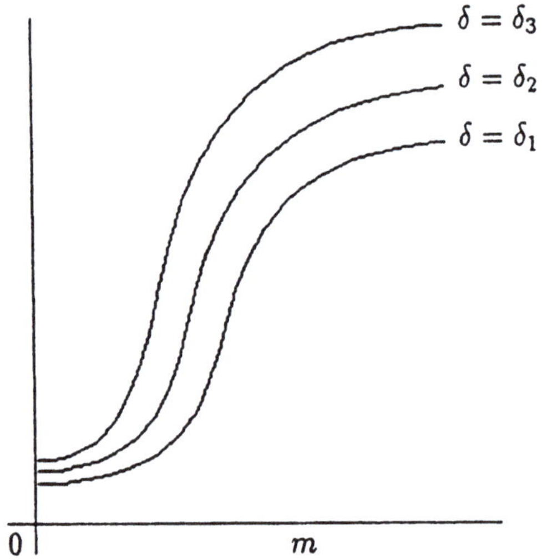

Figure 1. $\max_{\mathbf{x}} \bar{U}(\mathbf{x}, m)$ *vs* m for different δ , $\delta_1 < \delta_2 < \delta_3$

For the case when δ is such that $max_{\mathbf{x}}\bar{U}(\mathbf{x}, m) - m = 0$ has a single root, say m_0, it is not difficult to see that $max_{\mathbf{x}}\bar{u}(\mathbf{x}) \leq m_0$. Let δ be such that $max_{\mathbf{x}}\bar{U}(\mathbf{x}, m) - m = 0$ has three roots, say m_1, m_2, and m_3, where $m_1 < m_2 < m_3$ [See Figure(2)], then for those δ, either $0 \leq max_{\mathbf{x}}\bar{u}(\mathbf{x}) \leq m_1$ or $m_2 \leq max_{\mathbf{x}}\bar{u}(\mathbf{x}) \leq m_3$. We note that m_1 is of $O(1)$ while m_3 is of $O(e^{\alpha})$.

Let $\delta^{U cr}$ and $\delta_{U cr}$ be the largest and the smallest value of δ such that the line m intersects the graph $max_{\mathbf{x}}\bar{U}(\mathbf{x}, m) vs m$ at exactly two points. Figures

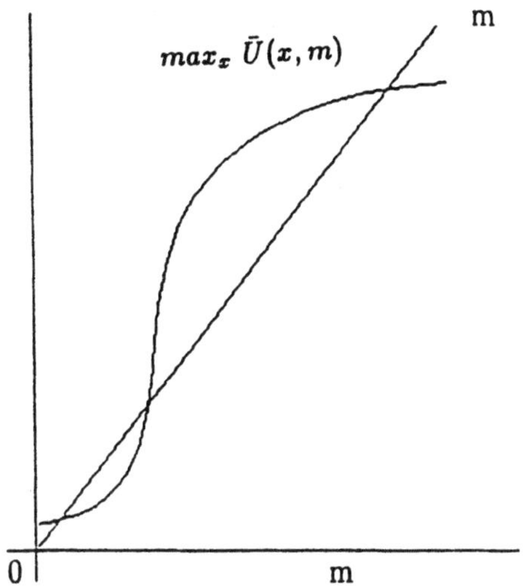

Figure 2. $\max_{\mathbf{x}} \bar{U}(\mathbf{x}, m)$ *vs* m

(3) and (4) illustrate these conditions. Let δ be such that $\delta_{U_{cr}} < \delta < \delta^{U_{cr}}$, for these δ, since $\bar{U}(\mathbf{x}, m)$ is an upper solution to $\bar{u}(\mathbf{x})$, if the initial condition $H(\mathbf{x})$ is sufficiently small then $u(\mathbf{x}, t)$ is small all the time. We note that this $\delta^{U_{cr}}$ is a lower bound of critical values δ^{cr}, where the steady state solution $\bar{u}$ undergoes a rapid transition from being $O(1)$ to $O(e^{\alpha})$. On the other hand, $\delta_{U_{cr}}$ is an upper bound of critical values δ_{cr}, where the steady state solution is $O(1)$ for $\delta < \delta_{cr}$.

4. Fundamental-Mode Approximation

We consider the following IBVP

$$\frac{\partial u}{\partial t} = \nabla^2 u + \delta E(\mathbf{x}) F(u) . \tag{15}$$

$$u(\mathbf{x}, 0) = H(\mathbf{x}) \geq 0 , \quad u(\mathbf{x}, t) = 0 \text{ on } \partial D . \tag{16}$$

Let φ_n and λ_n be the normalized eigenfunctions and eigenvalues of the Laplacian operator subject to the Dirichlet boundary condition and let $H(\mathbf{x}) = \Sigma_{i=1}^{i=\infty} C_i \varphi_i(\mathbf{x})$.

We adopt the following approximation procedure which can be attributed to Galerkin. Let

$$s_N(\mathbf{x}, t) = \Sigma_{i=1}^{i=N} A_i^{(N)}(t) \varphi_i(\mathbf{x}) , \tag{17}$$

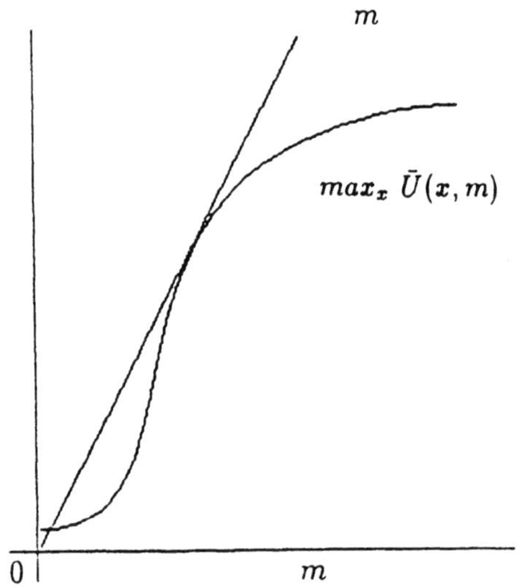

Figure 3. $\max_{\mathbf{x}} \bar{U}(\mathbf{x}, m)$ *vs* m for $\delta = \delta_{Ucr}$

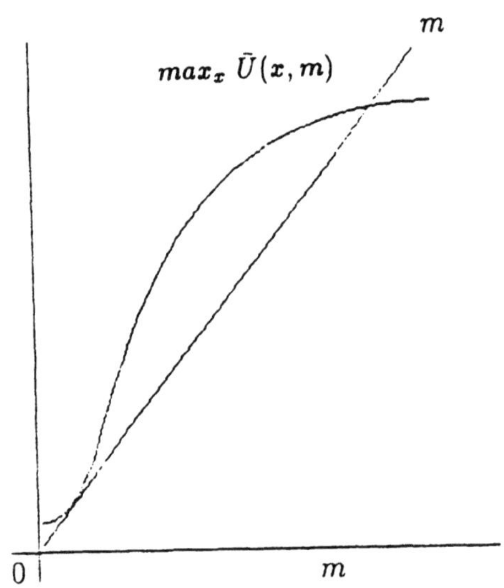

Figure 4. $\max_{\mathbf{x}} \bar{U}(\mathbf{x}, m)$ *vs* m, for $\delta = \delta^{Ucr}$

where $A_i^{(N)}(t)$ is the solution of the integral equation

$$\frac{dA_i^{(N)}}{dt} = -\lambda_i^2 A_i^{(N)} + \delta \int_D E(\mathbf{x}) F(\Sigma_{i=1}^{i=N} A_i^{(N)} \varphi_i(\mathbf{x})) \varphi_i(\mathbf{x}) dv(\mathbf{x}) , \quad (18)$$

$$A_i^{(N)}(0) = C_i\,, \tag{19}$$

for $1 \leq i \leq N$. Equations (18) and (19) constitute N integral equations with N unknowns.

If $\Sigma_{i=1}^{\infty} a_i(t)\varphi_i(\mathbf{x})$ is the eigenfunction expansion of $u(\mathbf{x}, t)$ then $A_i^{(N)}(t) \to a_i(t)$ as $N \to \infty$, and so $s_N(\mathbf{x}, t)$ converges to $u(\mathbf{x}, t)$ as $N \to \infty$ in the mean square.

Inspired by the behaviour of $max_{\mathbf{x}} u(\mathbf{x}, t)$ described in the previous section, which is similar to that of the amplitude of the fundamental mode described in Equation (20), we seek the first approximation $s_1(\mathbf{x}, t)$. In this first eigenfunction approximation $s_1(\mathbf{x}, t) = A(t)\varphi_1(\mathbf{x})$, where $A(t)$ is obtained from the integral equation

$$\frac{dA}{dt} = -\lambda_1^2 A + \delta \int_D E(\mathbf{x}) F(A\varphi_1(\mathbf{x}))\varphi_1(\mathbf{x}) dv(\mathbf{x})\,, \tag{20}$$

$$A(0) = C_1\,. \tag{21}$$

Let

$$I(A) = \int_D E(\mathbf{x}) F(A\varphi_1(\mathbf{x}))\varphi_1(\mathbf{x}) dv(\mathbf{x})\,. \tag{22}$$

The equilibrium values of A can be obtained graphically from the intersection of the straight line $\frac{\lambda_1^2 A}{\delta}$ vs A and the S-shape curve of the graph $I(A)$ vs A.

Let δ, $\delta_{cr} < \delta < \delta^{cr}$ for some critical values δ_{cr} and δ^{cr}, be such that there are three possible steady state solutions for Equation (21), say A_1, A_2, and A_3, where A_1 is of $O(1)$, A_3 is of $O(e^{\alpha})$ and $A_1 < A_2 < A_3$. We note that the middle solution A_2 is unstable while the other two are stable.

The critical value δ^{cr}, where the steady state of (20) is $O(e^{\alpha})$ for $\delta > \delta^{cr}$, is obtained when the straight line $\frac{\lambda_1^2 A}{\delta}$ vs A is tangent to the lower portion of the S-shape curve. On the otherhand, the critical value δ_{cr}, where the steady state of (20) $O(1)$ for $\delta < \delta_{cr}$, is obtained when the straight line $\frac{\lambda_1^2 A}{\delta}$ vs A is tangent to the upper portion of the S-shape curve [See Figure (5)].

If for some δ, $< \bar{s}_1, \bar{s}_1 >$ is in the same order of magnitude as $< \bar{s}_N, \bar{s}_N >$ for any $N > 1$, then for that δ, $\bar{s}_1$ approximates the steady state solution $\bar{u}$. Here $< f(\mathbf{x}), g(\mathbf{x}) >$ is defined as $\int_D f(\mathbf{x})g(\mathbf{x})d\mathbf{x}$. This argument is based on the convergence of $\bar{s}_N$ to $\bar{u}$ in the mean square. Further if for any $N > 1$, the smallest solution $\bar{s}_N(\mathbf{x}, \delta)$ changes its order of magnitude near the value of δ^{cr}, then critical parameter δ^{cr} obtained from the single mode approximates the critical parameter of the solution u. On the otherhand, if for any $N > 1$ the largest solution $\bar{s}_N(\mathbf{x}, \delta)$ changes its order of magnitude near the value of δ_{cr}, then critical parameter δ_{cr} obtained from the single mode approximates the critical parameter of the solution $\bar{u}$.

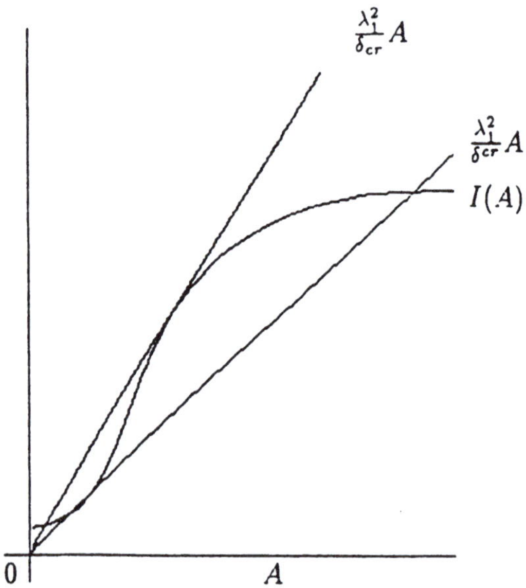

Figure 5. $\frac{\delta_{cr}}{\lambda_1^2}A$, $\frac{\delta^{cr}}{\lambda_1^2}A$ and $I(A)$ *vs* A

For the sphere configuration of a unit sphere, assuming the spherical symmetry with $\alpha = 10$, $\gamma = 0.2$, and $E(r) = e^{-(1-r)}$, we obtain $< \bar{s}_1, \bar{s}_1 >$ is in the same order of magnitude as $< \bar{s}_N, \bar{s}_N >$, for N=2,3,4, ... ,15, and δ =1,2,3,4,5,6,7,8,9,and 10. Here $s_1 = A\varphi(\mathbf{x})$ has been taken to be the smallest solution in the case of multiple solutions. [See Table (1) as examples.]

	$< \bar{s}_N, \bar{s}_N >$			
	N=1	N=5	N=10	N=15
$\delta = 5.0$	0.518896	0.514249	0.514244	0.514246
$\delta = 6.0$	1.068668	1.061605	1.061586	1.061587
$\delta = 7.0$	2.847950	3.058957	3.058752	3.058781
$\delta = 8.0$	$0.568998\ 10^6$	$0.572069\ 10^6$	$0.572078\ 10^6$	$0.572079\ 10^6$
$\delta = 9.0$	$0.866649\ 10^6$	$0.872309\ 10^6$	$0.872333\ 10^6$	$0.872335\ 10^6$
$\delta = 10.0$	$0.124418\ 10^7$	$0.125334\ 10^7$	$0.125339\ 10^7$	$0.125339\ 10^7$

Table (1) The magnitude of the smallest steady-state solution $\bar{s}_N$ for different δ with $\alpha = 10$, $\gamma = 0.2$.

For $\alpha = 10$ and $\gamma = 0.2$, the critical parameter δ^{cr} calculated from the fundamental-mode is $\delta^{cr} = 7.256$. It is shown in Table (2), that for δ near δ^{cr}, the magnitude of the smallest steady-state $< \bar{s}_1, \bar{s}_1 >$ is in the same order of magnitude as $< \bar{s}_N, \bar{s}_N >$ for N=5 and 10, except for $\delta = 7.2$ which is very close to the critical point $\delta^{cr} = 7.256$ indicating the critical parameter δ^{cr} obtained from the single mode approximates the critical parameter of the solution u with a relatively small error. On the otherhand, the critical parameter δ_{cr} for the above example is $\delta_{cr} = 3.494$. It is shown in Table (3), that for δ near δ_{cr}, the magnitude of the largest steady-state $< \bar{s}_1, \bar{s}_1 >$ is in the same order of magnitude as $< \bar{s}_N, \bar{s}_N >$ for N=5 and 10. This indicates that the critical parameter δ_{cr} obtained from the single mode approximates the critical parameter of the solution u with a relatively small the error. We repeat similar calculations for different α and γ and obtain similar results.

Thus, it is not only that the first mode is dominant but also that the critical values δ_{cr} and δ^{cr} obtained by using a single mode approximate the critical values δ of u.

	$< \bar{s}_N, \bar{s}_N >$		
$\delta^{cr} = 7.256$	N=1	N=5	N=10
$\delta = 7.0$	2.847926	3.058751	3.058752
$\delta = 7.1$	3.360060	3.915313	3.914835
$\delta = 7.2$	4.243676	$0.385138\ 10^6$	$0.385141\ 10^6$
$\delta = 7.3$	$0.404318\ 10^6$	$0.406070\ 10^6$	$0.406072\ 10^6$
$\delta = 7.4$	$0.425608\ 10^6$	$0.427681\ 10^6$	$0.427685\ 10^6$

Table (2) The magnitude of the smallest steady-state solution $\bar{s}_N$ for δ near δ^{cr} with $\alpha = 10$, $\gamma = 0.2$.

	$< \bar{s}_N, \bar{s}_N >$		
$\delta_{cr} = 3.494$	N=1	N=5	N=10
$\delta = 3.2$	0.138690	0.138178	0.138178
$\delta = 3.3$	0.150413	0.149806	0.149807
$\delta = 3.4$	0.162888	0.162175	0.162175
$\delta = 3.5$	$0.348079\ 10^4$	$0.406508\ 10^4$	$0.406555\ 10^4$
$\delta = 3.6$	$0.701667\ 10^4$	$0.712715\ 10^4$	$0.712740\ 10^4$

Table (3) The magnitude of the largest steady-state solution $\bar{s}_N$ for δ near δ_{cr} with $\alpha = 10$, $\gamma = 0.2$.

5. Numerical Results

Let δ^{cr} be the critical value of δ obtained from the single mode (18). Suppose that A_1 is the smallest A satisfying $I(A_1) = \frac{\lambda_1^2 A_1}{\delta^{cr}}$; then $\frac{d}{dA}\left\{I(A) - \frac{\lambda_1^2 A}{\delta^{cr}}\right\}_{A=A_1} = 0$. We may thus calculate δ^{cr} and the corresponding smallest steady state solution of A from

$$I(A_1) - A_1 I'(A_1) = 0 \quad \text{and} \quad \delta^{cr} = \frac{\lambda_1^2 A_1}{I(A_1)}, \tag{23}$$

noting that $I(A) - \frac{\lambda_1^2 A}{\delta^{cr}}$ attains a minimum at $A = A_1$.

Similarly for the critical value δ_{cr}, let A_2 is the largest A satisfying $I(A_2) = \frac{\lambda_1^2 A_2}{\delta_{cr}}$, then $I(A_2) - A_2 I'(A_2) = 0$, and $\delta_{cr} = \frac{\lambda_1^2 A_2}{I(A_2)}$. Here $I(A) - \frac{\lambda_1^2 A}{\delta^{cr}}$ attains a maximum at $A = A_2$.

We carry out some computations by taking D to be a sphere with unit radius. Assuming the solution to depend on the radius only, we have the eigenvalues $\{i^2\pi^2\}$ and the eigenfuctions $\{\frac{1}{\sqrt{2\pi}} \frac{\sin(i\pi r)}{r}\}$. For several values of α, and γ, in Table (4) we summerize the critical values δ^{cr} and the corresponding smallest steady state of A, whereas the critical values δ_{cr} and the corresponding smallest steady state of A are presented in in Table (5).

	$\gamma = 0.1$		$\gamma = 0.2$		$\gamma = 0.3$	
	δ^{cr}	A	δ^{cr}	A	δ^{cr}	A
$\alpha = 10$	6.654	1.932	7.256	2.187	8.033	2.952
$\alpha = 12$	6.473	1.647	7.018	1.998	7.696	2.551
$\alpha = 14$	6.354	1.581	6.864	1.888	7.488	2.351
$\alpha = 16$	6.268	1.532	6.756	1.816	7.345	2.226
$\alpha = 18$	6.204	1.498	6.675	1.762	7.240	2.138
$\alpha = 20$	6.155	1.474	6.613	1.724	7.160	2.076

Table (4): Critical values δ^{cr} and the corresponding smallest steady state solutions A for a sphere configuration of unit radius.

	$\gamma = 0.1$		$\gamma = 0.2$		$\gamma = 0.3$	
	δ_{cr}	A	δ_{cr}	A	δ_{cr}	A
$\alpha = 10$	1.344	0.134	3.494	0.419	6.326	1.106
$\alpha = 12$	0.556	0.053	1.978	0.207	4.403	0.581
$\alpha = 14$	0.225	0.021	1.098	0.108	2.996	0.342
$\alpha = 16$	0.089	0.009	0.603	0.058	2.015	0.212
$\alpha = 18$	0.035	0.004	0.335	0.032	1.348	0.135
$\alpha = 20$	0.013	0.002	0.196	0.019	0.898	0.087

Table (5): Critical values δ_{cr} and the corresponding smallest steady state solutions A for a sphere configuration of unit radius.

6. Concluding Remarks

We have considered a model for a microwave heating. ¿From a procedure based on an eigenfunction expansion, an ordinary differential equation for the amplitude of its fundamental mode is presented. It is shown that besides the fundamental mode dominance, the critical phenomena of the solution can be deduced from merely this ODE. Numerical results for the critical values δ_{cr} and δ^{cr} and corresponding smallest amplitudes of the solution, say A_{cr} and A^{cr} respectively, are carried for a sphere configuration with unit radius. A_{cr} is an approximated upper bound of the magnitude of the initial data for which the solution is of $O(1)$, whereas A^{cr} is an approximated lower bound of the magnitude of the initial data for which the solution is of $O(e^{\alpha})$. Numerical results show that larger α corresponds to smaller δ_{cr} and δ^{cr}. We note that δ^{cr} and δ_{cr} are the critical values of δ, where the steady state solution is $O(1)$, for $\delta < \delta_{cr}$, and $O(e^{\alpha})$, for $\delta > \delta^{cr}$.

Acknowledgment

The author is very grateful to Prof. K.K. Tam for detailed comments on the manuscript of this paper. She would like to thank Prof. B. van Groesen and Dr. Edi Soewono for some corrections and improvement. She acknowledges the financial support of P4M Institut Teknologi Bandung, Indonesia. This research was partly done during the author visit to Twente University in the Netherlands.

References

1. Andonowati. *A study of some problems arising from combution and the microwave heating.* Ph.D. thesis, McGill University, Montreal, Canada (1995).
2. C. J. Coleman. *On the microwave hotspot problem.* J. Austral. Math. Soc. B. 33:1-8 (1987).
3. D. A. Frank-Kamenestskii. *Diffusion and Heat Transfer in Chemical Kinetics.* Plenum Press. New York. (1959).
4. J. M. Hill and A. H. Pincombe. *Some similarity temperature profiles for the microwave heating of a half-space.* J. Austral. Math. Soc. B. 33:290-320 (1992).
5. A. Lacey and G.C. Wake. *Therman ignition with variable thermal conductivity.* IMA J. Appl. Math 28:23-39 (1982).
6. N. F. Smyth. *The effect of conductivity on hotspots.* J. Austral. Math. Soc. B. 33:403-413 (1992).
7. K. K. TAM. *Criticality dependence on data and parameters for a problem in combustion theory.* J. Austral. Math. Soc. B. 27:416-441 (1986).
8. K. K. TAM. *Criticality dependence on data and parameters for a problem in combustion theory, with temperature-dependent conductivity.* J. Austral. Math. Soc. B. 31:76-80 (1989).

ON THE CONVERGENCE OF SAMPLED DATA NONLINEAR SYSTEMS

P. ASTUTI
Jurusan Matematika
Institut Teknologi Bandung
Jln Ganesha 10
Bandung 40132, INDONESIA

M. CORLESS
School of Aeronautics
Purdue University
West Lafayette, Indiana 47907, USA

AND

D. WILLIAMSON
Department of Engineering
The Faculties
Australian National University
ACT 0200, AUSTRALIA

Abstract. The effect of sampling on the convergence of closed loop continuous systems is investigated. The systems under consideration are nonlinear containing parameter uncertainties. It is shown that the discretized of a converging controller results in convergence of the sampled data closed loop system; i.e., a nonlinear continuous-time system subject to a discrete controller; if the sampling time is in an open set and the continuous plant satisfies certain Lipschitz type conditions. The results are derived for both global and local convergences.

1. Introduction

Sampled data control systems consist of continuous-time plants controlled by discrete controllers. Recently, research interest on the effect of sampling times on sampled data nonlinear control systems has increased due to the advance development of nonlinear control design. For exsample, conditions for the preservation of the linear equivalent and feedback linearizable properties have been investigated in [6], [5], [7], and [1]. In [8] a modification of sampled linearizing feedback to improve the performance of closed loop sampled data control systems was proposed. In another report, [9] applied a discrete event controller design methodology to improve a performance of a sampled data control system.

This paper investigates the effect of sampling on the convergence property of the closed loop systems. The effect of sampling on the stability property of the closed loop systems has been investigated in [2].

The organization of this paper is as follows. Section 2 introduces sampled data systems. Section 3 investigates a class of nonlinear systems for which the discrete implementation of the continuous controllers can preserve the global convergence. Then, Section 4 investigates the effect of sampling on local convergence. Finally, conclusion is in Section 5.

2. Sampled Data Systems

Consider a nonlinear uncertain system described by equation

$$\dot{\mathbf{x}} = \mathbf{F}(\mathbf{x}(t), \mathbf{u}(t), \delta(t, \mathbf{x}(t))) \tag{1}$$

where $t \in \mathbb{R}$ is the time, $\mathbf{x} \in \mathbb{R}^n$ is the state, and $\mathbf{u} \in \mathbb{R}^m$ is the control input. All the uncertainties in the system is characterized by the uncertain term δ which is contained in a known set Δ. For example Δ is of the form an interval $\Delta = [-\rho, \rho]$ for some positive constant ρ. Assume that $\mathbf{F}$ is continuous such that a solution of equation (1) exists for any initial state $\mathbf{x}_0$ at time t_0 and any continuous input $\mathbf{u}$ [10]. For simplicity we will assume $t_0 = 0$.

Suppose a continuous-time controller for (1) is described by equation

$$\mathbf{u}(t) = \mathbf{p}(\mathbf{x}(t)) \tag{2}$$

where $\mathbf{p}$ is continuous. Then the corresponding closed loop system is described by

$$\dot{\mathbf{x}}(t) = (\mathbf{x(t)}, \mathbf{p}(\mathbf{x(t)}), \delta(\mathbf{t}, \mathbf{x(t)})) \tag{3}$$

Our first assumption requires that the closed loop sytem be *globally uniformly exponentially convergent* (GUEC) to a ball.

Assumption 1 *There exists a continuously differentiable function* $V : \mathbb{R}^n \longrightarrow \mathbb{R}$ *and a positive scalars* $\omega_1, \omega_2, \alpha, V^*$ *such that for every* $\mathbf{x} \in \mathbb{R}^n$ *and* $t \in \mathbb{R}$

i. $\omega_1 \parallel \mathbf{x} \parallel^2 \leq V(\mathbf{x}) \leq \omega_2 \parallel \mathbf{x} \parallel^2$

ii. $\nabla V(\mathbf{x}).\mathbf{F}(\mathbf{x}, \mathbf{p}(x), \delta(t, \mathbf{x})) \leq -2\alpha(V(\mathbf{x}) - V^*)$ *for all* $V(\mathbf{x}) > V^*$

where $\nabla V(\mathbf{x})$ *denotes the gradient of* V *at* $\mathbf{x}$.

□

Note that if the closed loop system (3) satisfies Assumption 1, it can be shown that every solution of the closed loop system (3) satisfies

$$\| \mathbf{x}(t) \| \leq \beta \| \mathbf{x}(0) \| e^{-\alpha t} + r_c \quad \text{for} \quad t \geq 0 \tag{4}$$

where $\beta = \sqrt{\frac{\omega_2}{\omega_1}}$ and $r_c = \sqrt{\frac{V^*}{\omega_1}}$[3]. This implies that the continuous controller (2) results in the system (1) GUEC to the ball with radius r_c,

$$B(r_c) = \{\mathbf{x} \in \mathbb{R}^n | \, \| \mathbf{x} \| \leq r_c\}$$

and rate α. In other words, the continuous controller (2) drives the system (1) to go to the ball $B(r_c)$ exponentially as the time goes to infinity.

In this paper we will analyse the effect of discretization of the controller (2). In particular, we derive sufficient conditions under which the discretization of the continuous controller (2) results in the system (1) asymptotically convergence to a ball.

To begin with, we introduce a sampled data closed loop system. Let T be a sampling time and define

$$t_0 = 0$$
$$t_k = t_{k-1} + T \quad \text{for} \quad k = 1, 2, \ldots$$

i.e.

$$t_k = kT$$

The discretization of the continuous controller (2) is obtained by holding the control input constant over each interval $[t_k, t_{k+1})$, i.e.

$$\mathbf{u}(t) = \mathbf{p}(\mathbf{x}(t_k)) \quad \text{for} \quad t_k \leq t < t_{k+1} \tag{5}$$

The *sampled data closed loop system* resulting from (1) and (5) is described by equation

$$\dot{\mathbf{x}}(t) = \mathbf{F}(\mathbf{x}(t), \mathbf{p}(\mathbf{x}(t_k)), \delta(t, \mathbf{x}(t))) \quad \text{for} \quad t_k \leq t < t_{k+1} \tag{6}$$

To simplify the notation we will denote $\mathbf{x}_t = \mathbf{x}(t)$ and $\mathbf{x}_k = \mathbf{x}(t_k)$.

3. Global Convergence

In this section we present sufficient conditions which guarantees global asymptotic convergence to a ball of the sampled data closed loop system (6). The conditions are stated in the following assumption which are natural conditions for some systems. For example, the control system of an inverted pendulum which contains uncertainties as described in [3] satisfies this assumption.

Assumption 2 *There exists positive scalars* $\beta_1, \beta_2, \beta_3$ *and* β_4 *such that for every* $\mathbf{x}, \mathbf{y} \in \mathbb{R}^n$ *and* $t \in \mathbb{R}$

i. $\| \mathbf{F}(\mathbf{x}, \mathbf{p}(\mathbf{x})\delta(t,\mathbf{x})) \| \leq \beta_1 \| \mathbf{x} \| + \beta_2$

ii. $\| \mathbf{F}(\mathbf{x}, \mathbf{p}(\mathbf{y}), \delta(t,\mathbf{x})) - \mathbf{F}(\mathbf{x}, \mathbf{p}(\mathbf{x}), \delta(t,\mathbf{x})) \| \leq \beta_3 \| \mathbf{x} - \mathbf{y} \|$

iii. $\| \nabla V(\mathbf{x}) \| \leq \beta_4 \| \mathbf{x} \|$ □

The result for a sampled data nonlinear system satisfying this assumption is stated in the following theorem.

Theorem 1 *Consider the sampled data closed loop system (6) and suppose Assumptions 1 and 2 hold. Then every solution is global. In addition, suppose for* $T > 0$, $\mathbf{x}_t(T)$ *be the solution of (6) with the sampling time* T *and* $\mathbf{x}_t(0)$ *is the solution of (3). Then, there exists a positive number* $T^* > 0$ *and a continuous nondecreasing function* $r_d : [0, T^*) \longrightarrow \mathbb{R}$ *with* $r_d(0) = r_c = \sqrt{\frac{V^*}{\omega_1}}$ *such that for every sampling time* $T \in [0, T^*)$ *the solution* $\mathbf{x}_t(T)$ *goes to the ball* $B(r_d(T))$ *as the time* t *goes to infinity.* □

Proof :

Using induction we shall show that every solution $\mathbf{x}(.)$ of (6) is global and establish a bound on the norm $\| \mathbf{x}_t - \mathbf{x}_k \|$ for all $t \in [t_k, t_{k+1}]$, $k = 0, 1, \ldots$.

Suppose $\mathbf{x}_t$ is a solution of the sampled data system (6) which is defined on the interval $[0, t_k]$, for some $k \in \{0, 1, \ldots\}$. Clearly it is true for $k = 0$. For this, the solution $\mathbf{x}_t$ can be extended over the interval $[0, t_k + \theta_k]$ for some positive number $\theta_k < T$. For every $t \in [t_k, t_k + \theta_k]$, the solution $\mathbf{x}_t$ satisfies

$$\| \mathbf{x}_t - \mathbf{x}_k \| \leq \int_{t_k}^{t} \| \mathbf{F}(\mathbf{x}_\tau, \mathbf{p}(\mathbf{x}_\tau), \delta(\tau, \mathbf{x}_\tau)) \| + \| \Delta\mathbf{F}(\mathbf{x}_\tau, \mathbf{x}_k) \| \, d\tau$$

where $\Delta\mathbf{F}(\mathbf{x}_t, \mathbf{x}_k) = \mathbf{F}(\mathbf{x}_t, \mathbf{p}(\mathbf{x}_k), \delta(t, \mathbf{x}_t)) - \mathbf{F}(\mathbf{x}_t, \mathbf{p}(\mathbf{x}_t), \delta(t, \mathbf{x}_t))$. By applying Assumption 2 and the Granwall inequality, we finally obtain

$$\| \mathbf{x}_t - \mathbf{x}_k \| \leq \frac{\beta_1 \| \mathbf{x}_k \| + \beta_2}{\alpha_1} (e^{\alpha_1 (t - t_k)} - 1) \tag{7}$$

where $\alpha_1 = \beta_1 + \beta_3$.

Consider that the right hand side of the inequality (7) is bounded above for $t \in [t_k, t_{k+1}]$. Hence the continuity of $\mathbf{x}_t$ implies that the solution of (6) can be extended to the interval $[0, t_{k+1}]$. Thus, any solution of (6) is global and satisfies inequality (7) for all $t \in [t_k, t_{k+1}]$, $k = 0, 1, \ldots$

Define $\eta(t) = V(\mathbf{x}_t)^{\frac{1}{2}}$ and $\eta^* = (V^*)^{\frac{1}{2}}$ where V is the function satisfying Assumption 1. Then, whenever $\eta(t) > \eta^*$, η is differentiable and satisfies

$$\begin{aligned} \dot{\eta}(t) &= \tfrac{1}{2} V(\mathbf{x}_t)^{-\frac{1}{2}} \nabla V(\mathbf{x}_t) \dot{\mathbf{x}}_t \\ &= \tfrac{1}{2} V(\mathbf{x}_t)^{-\frac{1}{2}} [\nabla V(\mathbf{x}_t) \mathbf{F}(\mathbf{x}_t, \mathbf{p}(\mathbf{x}_t), \delta(t, \mathbf{x}_t)) + \nabla V(\mathbf{x}_t) \Delta\mathbf{F}(\mathbf{x}_t, \mathbf{x}_k)] \end{aligned}$$

Now, taking into account the Assumptions 1 and 2 we obtain

$$\begin{array}{rcl}\dot{\eta}(t) & \leq & \frac{1}{2}V(\mathbf{x}_t)^{-\frac{1}{2}}[-2\alpha(V(\mathbf{x}_t)-V^*)+\beta_3\beta_4 \parallel \mathbf{x}_t \parallel \parallel \mathbf{x}_t-\mathbf{x}_k \parallel] \\ & \leq & -\alpha(V(\mathbf{x}_t)^{\frac{1}{2}}-V(\mathbf{x}_t)^{-\frac{1}{2}}V^*)+\frac{\beta_3\beta_4}{2\sqrt{\omega_1}}(\frac{\beta_1\|\mathbf{x}_k\|+\beta_2}{\alpha_1})(e^{\alpha_1(t-t_k)}-1) \\ & \leq & -\alpha(\eta(t)-\eta^*)+\frac{\beta_3\beta_4(\beta_1\|\mathbf{x}_k\|+\beta_2)}{2\alpha_1\sqrt{\omega_1}}(e^{\alpha_1(t-t_k)}-1)\end{array}$$

Hence, if we define $\eta_1(t) = \eta(t) - \eta^*$, then whenever $\eta_1(t) > 0$, it is differentiable where its derivative satisfies

$$\dot{\eta}_1(t) \leq -\alpha\eta_1(t) + \frac{\beta_3\beta_4(\beta_1 \parallel \mathbf{x}_k \parallel + \beta_2)}{2\alpha_1\sqrt{\omega_1}}(e^{\alpha_1(t-t_k)}-1)$$

According to a Lemma in [4] the function η_1 satisfies

$$\eta_1(t) \leq \eta_2(t) \quad t \in [t_k, t_{k+1}]$$

where $\eta_2(t)$ is the solution of equation

$$\begin{array}{rcl}\dot{\eta}_2(t) & = & -\alpha\eta_2(t)+\frac{\beta_3\beta_4(\beta_1\|\mathbf{x}_k\|+\beta_2)}{2\alpha_1\sqrt{\omega_1}}(e^{\alpha_1(t-t_k)}-1) \quad \text{for} \quad t\in[t_k,t_{k+1}] \\ \eta_2(t_k) & = & \eta_1(t_k)=\eta(t_k)-\eta^*\end{array}$$

It can be shown that solution η_2 is of the form

$$\begin{array}{rcl}\eta_2(t) & = & e^{-\alpha(t-t_k)}(\eta(t_k)-\eta^*) \\ & & +\left(\frac{\beta_1\beta_3\beta_4\|\mathbf{x}_k\|}{2(\alpha+\alpha_1)\alpha\alpha_1\sqrt{\omega_1}}+\frac{\beta_2\beta_3\beta_4}{2(\alpha+\alpha_1)\alpha\alpha_1\sqrt{\omega_1}}\right)O(t-t_k)\end{array}$$

for $t \in [t_k, t_{k+1}]$ where

$$O(t) = \alpha(e^{\alpha_1 t}-1)+\alpha_1(e^{-\alpha t}-1).$$

So, by taking into account the Assumption 1 part i we have that for every $t \in [t_k, t_{k+1}]$

$$\eta(t) \leq \eta(t_k)\lambda(t-t_k)+O_1(t-t_k) \tag{8}$$

where

$$\lambda(t) = e^{-\alpha t}+\gamma_1 O(t)$$

$$O_1(t) = (1-e^{-\alpha t})\eta^*+\gamma_2 O(t)$$

$$\gamma_1 = \frac{\beta_1\beta_3\beta_4}{2(\alpha+\alpha_1)\alpha\alpha_1\omega_1}; \quad \gamma_2 = \frac{\beta_2\beta_3\beta_4}{2(\alpha+\alpha_1)\alpha\alpha_1\sqrt{\omega_1}}.$$

Note that :

i. the function λ satisfies $\lambda(0) = 1$ and $\lambda'(0) = -\alpha < 0$. The continuity of the function λ guarantees the existence of a positive number T_1 such that $\lambda(t) < 1$ for all $t \in (0, T_1)$. Denote

$$T^* = \sup\{T > 0 | \lambda(t) < 1 \text{ for all } t \in (0, T)\}.$$

ii.

$$\lim_{t \longrightarrow 0^+} \frac{O_1(t)}{1 - \lambda(t)} = \eta^*$$

So, we can define a radius function $r_d : [0, T^*) \longrightarrow \mathbb{R}$ with

$$r_d(t) = \begin{cases} r_c = \frac{\eta^*}{\sqrt{\omega_1}} & t = 0 \\ \sup\left\{ \frac{O_1(\tau)}{\sqrt{\omega_1}(1-\lambda(\tau))} \mid \tau \in (0, t] \right\} & t \in (0, T^*) \end{cases}$$

which is a continuous and nondecreasing function on the interval $[0, T^*)$.

If we can show that for every $T \in [0, T^*)$ the solution $\mathbf{x}_t(T)$ goes to the ball $B(r_d(T))$ as the time t goes to infinity than we will get the theorem.

Note that it is clear the above situation holds for $T = 0$. So, let $T \in (0, T^*)$ be the sampling time for the equation (6). By using induction and the equation (8) we can derive for all $k = 1, 2, \ldots$

$$\eta(t_k) \leq \eta(0)\lambda(T)^k + O_1(T) \sum_{i=0}^{k-1} \lambda(T)^i$$

In addition, for all $t \in [t_k, t_{k+1}]$ we obtain

$$\begin{aligned} \eta(t) &\leq \eta(t_k)\lambda(t - t_k) + O_1(t - t_k) \\ &\leq \eta(0)\lambda(T)^k\lambda(t - t_k) + O_1(T) \textstyle\sum_{i=0}^{k-1} \lambda(T)^i\lambda(t - t_k) + O_1(t - t_k) \\ &\leq \eta(0)\lambda(T)^k\lambda(t - t_k) + \frac{O_1(T)}{1-\lambda(T)}\lambda(t - t_k) + \frac{O_1(t-t_k)}{1-\lambda(t-t_k)}(1 - \lambda(t - t_k)) \\ &\leq \eta(0)\lambda(T)^k\lambda(t - t_k) + \sqrt{\omega_1}r_d(T)\lambda(t - t_k) + \sqrt{\omega_1}(1 - \lambda(t - t_k)) \\ &\leq \eta(0)\lambda(T)^k\lambda(t - t_k) + \sqrt{\omega_1}r_d(T) \end{aligned}$$

Since, for $t \longrightarrow \infty$ implies $k \longrightarrow \infty$ and $\lim_{k \longrightarrow \infty} \lambda^k(T) = 0$, the last inequality above implies that

$$\lim_{t \longrightarrow \infty} \eta(t) \leq \sqrt{\omega_1}r_d(T)$$

As a result,

$$\lim_{t \longrightarrow \infty} \| \mathbf{x}_t(T) \| \leq \lim_{t \longrightarrow \infty} \frac{\eta(t)}{\sqrt{\omega_1}} \leq r_d(T)$$

That is the solution $\mathbf{x}_t(T)$ goes to the ball $B(r_d(T))$ as the time t goes to infinity. □

Remarks :

1. Considering the proof of Theorem 1, one can conclude that one way to guarantee the discrete implementation of a converging continuous controller for (1) also resulting in convergence to a ball, the sampling time T should be less than a certain number, T^*. This bound of sampling time can be determined as the smallest positif number satisfying the equation
$$\lambda(T) = 1$$
2. The above theorem can be viewed as describing the continuity of the radius of the ball of convergence with respect to sampling time T. In particular, for every positif number $r > r_c$, one can find a positif number T such that the resulting closed loop data system with sampling time T asymptotically converges to the ball $B(r)$. This sampling time can be determined as the solution of inequalities
$$\lambda(T) < 1 \qquad \text{and} \qquad r_d(T) \leq 1$$
3. Suppose the continuous controller (2) results in an exponentially stable at 0 of the closed loop system (ie. $r_c = 0$) and the Assumption 2 holds for $\beta_2 = 0$. Then, we obtain the closed loop data system is asymptotically stable at 0 provided the sampling time T satisfies the inequality
$$\lambda(T) < 1.$$
Consider that this is the result stated in [2] (see Theorem 1). This means the above theorem is a generalization of Theorem 1 in [2]. □

The above result can also be stated as the following corollary.

Corollary 1 *Consider the sampled data closed loop system (6) and suppose Assumptions 1 and 2 hold and the sampling time T satisfies*
$$\lambda(T) < 1$$
where
$$\lambda(T) = e^{-\alpha T} + \gamma_1 O(T)$$
$$O(T) = \alpha(e^{\alpha_2 T} - 1) + \alpha_2(e^{-\alpha T} - 1); \quad \gamma_1 = \frac{\beta_1\beta_3\beta_4}{2(\alpha + \alpha_1)\alpha\alpha_1\omega_1}.$$
Then, the solution of the closed loop data System (6) goes to the ball $B(r_d(T))$ where
$$r_d(T) = \begin{cases} r_c = \sqrt{\frac{V^*}{\omega_1}} & T = 0 \\ \sup\left\{ \frac{(1-e^{-\alpha t})\sqrt{V^*} + \gamma_2 O(t)}{\sqrt{\omega_1}(1-\lambda(\tau))} \mid \tau \in (0, T] \right\} & T > 0 \end{cases}$$

and

$$\gamma_2 = \frac{\beta_2\beta_3\beta_4}{2(\alpha+\alpha_1)\alpha\alpha_1\sqrt{\omega_1}}.$$

4. Local Convergence

In this section we only assume a local version of Assumptions 1-2 and obtain conditions which guarantee local asymptotic convergence of the closed loop system under discritized controller. So, suppose the requirements of Assumption 1-2 are only satisfied by $\mathbf{x}, \mathbf{y}$ in a closed ball,

$$B(r) = \{\mathbf{x} \in \mathbb{R}^n | \parallel \mathbf{x} \parallel \leq r\}$$

where $r > 0$. That is, suppose the following assumption holds.

Assumption 3 *There exists a continuously differentiable function $V : \mathbb{R}^n \longrightarrow \mathbb{R}$ and a positive scalars $r, \omega_1, \omega_2, \alpha, V^*$, $\beta_1, \beta_2, \beta_3$ and β_4 such that for every $\mathbf{x} \in \mathbb{R}^n$ and $t \in \mathbb{R}$*

i. $\omega_1 \parallel \mathbf{x} \parallel^2 \leq V(\mathbf{x}) \leq \omega_2 \parallel \mathbf{x} \parallel^2$

ii. $\nabla V(\mathbf{x}).\mathbf{F}(\mathbf{x}, \mathbf{p}(x), \delta(t, \mathbf{x})) \leq 2\alpha(V(\mathbf{x}) - V^*)$ *for all* $V(\mathbf{x}) > V^*$ *where* $\nabla V(\mathbf{x})$ *denotes the gradient of* V *at* $\mathbf{x}$.

And for every $\mathbf{x}, \mathbf{y} \in B(r)$ *and* $t \in \mathbb{R}$

iii. $\parallel \mathbf{F}(\mathbf{x}, \mathbf{p}(\mathbf{x})\delta(t, \mathbf{x})) \parallel \leq \beta_1 \parallel \mathbf{x} \parallel + \beta_2$

iv. $\parallel \mathbf{F}(\mathbf{x}, \mathbf{p}(\mathbf{y}), \delta(t, \mathbf{x})) - \mathbf{F}(\mathbf{x}, \mathbf{p}(\mathbf{x}), \delta(t, \mathbf{x})) \parallel \leq \beta_3 \parallel \mathbf{x} - \mathbf{y} \parallel$

v. $\parallel \nabla V(\mathbf{x}) \parallel \leq \beta_4 \parallel \mathbf{x} \parallel$ □

Then, we have the following result.

Theorem 2 *Consider the sampled data closed loop system (6) and suppose Assumption 3 holds and the sampling time T satisfies*

$$\lambda(T) < 1; \ r_d(T) < r$$

where $\lambda(T)$ and $r_d(T)$ are defined in Corollary 1. If the initial state satisfies

$$\parallel \mathbf{x}(0) \parallel < r\sqrt{\frac{\omega_1}{\omega_2}}$$

then the solution of the closed loop data System (6) goes to the ball $B(r_d(T))$. □

Proof : Recalling Theorem 1 and its proof, it is sufficient to show that the hypotheses of the present theorem imply that

$$\parallel \mathbf{x}(t) \parallel \leq r \quad \text{for all} \quad t \geq 0 \tag{9}$$

Recalling definition of $\eta(t)$, it should be clear that (9) holds if

$$\eta(t) \leq r\sqrt{\omega_1} \quad \text{for all} \quad t \geq 0 \tag{10}$$

We first note that

$$\eta(0) \leq \sqrt{\omega_2} \parallel \mathbf{x}(0) \parallel < \sqrt{\omega_2} r \sqrt{\frac{\omega_1}{\omega_2}} = r\sqrt{\omega_1}$$

i.e., $\eta(0) < r\sqrt{\omega_1}$. We now show that for any $k = 0, 1, \ldots$

$$\eta(t_k) < r\sqrt{\omega_1} \tag{11}$$

implies that

$$\eta(t) < r\sqrt{\omega_1} \quad \text{for all} \quad t_k \leq t \leq t_{k+1} \tag{12}$$

Then, (10) follows by induction and the theorem is proven.

So, suppose on the contrary that (11) holds but (12) is false. Let k be the smallest positif integer such that the equation (11) holds but (12) does not. Since $\eta(.)$ is continuous, there exists $t^* \in (t_k, t_{k+1}]$, such that

$$\eta(t) \leq r\sqrt{\omega_1}, \quad t_k \leq t \leq t^*$$

$$\eta(t^*) = r\sqrt{\omega_1}$$

Recalling the development of (8) in the proof of Theorem 1 it follows that

$$\begin{aligned}
\eta(t^*) &\leq \eta(t_k)\lambda(t^* - t_k) + O_1(t^* - t_k) \\
&< r\sqrt{\omega_1}\lambda(t^* - t_k) + \tfrac{O_1(t^* - t_k)}{1 - \lambda(t^* - t_k)}(1 - \lambda(t^* - t_k)) \\
&\leq r\sqrt{\omega_1}\lambda(t^* - t_k) + r_d(T)\sqrt{\omega_1}(1 - \lambda(t^* - t_k)) \\
&\leq r\sqrt{\omega_1}\lambda(t^* - t_k) + r\sqrt{\omega_1}(1 - \lambda(t^* - t_k)) \\
&= r\sqrt{\omega_1}
\end{aligned}$$

which yields a contradiction. Hence (12) holds and the theorem is proven. □

5. Conclusions

This paper has shown that the discrete implementation of a converging controller will keep the converging property of the original controller if some conditions hold. The conditions deal with the plan as well as the sampling time. In addition, for a certain class of systems, the conditions also deal with the initial state. It was also shown that the ball of convergence of the sampled data closed loop system is a continuous function of the sampling time.

Acknowledgements

P. Astuti's research was partly done during her visit to P4M-ITB (1st July - 30th Sept 1996), and partially supported by Advanced Study Program, P4M-ITB.

References

1. Arapostathis, A., B. Jakubczyk, H.-B. Lee, S.I. Marcus, and E.D. Sontag (1989) The effect of sampling of linear equivalence and feedback linearization. *Systems and Control letters*, 13 : 373-381.
2. Astuti, P., D. Williamson, and M. Corless (1993) Indirect digital control fo nonlinear systems with application to robotics. In *Proc. of the European Control Conference*, Groningen, the Netherlands.
3. Corless, M. (1993) Control of Uncertain Nonlinear Systems. *Trans. of the ASME*, Vol 115, pp 362 - 372.
4. Corless, M. and L. Glielmo (1991) A new converse lyapunov result on exponential stability. In *Proc. IFAC Symposium on Design Methods for Control Systems*, Zurich.
5. Grizzle, J.W. and P.V. Kokotivic (1988) Feedback linearization of sampled-data systems. *IEEE Trans. Automatic Control*, 33 : 857-859.
6. Jakubczyk, B. and E.D. Sontag (1987) The effect of sampling on feedback linearization, In *Proc. of the 26th IEEE Conference on Decision and Control*, pp. 1374-1379, los Angeles, CA.
7. Lee, H.-G., A. arapotathis, and S.I. Marcus (1987) Remarks on discretization and linear equivalence of continuous time nonlinear systems. In *Proc. of the 26th IEEE Conference on Decision and Control*, pp. 1783-1785, Los Angeles, CA.
8. Monaco, S., D. Normand-Cyrot, and S. Stornelli (1986) On the linearizing feedback in nonlinear sampled data control schemes. In *Proc. of the 25th IEEE Conference on Decision and Control*, pp. 2056-2060, Athens.
9. Stiver, J.A, P.J. Antsaklis, and M.D. Lemmon (1994) Digital control from a hybrid perspective. In *Proc. of the 33th Conference on Decision and Control*, pp. 4241-4246, Florida.
10. Vidyasagar, M (1993) *Nonlinear Systems Analysis*. Pentice-Hall, new Jersey, 2nd endition, 1993.

NONLINEAR GALERKIN METHOD FOR HAMILTONIAN SYSTEMS

Comparison of two approximation techniques

F.P.H. VAN BECKUM

Department of Applied Mathematics, University of Twente
P.O.Box 217, 7500 AE Enschede, the Netherlands

M. MUKSAR

Pend. Matematika FPMIPA IKIP Malang
Surabaya 6, Malang 65145, Indonesia

AND

E. SOEWONO

Department of Mathematics & Center Mathematics
Institut Teknologi Bandung
Ganesha 10, Bandung 40132, Indonesia

Abstract. In its simplest form, the Galerkin method is the truncation of a differential equation by projection on a set of (spatial) base functions, in the present case: truncation to a certain number n of Fourier modes. But rather than neglecting all other modes completely, the nonlinear modification consists in taking some of the effects of the higher modes into account in the calculation of the first n modes. Specifically, if in the dynamic equations for the higher modes the time-derivative is set equal to zero, the equations simplify to quasi-stationary relations from which the higher modes can be solved, as function of the lower modes.
In dissipative systems this procedure is motivated by the very fast decay of higher modes. In the present paper, however, we apply the idea on Hamiltonian, i.e. conservative, systems. With the Korteweg-de Vries equation as an example, we will consider the direct truncation and a Nonlinear Galerkin method, both in Hamiltonian formulation. The accuracy of both methods is analysed in terms of negative powers of n, and comparisons are visualised graphically.

1. Introduction

A description of the Nonlinear Galerkin method for dissipative systems can be found in a number of papers, in particular by Temam and others; see [3], [4]. For conservative (Hamiltonian) systems we refer to [2] where the formalism is described and to [1] where numerical results are presented. The present paper in an extension of [1] in as much as it gives an estimate of the error as the solution evolves in time. Moreover, the different ways the error grows from zero in the two methods, which became appearent in numerical results by zooming in for small t, can be understood from the error analysis given.

The following procedure is summarized from Ref.[1], [2]. Let V be the space of 2π-periodic C^1-functions in the spatial variable x, having zero total mass, i.e. $\int u\,dx = 0$. (Here, and in the sequel, spatial integrals are supposed to run over one period.) V is equipped with the usual L_2-innerproduct. For functions $u(x,t)$ which for each t are element of V, consider the Korteweg-de Vries (KdV) equation in the following standard form

$$u_t + 6uu_x + u_{xxx} = 0 \quad \text{or} \quad u_t = \partial_x(-u_{xx} - 3u^2) \tag{1}$$

which can be written as a Hamiltonian system:

$$u_t = \Gamma\,\delta H(u) \tag{2}$$

where the operator Γ is partial differentiation ∂_x, the symbol δ denotes variational derivation, and the functional H (on V) is the Hamiltonian energy of the system:

$$H(u) = \int (\tfrac{1}{2}u_x^2 - u^3)\,dx\,. \tag{3}$$

Remark: The Hamiltonian structure (2) guarantees the conservation of energy, of total mass and of momentum ($\equiv \int \frac{1}{2}u^2\,dx$ for KdV), as well as the existence of solitons, provided the operator Γ is antisymmetric and the Hamiltonian integrand is not explicitly depending on x.

1.1. PROJECTION IN V_N AND $V_N^\perp$

For a given positive integer n, and $N \equiv 2n$, define V_N as the span of the first n Fourier modes, i.e. V_N consists of all functions u that are linear combinations of $\cos kx$ and $\sin kx$, $k = 1,...,n$. Then decompose V as $V = V_N + V_N^\perp$, where $V_N^\perp$ is the orthogonal complement of V_N in V (i.e. $V_N^\perp$ contains all higher modes). Note that the constant function is not involved because of the condition of zero total mass $\int u\,dx = 0$. Therefore, any $u \in V$ can uniquely be written as $u = v + w$, which $v \in V_N$ and $w \in V_N^\perp$.

Define the projection operator $P : V \to V_N$ by $Pu = v$ and the complementary projection operator $Q : V \to V_N^{\perp}$ with $Qu = w$, so u can be written as

$$u = v + w = Pu + Qu. \tag{4}$$

In terms of v and w the Hamiltonian (3) can be written as

$$\bar{H}(v,w) \equiv H(v+w) = \int \{\tfrac{1}{2}(v_x^2 + w_x^2) - (v^3 + 3v^2w + 3vw^2 + w^3)\}\, dx. \tag{5}$$

The dynamics for the two components v and w can be found from (2) by projection onto V_N and $V_N^{\perp}$, respectively. Since the operator Γ commutes with the projection operators P and Q, we have

$$\begin{aligned} v_t &= \Gamma P \delta_v \bar{H}(v,w) &&= \partial_x(-v_{xx} - 3P(v+w)^2) && (6)\\ w_t &= \Gamma Q \delta_w \bar{H}(v,w) &&= \partial_x(-w_{xx} - 3Q(v+w)^2)\,. && (7) \end{aligned}$$

Remark 1. Note how the two equations (6), (7) are coupled through the nonlinear term $(v+w)^2$. When starting within V_N, i.e. with $w = 0$, the square v^2 in (7) generates modes higher than n; and conversely, when starting with in $V_N^{\perp}$, i.e. $v = 0$, the square w^2 in (6) generates modes lower than n.

Remark 2. Note that (6) and (7) still describe the full dynamics; no approximation has been made yet. But we will do so in the next subsection.

1.2. DIRECT FOURIER TRUNCATION

The simplest approximation scheme is obtained by neglecting all w-terms in (6), (7) and will be referred to as Direct n modes Fourier Truncation. The w-equation (7) disappears, and the resulting equation for v is still of Hamiltonian structure:

$$v_t = \Gamma\, \delta H_1(v) = \partial_x(-v_{xx} - 3Pv^2)\,, \tag{8}$$

where the Hamiltonian H_1 is found from (5) by neglecting all w-terms:

$$H_1(v) = \int (\tfrac{1}{2}v_x^2 - v^3)\, dx. \tag{9}$$

1.3. SINGULAR PERTURBATION

In the following, an inprovement to the Direct Fourier Truncation will be considered. First, we will see that the operator $\Gamma \equiv \partial_x$ acts with different weights on V_N and $V_N^{\perp}$. The effect of the differential operator ∂_x on the

k-th Fourier mode is essentially a multiplication by k. Since $1 \leq k \leq n$ in V_N, and $k \geq n+1$ in $V_N^\perp$, we find

$$\begin{aligned} & ||v||^2 \leq ||\partial_x Pu||^2 = ||\partial_x v||^2 \leq n^2 ||v||^2 \\ \text{and} \quad & ||\partial_x Qu||^2 = ||\partial_x w||^2 \geq (n+1)^2 ||w||^2, \end{aligned} \tag{10}$$

respectively. We conclude that the operators $\partial_x P$ and $\partial_x Q$ satisfy

$$1 \leq ||\partial_x P|| \leq n \quad \text{and} \quad ||\partial_x Q|| \geq n+1. \tag{11}$$

Now let $\varepsilon \equiv \frac{1}{n+1}$. Then, defining new operators $\Gamma_1 \equiv \partial_x P$ and $\Gamma_2 \equiv \varepsilon \partial_x Q$, the equations in (6), (7) can be written as

$$v_t = \Gamma_1 \, \delta_v \bar{H}(v,w) = \Gamma_1(-v_{xx} - 3(v+w)^2) \tag{12}$$
$$\varepsilon w_t = \Gamma_2 \, \delta_w \bar{H}(v,w) = \Gamma_2(-w_{xx} - 3(v+w)^2), \tag{13}$$

where the new operators Γ_1 and Γ_2 have *the same lower bound:* $||\Gamma_{1,2}|| \geq 1$. For n large, $\varepsilon \equiv \frac{1}{n+1}$ small, the second equation (13) can be viewed as a singular perturbation. From this observation we may conclude that for KdV, the small scale contributions w act as singular perturbation on the large scale v.
Note that the equations (12), (13) are still equivalent to the full system (6), (7) : no approximation has been made yet; but this is done in the following subsection.

1.4. NONLINEAR GALERKIN METHOD

Rather than completely neglecting all w-tems, we now approximate (13) by its *limit equation* for $\varepsilon = 0$, i.e.: $0 = \partial_x(-w_{xx} - 3Q(v+w)^2)$. Then, upon integrating (the constant of integration is 0 in $V_N^\perp$), we get:

$$0 = -w_{xx} - 3Q(v+w)^2 \tag{14}$$

When eliminating w from (14) and (12) we get a dynamic equation for v only. However, (14) is nonlinear in w and for the ease of computation we would rather replace (14) by the linear equation

$$0 = -w_{xx} - 3Qv^2. \tag{15}$$

(In Ref.[1]) it is shown that solutions of (14) and its linearisation (15) are essentially of the same order in ε.)
Therefore we restart the approach by first replacing the Hamiltonian (5) by the approximation

$$H_0(v,w) = \int \{\tfrac{1}{2}(v_x^2 + w_x^2) - (v^3 + 3v^2 w)\} \, dx. \tag{16}$$

with which (12), (13) is replaced by

$$v_t = \Gamma_1 \delta_v H_0(v,w) = \Gamma_1(-v_{xx} - 3(v^2 + 2vw)) \tag{17}$$

$$\varepsilon w_t = \Gamma_2 \delta_w H_0(v,w) = \Gamma_2(-w_{xx} - 3v^2). \tag{18}$$

Following the same procedure, ε is set equal to zero and the w-equation becomes (15), which can be solved explicitly to give

$$w \equiv W_0(v) = 3(-\partial_x^2)^{-1} Q v^2. \tag{19}$$

This enables us to express the Hamiltonian in terms of v alone as:

$$K_0(v) \equiv H_0(v, W_0(v)) = \int \{\tfrac{1}{2}v_x^2 - v^3 - \tfrac{9}{2}v^2(-\partial_x^2)^{-1}Qv^2\}\, dx\,, \tag{20}$$

and the dynamic equation for v becomes

$$v_t = \Gamma_1 \delta_v K_0(v) = \partial_x(-v_{xx} - 3P(v^2 + 6v(-\partial_x^2)^{-1}Qv^2))\,. \tag{21}$$

Remark. The additional term $\int \frac{9}{2}v^2(-\partial_x^2)^{-1}Qv^2\, dx$ in K_0 as compared to H_1 in the direct truncation (11), represents the contribution of the smaller scales to the larger scales.

2. Convergence

Throughout this section, $u = v + w$ denotes the exact solution of the full system (6), (7), whereas $v_0 + w_0$ will be used for the solution of the approximating equations. The differences (i.e. the errors we make by using an approximating scheme instead of the full system) will be denoted by capitals: $V \equiv v_0 - v, \quad W \equiv w_0 - w\,.$

We will give estimates for these errors over a finite time interval $[0, T]$. We write the L_2-norm as $||\,.\,||$, and use $|||\,.\,|||$ to denote $sup_{[0,T]}\{||\cdot||\}\,.$

An important inequality is derived directly from (7): since $v + w$ and w_t are bounded, by integrating three times we get $||w|| \leq \varepsilon^3||w_t|| + 3\varepsilon^2||u||^2$, and then, taking the supremum over t and defining $A = |||u_t|||, B = |||u|||$:

$$||w|| \leq (\varepsilon A + 3B^2)\,\varepsilon^2. \tag{22}$$

2.1. DIRECT TRUNCATION

Let v_0 be a solution to (8), and $v+w$ the corresponding solution to (6), (7), i.e. starting from the same initial condition: $v|_{t=0} = v_0|_{t=0}$ and $w|_{t=0} = 0$. Subtracting (6) from (8), we find that the error $V \equiv v_0 - v$ satisfies

$$V_t = (-V_{xx} - 3P(V(V + 2v) - 2vw - w^2))_x. \tag{23}$$

Multiplying (23) by $2V$, and integrating over one spatial period, we get

$$\partial_t(\int V^2\,dx) = -6\int V^2 v_x\,dx - 6\int P(2vw + w^2)V_x\,dx \tag{24}$$

(Note that here (and later) P can be left out because of orthogonality.) Now rename $\zeta(t)^2 \equiv \int V^2\,dx \equiv ||V||^2$, $K \equiv 6\sup_{x,t}\{|u_x|\}$ and $f(t) \equiv 6\int P(2vw + w^2)V_x\,dx$. Then (24) can be estimated as

$$\partial_t(\zeta^2) \le K\zeta^2 + |f|\,. \tag{25}$$

Multiplying (25) by e^{-Kt} and then integrating over $(0,t)$, we have

$$\zeta^2 e^{-Kt} \le \int_0^t e^{-Ks}\,|f|\,ds \quad \text{or} \quad \zeta^2 \le \int_0^t e^{K(t-s)}\,|f|\,ds\,. \tag{26}$$

Applying Schwarz' inequality on the integral f and using the fact that $||V_x|| \le \frac{1}{\varepsilon}||V||$, we get from (26), for $t \in [0,T]$,

$$|||V||| \le 6K_0(\frac{2B|||w||| + |||w|||^2}{\varepsilon}), \tag{27}$$

where $K_0 = e^{KT}$. From (27) and (22) we conclude:

$$|||v - v_0||| \le C\varepsilon\,, \tag{28}$$

where $C = 36K_0B^3 + 12K_0AB\varepsilon + O(\varepsilon^2)$.
This shows that the rate of convergence of the direct truncation scheme is $O(\varepsilon)$.

2.2. NONLINEAR GALERKIN METHOD

Let v_0 be a solution to (21) and let $w_0 \equiv W_0(v_0)$ be given by (19). Let $v+w$ be the corresponding solution to (6), (7), i.e. starting from the same initial condition: $v|_{t=0} = v_0|_{t=0}$ and $w|_{t=0} = w_0|_{t=0}$.

The analysis follows essentially the same lines as before. Subtracting (6), (7) from (17), (15), we have, with $V \equiv v_0 - v$ and $W \equiv w_0 - w$:

$$V_t = (-V_{xx} - 3P(V(V+2v) + 2vW + 2Vw + 2VW - w^2))_x \tag{29}$$
$$-w_t = (-W_{xx} - 3Q(V(V+2v) - 2vw - w^2))_x. \tag{30}$$

Remark. Expression (29) looks much more complicated than (23): instead of $-2vw$ in (23) we now have three terms: $2vW + 2Vw + 2VW$. Note, however, the gain we can expect. The factor v in $-2vw$ in (23) is the one that created the constant B which is dominating the estimate (27) or (28). Now, since all three terms replacing $-2vw$ are at least one order in ε smaller than $-2vw$, we can expect that the error estimate will be one order better.

From (30), integrating three times and using (22), we easily find

$$||W|| \leq \varepsilon^3 A+3\varepsilon^2(||V||^2+2B||V||+2B\varepsilon^2(\varepsilon A+3B^2)+\varepsilon^4(\varepsilon A+3B^2)^2). \quad (31)$$

Now, multiplying (29) by $2V$ and integrating over one period we have

$$\partial_t(\int V^2 dx) = -6\int V^2 u_x\, dx + 6\int P(2vW + 2VW - w^2)V_x\, dx. \quad (32)$$

With $\zeta(t) = ||V||$, $K = 6\sup_{t,x}\{|u_x|\}$, $f(t) = 6\int P(2vW+2VW-w^2)V_x\,dx$ and using the same argument as before we get $(\zeta^2)_t \leq K\zeta^2 + |f|$ and then:

$$\zeta^2 e^{-Kt} \leq \int_0^t e^{-Ks}|f|ds \quad \text{or} \quad \zeta^2 \leq \int_0^t e^{K(t-s)}|f|ds, \quad (33)$$

By applying Schwarz' inequality on the integral f, and using (22) we find for $t \in [0,T]$, with $K_0 = e^{KT}$:

$$|||V||| \leq 6K_0((\frac{(2B + 2|||V|||)|||W|||}{\varepsilon}) + \varepsilon^3((A\varepsilon + B)^2)), \quad (34)$$

Now the inequality for $|||V|||$ that we can derive by substituting (31) into (34) is an implicit one:

$$\begin{aligned}-|||V||| + 6K_0(\frac{1}{\varepsilon}(2B + 2|||V|||)(\varepsilon^3 A + 3\varepsilon^2(|||V|||^2 + 2B|||V||| +\\ 2B\varepsilon^2(\varepsilon A + 3B^2) + \varepsilon^4(\varepsilon A + 3B^2)^2)) + \varepsilon^3((A\varepsilon + B)^2)) \geq 0, \quad (35)\end{aligned}$$

which needs further analysis. Denote the left hand side by

$$F(|||V|||) = a_0 + a_1|||V||| + a_2|||V|||^2 + a_3|||V|||^3, \quad (36)$$

then the inequality (35) implies that $|||V|||$ is bounded by the first positive

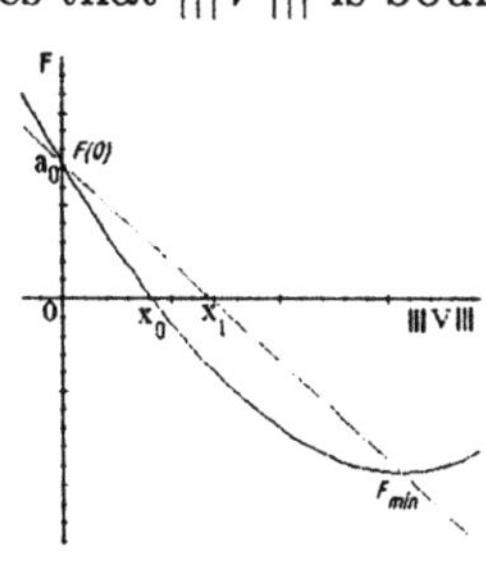

zero x_0 of the cubic polynomial F. However, it is not easy to express the value of x_0 in terms of ε. This is more easily done for x_1, the zero of the chord connecting $F(0)$ and F_{min}; see figure. The estimate $x_0 < x_1$ can be proved from the observation that $a_1 < 0$ and $a_0, a_2, a_3 > 0$, at least for ε sufficiently small. Now calculating the point x_1 and expressing it in ε, we arrive at the following estimate for $|||V|||$:

$$|||V||| = |||v - v_0||| \leq C\varepsilon^2 , \tag{37}$$

where $C = 18K_0A^2 + (36\sqrt{3}A^2B + 18\sqrt{3}A^3)K_0^{\frac{3}{2}}\varepsilon^{\frac{1}{2}} + O(\varepsilon)$.
This shows that the rate of convergence of Nonlinear Galerkin scheme is $O(\varepsilon^2)$.

3. Numerical results

For the numerics, we choose an initial condition of the form $u(x,0) = a\cos x$. In this way, we control the nonlinearity with one parameter, the amplitude a. We compare the error in NLG with the error in Direct Truncation, both run with the same n. For the "exact" solution in V, to be used for reference, we make a Direct Truncation run with $m = n^3$ modes.
In the figure the error is shown for $a = 1$ and $a = 0.5$, respectively, both for a two and for a four modes run.

Figure 1. Error of Direct Truncation and NLG, for initial condition $u = \cos x$. Left: run with 2 modes, right: run with 4 modes.

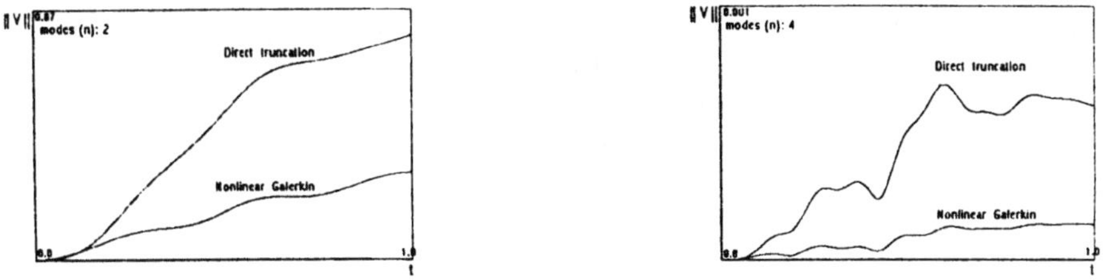

Figure 2. Error of Direct Truncation and NLG, for initial condition $u = \frac{1}{2}\cos x$. Left: run with 2 modes, right: run with 4 modes.

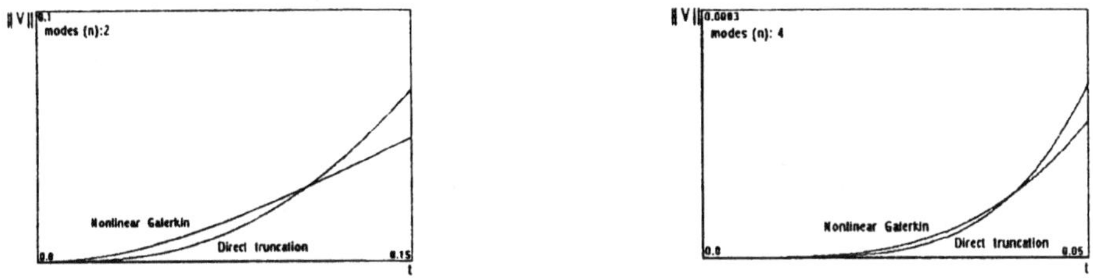

Figure 3. Zoom of figure 1 near $t = 0$. Note that at first instance, Direct Truncation is more accurate; later NLG performes better.

4. Discussion

1. The behaviour near $t = 0$, shown in figure 3, seems unexpected. The explanation is found in (23) and (29). DT starts with a zero $V_N^{\perp}$-component, whereas NLG starts with $w \equiv W_0(v) = 3(-\partial_x^2)^{-1}Qv^2$; see (19). Therefore, at $t = 0$ the righthand side of (23) is zero, whereas that of (29) is not necessarily.

2. It should be noted that the error estimates derived in this paper are for a finite time interval. When dynamical systems are considered over longer times, the estimates will be of limited use. Nonetheless, over the time periods shown in the graphs the application of the NLG reduces the error by a factor 2 to 4.

3. Computationally, the advantage of Hamiltonian formulation lies in the conservation properties. In this respect, even standard Runge-Kutta time integration for DT performs much better than all existing methods for KdV; see [1].

4. The advantage of NLG over DT is mainly due to the stability properties, as is noticed in [1]. Applying an explicit time integrator (like runge-Kutta) on an n modes Direct Truncated system, an increase of n would cubicly decrease the allowable time step, whereas an $n + n$ modes NLG enjoys the same stability condition as the n modes DT itself, with much higher accuracy.

Acknowledgement Part of the research is sponsored by (1) The Commission of the European Communities, Directorate General XII-B, Joint Project CI1*-CT93-0018 between the Department of Mathematics, Institut Teknologi Bandung, Indonesia, and the Faculty of Applied Mathematics, University of Twente, The Netherlands, and (2) Hibah Tim URGE Project No. 007/HTPP/URGE/1995, DIKTI.

References

1. Beckum, F.P.H. van (1995): *Hamiltonian-Consistent Discretisation of Wave Equations* (Thesis).
2. Groesen, E. van, and Jager, E.M. de (1994): *Mathematical structures in continuous dynamical systems.* Studies in Mathematical Physics, North-Holland, Elsevier, Amsterdam.
3. Jauberteau, F., Rosier, C., and Temam, R. (1990): *A nonlinear Galerkin method for the Navier-Stokes equations.* J. Comput. Met. **80**, 245-260.
4. Gottlieb, D., and Temam, R. (1993): *Implementation of the Nonlinear Galerkin Method with pseudospectral (collocation) discretisations.* Appl. Numer. Math. **12**. 119-134.

ON INTERNAL RESONANCES FOR A WEAKLY NONLINEAR BEAM EQUATION

G.J. BOERTJENS AND W.T. VAN HORSSEN
Department of Technical Mathematics and Informatics, Delft University of Technology, P.O. Box 5031, 2600 GA Delft, The Netherlands.

Abstract. In this paper an initial-boundary value problem for a weakly nonlinear Rayleigh beam equation will be studied. It will be shown that the analysis for this problem differs substantially from the Rayleigh wave equation as for instance presented in [3] or [5]. The initial-boundary value problem can be regarded as a simple model describing wind-induced oscillations of flexible structures like suspension bridges or overhead transmission lines. Using a two-timescales perturbation method approximations for solutions of this initial-boundary value problem will be constructed.

1. Introduction

Flexible structures like tall buildings, suspension bridges or iced transmission lines with bending stiffness, are subjected to oscillations due to wind-forces or other causes (e.g. earthquakes). These oscillations can be described by nonlinear second and fourth order PDE's. Asymptotic methods can be used to construct approximations for solutions of this type of equations. In [3] an asymptotic theory for second order PDE's is presented for models describing the growth of wind-induced oscillations of iced overhead transmission lines. However, the use of perturbation methods for fourth order PDE's has not yet been examined extensively.
In this paper we consider the following initial-boundary value problem:

$$w_{tt} + w_{xxxx} + p^2 w = \epsilon\left(w_t - \frac{1}{3}w_t^3\right), \qquad 0 < x < \pi,\ t > 0, \tag{1}$$

$$w(0,t) = w(\pi,t) = 0, \qquad t > 0, \tag{2}$$

$$w_{xx}(0,t) = w_{xx}(\pi,t) = 0, \qquad t > 0, \tag{3}$$

$$w(x,0) = w_0(x), \quad w_t(x,0) = w_1(x), \qquad 0 < x < \pi, \tag{4}$$

where ϵ and p are constants with $0 < \epsilon \ll 1$ and $0 < p^2 < 10$, $w=w(x,t)$ is the vertical displacement of the beam, x is the co-ordinate along the beam, $w_0(x)$ is the initial displacement of the beam in vertical direction and $w_1(x)$ is the initial velocity of the beam in vertical direction. All functions are sufficiently smooth. The first two terms on the left hand side are the linear part of the beam equation, p^2w represents a linear restoring force and the right hand side is a so-called Rayleigh perturbation, which describes an external (wind)force (see [3]). The boundary conditions describe a simply supported beam.
In [6] a survey of literature on oscillations of suspension bridges is given. The initial-boundary problem (1)-(4) can be considered as a simple model for nonlinear oscillations of a suspension bridge, where the stays of the bridge are treated as two-sided springs, as can be seen in Fig. 1.

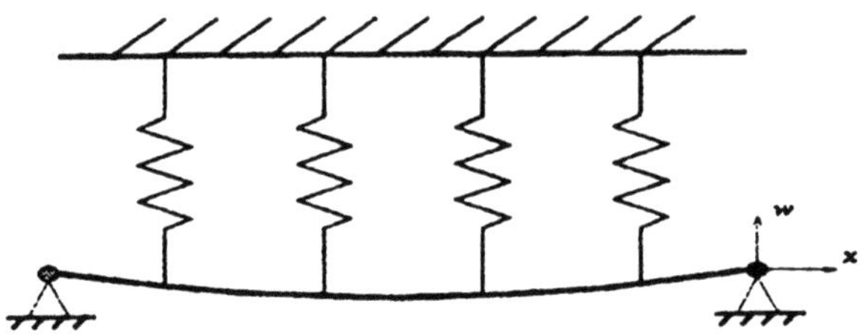

Figure 1. A simple model of a suspension bridge

In this paper formal approximations, i.e. functions that satisfy the differential equation and the initial and boundary conditions up to some order in ϵ, will be constructed for the initial-boundary value problem (1)-(4). The interaction between the different oscillation modes will be considered. It will be shown that for some specific values of p^2 'extra' interactions occur that cause complicated internal resonances. Resonances of this type are for example analysed in [4] and [8]. For these p^2-values truncation to one or two modes, as performed in [7] or [1], is not valid. We will show that for instance for $p^2=9$ energy transfer occurs between modes 1 and 3. To our knowledge these mode interactions for weakly nonlinear beam equations have not yet been studied thoroughly.
The outline of this paper is as follows. In section 2 we apply a two-timescales perturbation method to the initial-boundary value problem (1)-(4), and construct formal approximations for solutions of this initial-boundary value problem. We show that for most p^2-values mode interactions occur only between modes with non-zero initial energy (up to $\mathcal{O}(\epsilon)$). For some specific values, $p^2=\frac{693}{152}$ and $p^2=9$, also modes with zero initial energy are excited. We consider the cases $p^2 \in]0,10[\backslash\{\frac{693}{152},9\}$ and $p^2=9$. To understand the

transition from $p^2 \neq 9$ to $p^2 = 9$ we study the detuning case $p^2 = 9 + \epsilon\alpha$ with $\alpha > 0$. In section 3 we make some remarks on the relation of these results to other literature on this subject.

2. The perturbation method and formal approximations

In this section we want to construct formal approximations of solutions for the initial-boundary value problem (1)-(4). When straightforward ϵ-expansions are used to approximate solutions, secular terms occur in the approximations. To avoid these secular terms we use a two-timescales perturbation method. The boundary conditions imply that w can be written as a Fourier sine-series: $w(x,t) = \sum_{m=1}^{\infty} q_m(t) sin(mx)$. Substituting this series into (1), we obtain the following system of equations:

$$\sum_{m=1}^{\infty} \left(\ddot{q}_m + (m^4 + p^2) q_m\right) sin(mx) = \tag{5}$$

$$\epsilon \left(\sum_{m=1}^{\infty} \dot{q}_m sin(mx) - \frac{1}{3} \sum_{m,k,j=1}^{\infty} \dot{q}_m \dot{q}_k \dot{q}_j sin(mx) sin(kx) sin(jx) \right).$$

Multiplying (5) with $sin(nx)$ and integrating with respect to x from 0 to π gives us the following equations for each q_n (see [2]):

$$\ddot{q}_n + (n^4 + p^2) q_n = \tag{6}$$

$$\epsilon \left[\dot{q}_n + \frac{1}{4} \left(- \sum_{n=m+k-j} + \sum_{n=-m-k+j} + \frac{1}{3} \sum_{n=m+k+j} \right) \dot{q}_m \dot{q}_k \dot{q}_j \right],$$

for $n = 1,2,3,\ldots$, where q_n must satisfy the initial conditions given by (4):

$$q_n(0) = \frac{2}{\pi} \int_0^{\pi} w_0(x) sin(mx) dx, \; \dot{q}_n(0) = \frac{2}{\pi} \int_0^{\pi} w_1(x) sin(mx) dx \; .$$

We introduce the two time-scales: $t_0 = t$ and $t_1 = \epsilon t$, and assume that q_n can be expanded in a formal powerseries in ϵ, that is, $q_n(t) = q_{n,0}(t_0, t_1) + \epsilon q_{n,1}(t_0, t_1) + \epsilon q_{n,2}(t_0, t_1) + \cdots$. We substitute this into equation (6), and collect equal powers of ϵ. The $\mathcal{O}(\epsilon^0)$-problem becomes:

$$\frac{\partial^2}{\partial t_0^2} q_{n,0} + (n^4 + p^2) q_{n,0} = 0, \; t > 0, \tag{7}$$

$$q_{n,0}(0,0) = \frac{2}{\pi} \int_0^{\pi} w_0(x) sin(mx) dx, \tag{8}$$

$$\frac{\partial}{\partial t_0} q_{n,0}(0,0) = \frac{2}{\pi} \int_0^{\pi} w_1(x) sin(mx) dx, \tag{9}$$

for n=1,2,3,.... The general solution for (7)-(9) is:

$$q_{n,0}(t_0,t_1) = A_{n,0}(t_1)cos(\omega_{n_p}t_0) + B_{n,0}(t_1)sin(\omega_{n_p}t_0), \tag{10}$$

where $\omega_{n_p}=(n^4+p^2)^{\frac{1}{2}}$ and $A_{n,0}$, $B_{n,0}$ satisfy the initial conditions (8)-(9):

$$A_{n,0}(0) = q_{n,0}(0,0),\ B_{n,0}(0) = \frac{1}{\omega_{n_p}}\frac{\partial}{\partial t_0}q_{n,0}(0,0).$$

Next we consider the $\mathcal{O}(\epsilon^1)$-problem:

$$\frac{\partial^2}{\partial t_0^2}q_{n,1} + (n^4+p^2)q_{n,1} = -2\frac{\partial^2}{\partial t_0 t_1}q_{n,0} + \frac{\partial}{\partial t_0}q_{n,0} \tag{11}$$
$$+\frac{1}{4}\left(-\sum_{n=m+k-j} + \sum_{n=-m-k+j} + \frac{1}{3}\sum_{n=m+k+j}\right)\frac{\partial}{\partial t_0}q_{m,0}\frac{\partial}{\partial t_0}q_{k,0}\frac{\partial}{\partial t_0}q_{j,0},$$

$$q_{n,1}(0,0) = 0, \quad \frac{\partial}{\partial t_0}q_{n,1}(0,0) = -\frac{\partial}{\partial t_1}q_{n,0}(0,0), \tag{12}$$

for n=1,2,3,.... We substitute (10) into (11) and get:

$$\frac{\partial^2}{\partial t_0^2}q_{n,1} + (n^4+p^2)q_{n,1} = \omega_{n_p}\left(2\frac{dA_{n,0}}{dt_1}sin(\omega_{n_p}t_0) - 2\frac{dB_{n,0}}{dt_1}cos(\omega_{n_p}t_0)\right)$$
$$+H_n + \frac{1}{4}\left(-\sum_{n=m+k-j} + \sum_{n=-m-k+j} + \frac{1}{3}\sum_{n=m+k+j}\right)H_mH_kH_j, \tag{13}$$

with $H_l = \omega_{l_p}(B_{l,0}cos(\omega_{l_p}t_0) - A_{l,0}sin(\omega_{l_p}t_0))$.
Since $cos(\omega_{n_p}t_0)$ and $sin(\omega_{n_p}t_0)$ are solutions of the homogeneous differential equation for $q_{n,1}$, we want the coefficients of $cos(\omega_{n_p}t_0)$ and $sin(\omega_{n_p}t_0)$ in the right hand side of (13) to be equal to zero (elimination of secular terms). This gives us equations for $A_{n,0}$ and $B_{n,0}$ (see [2]). Finding solutions for $A_{n,0}$, $B_{n,0}$ and thus for $q_{n,0}$ gives us a first order approximation of the exact solution of the initial-boundary value problem (1)-(4).
In [2] we examine the different possibilities for the solutions of the diophantic-like equations,

$$\begin{cases} n = m+k-j \vee n = -m-k+j \vee n = m+k+j, \\ \pm(n^4+p^2)^{\frac{1}{2}} = \pm(m^4+p^2)^{\frac{1}{2}} \pm (k^4+p^2)^{\frac{1}{2}} \pm (j^4+p^2)^{\frac{1}{2}}. \end{cases} \tag{14}$$

These equations need to be solved in order to determine the contributions of the sums in (13) to the equations for $A_{n,0}$ and $B_{n,0}$. Only specific combinations of m, k, j will give solutions to (14). In [2] it is shown that for some values of $p^2 \in]0, 10[$ there are additional contributions, i.e. for $p^2=\frac{693}{152}$ and for $p^2=9$. Also the case $p^2=0$ is considered in [2].

2.1. THE CASE $P^2 \neq \frac{693}{152}, 9$ AND $0 < P^2 < 10$

For $p^2 \neq \frac{693}{152}, 9$ the contributions are similar to the wave-equation case, described in [5]. We scale $A_{n,0}$, $B_{n,0}$ with ω_{n_p} and get the following equations:

$$\frac{dA_{n,0}}{dt_1} = \frac{1}{2}A_{n,0}\left(1+\frac{1}{16}\left(A_{n,0}^2+B_{n,0}^2\right)-\frac{1}{4}\sum_{k=1}^{\infty}\left(A_{k,0}^2+B_{k,0}^2\right)\right), \quad (15)$$

$$\frac{dB_{n,0}}{dt_1} = \frac{1}{2}B_{n,0}\left(1+\frac{1}{16}\left(A_{n,0}^2+B_{n,0}^2\right)-\frac{1}{4}\sum_{k=1}^{\infty}\left(A_{k,0}^2+B_{k,0}^2\right)\right), \quad (16)$$

for $n=1,2,3,\ldots$. From (15) and (16) we can see that if $A_{n,0}(0)=B_{n,0}(0)=0$ then $A_{n,0}(t_1)=B_{n,0}(t_1)=0$ $\forall t_1 > 0$. So, if we start with zero initial energy in the n-th mode, there will be no energy present up to $\mathcal{O}(\epsilon)$. We say the coupling between the modes is of $\mathcal{O}(\epsilon)$. This 'allows' truncation to those modes that have non-zero initial energy. We will see in subsection 2.2 that this does not hold for the case $p^2=9$. In that case mode 3 will be excited by mode 1 even if the third mode has zero initial energy. This means that an extra coupling of $\mathcal{O}(1)$ occurs between modes 1 and 3. Therefore we examine only the equations for modes 1 and 3 in subsection 2.2. To understand the transition from $p^2 \neq 9$ to $p^2=9$ better, we will examine (15)-(16) for $n=1$ and $n=3$ explicitly. We transform (15)-(16) for $n=1,3$ using:

$$A_{n,0} = r_n \cos(\phi_n) \;, \; B_{n,0} = r_n \sin(\phi_n) \;, \quad (17)$$

with r_n the amplitude and ϕ_n the phase of the oscillation, and get the following equations for r_1, r_3, ϕ_1, ϕ_3:

$$\dot{r}_1 = \frac{1}{2}r_1(1-\frac{3}{16}r_1^2-\frac{1}{4}r_3^2), \quad (18)$$

$$\dot{r}_3 = \frac{1}{2}r_3(1-\frac{1}{4}r_1^2-\frac{3}{16}r_3^2), \quad (19)$$

$$r_1^2\dot{\phi}_1 = 0, \quad (20)$$

$$r_3^2\dot{\phi}_3 = 0. \quad (21)$$

We can analyse (18)-(21) in the r_1, r_3-plane. Elementary analysis gives us the behaviour of the solutions of (18)-(21) as shown in Fig. 2.
We see that depending on the initial values the system will oscillate in only one mode for large times (up to $\mathcal{O}(\epsilon)$ on a time scale of order $1/\epsilon$).
In the next subsection we will consider the case $p^2=9$. The analysis of the

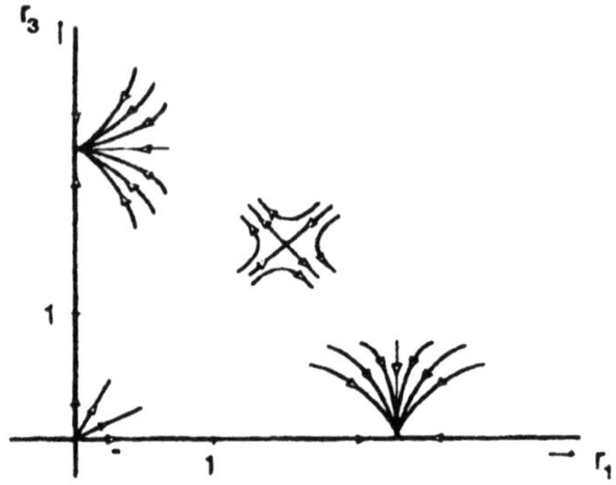

Figure 2. Phase plane for $p^2 \neq 9$

equations for $A_{1,0}, A_{3,0}, B_{1,0}, B_{3,0}$ will be brought back to an analysis in a $3d$-space with r_1, r_3 and $\psi=\phi_3-3\phi_1$. Looking at the above case in this $3d$-space will give us the same behaviour as given in Fig. 2 for every ψ, which means we have lines of critical points in the $3d$-space.

2.2. THE CASE $P^2=9$

For $p^2=9$ there are additional contributions in the equations for $A_{n,0}, B_{n,0}$ for $n=1,3$. These indicate a 3 : 1 resonance between modes 1 and 3. For $n=2,4,5,\ldots$ (15) and (16) still hold. For $A_{1,0}, B_{1,0}, A_{3,0}, B_{3,0}$ the equations, after scaling with ω_1, ω_3, respectively, are:

$$\frac{dA_{1,0}}{dt_1} = \frac{1}{2}A_{1,0}\left(1-\frac{3}{16}(A_{1,0}^2+B_{1,0}^2)-\frac{1}{4}(A_{3,0}^2+B_{3,0}^2)\right) + \frac{1}{16}\left((B_{1,0}^2-A_{1,0}^2)A_{3,0}-2A_{1,0}B_{1,0}B_{3,0}\right), \quad (22)$$

$$\frac{dB_{1,0}}{dt_1} = \frac{1}{2}B_{1,0}\left(1-\frac{3}{16}(A_{1,0}^2+B_{1,0}^2)-\frac{1}{4}(A_{3,0}^2+B_{3,0}^2)\right) + \frac{1}{16}\left(2A_{1,0}B_{1,0}A_{3,0}+(B_{1,0}^2-A_{1,0}^2)B_{3,0}\right), \quad (23)$$

$$\frac{dA_{3,0}}{dt_1} = \frac{1}{2}A_{3,0}\left(1-\frac{1}{4}(A_{1,0}^2+B_{1,0}^2)-\frac{3}{16}(A_{3,0}^2+B_{3,0}^2)\right) + \frac{1}{48}A_{1,0}\left(3B_{1,0}^2-A_{1,0}^2\right), \quad (24)$$

$$\frac{dB_{3,0}}{dt_1} = \frac{1}{2}B_{3,0}\left(1-\frac{1}{4}(A_{1,0}^2+B_{1,0}^2)-\frac{3}{16}(A_{3,0}^2+B_{3,0}^2)\right) + \frac{1}{48}B_{1,0}\left(B_{1,0}^2-3A_{1,0}^2\right). \quad (25)$$

We transform (22)-(25) using (17):

$$\dot{r}_1 = \frac{1}{2}r_1(1-\frac{3}{16}r_1^2-\frac{1}{4}r_3^2)-\frac{1}{32}r_1^2r_3cos(\phi_3-3\phi_1), \quad (26)$$

$$\dot{r}_3 = \frac{1}{2}r_3(1-\frac{1}{4}r_1^2-\frac{3}{16}r_3^2)-\frac{1}{96}r_1^3cos(\phi_3-3\phi_1), \quad (27)$$

$$r_1^2\dot{\phi}_1 = -\frac{1}{32}r_1^3r_3sin(\phi_3-3\phi_1), \quad (28)$$

$$r_3^2\dot{\phi}_3 = \frac{1}{96}r_1^3r_3sin(\phi_3-3\phi_1). \quad (29)$$

In [2] we analyse (26)-(29) in the (r_1, r_3, ψ)-space, with $\psi=\phi_3-3\phi_1$. The behaviour of the solutions of (26)-(29) in the (r_1, r_3, ψ) phase space is given in Fig. 3.

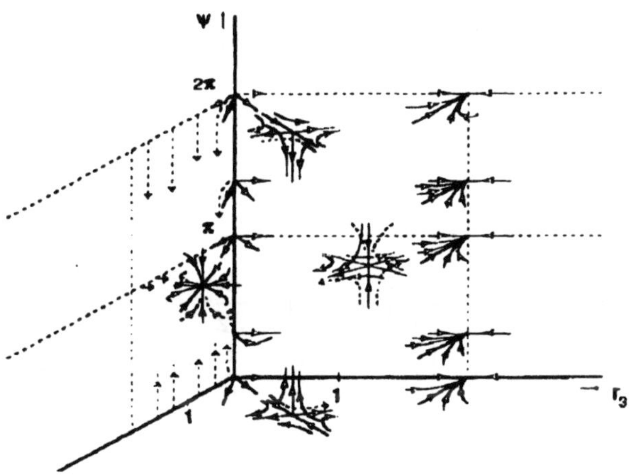

Figure 3. phase space for $p^2=9$

The system is 2π-periodic in ψ. In the plane $\psi=0$ there is a stable node $(0,\sqrt{16/3},0)$ and a saddle $(1.593,1.262,0)$. The point $(\sqrt{16/3},0,0)$ seems to attract, but as soon as $r_3=0$ a phase-jump occurs to the $(\psi=\pi)$-plane, imposed by the differential equations. The lines $(0,\sqrt{16/3},\psi)$ and $(0,0,\psi)$ are lines of singular points. We see that for large times the system will either oscillate in the third mode, with a phase depending on the initial values, or in a combined oscillation of the first and third mode, with a phase difference π.

2.3. THE CASE $P^2=9+\epsilon\alpha$

To understand the detuning from the case $p^2=9$ we consider $p^2=9+\epsilon\alpha$ with $\epsilon \ll 1$ and $\alpha \neq 0$ of $\mathcal{O}(1)$. The equations for $A_{1,0}, B_{1,0}, A_{3,0}, B_{3,0}$ now are:

$$\frac{dA_{1,0}}{dt_1} = \frac{1}{2}A_{1,0}\left(1 - \frac{3}{16}(A_{1,0}^2 + B_{1,0}^2) - \frac{1}{4}(A_{3,0}^2 + B_{3,0}^2)\right) + \frac{\alpha}{\sqrt{10}} B_{1,0} + \frac{1}{16}\left((B_{1,0}^2 - A_{1,0}^2)A_{3,0} - 2A_{1,0}B_{1,0}B_{3,0}\right), \tag{30}$$

$$\frac{dB_{1,0}}{dt_1} = \frac{1}{2}B_{1,0}\left(1 - \frac{3}{16}(A_{1,0}^2 + B_{1,0}^2) - \frac{1}{4}(A_{3,0}^2 + B_{3,0}^2)\right) - \frac{\alpha}{\sqrt{10}} A_{1,0} + \frac{1}{16}\left(2A_{1,0}B_{1,0}A_{3,0} + (B_{1,0}^2 - A_{1,0}^2)B_{3,0}\right), \tag{31}$$

$$\frac{dA_{3,0}}{dt_1} = \frac{1}{2}A_{3,0}\left(1 - \frac{1}{4}(A_{1,0}^2 + B_{1,0}^2) - \frac{3}{16}(A_{3,0}^2 + B_{3,0}^2)\right) + \frac{\alpha}{3\sqrt{10}} B_{3,0} + \frac{1}{48}A_{1,0}\left(3B_{1,0}^2 - A_{1,0}^2\right), \tag{32}$$

$$\frac{dB_{3,0}}{dt_1} = \frac{1}{2}B_{3,0}\left(1 - \frac{1}{4}(A_{1,0}^2 + B_{1,0}^2) - \frac{3}{16}(A_{3,0}^2 + B_{3,0}^2)\right) - \frac{\alpha}{3\sqrt{10}} A_{3,0} + \frac{1}{48}B_{1,0}\left(B_{1,0}^2 - 3A_{1,0}^2\right). \tag{33}$$

For $n=2,4,5,\ldots$ (15) and (16) (with an extra α-term as above) hold. We transform (30)-(33) using (17):

$$\dot{r}_1 = \frac{1}{2}r_1(1 - \frac{3}{16}r_1^2 - \frac{1}{4}r_3^2) - \frac{1}{32}r_1^2 r_3 cos(\phi_3 - 3\phi_1), \tag{34}$$

$$\dot{r}_3 = \frac{1}{2}r_3(1 - \frac{1}{4}r_1^2 - \frac{3}{16}r_3^2) - \frac{1}{96}r_1^3 cos(\phi_3 - 3\phi_1), \tag{35}$$

$$r_1^2\dot{\phi}_1 = -\frac{1}{2}\frac{\alpha}{\sqrt{10}}r_1^2 - \frac{1}{32}r_1^3 r_3 sin(\phi_3 - 3\phi_1), \tag{36}$$

$$r_3^2\dot{\phi}_3 = -\frac{1}{6}\frac{\alpha}{\sqrt{10}}r_3^2 + \frac{1}{96}r_1^3 r_3 sin(\phi_3 - 3\phi_1). \tag{37}$$

In [2] we analyse (34)-(37) in the (r_1, r_3, ψ)-space, with $\psi=\phi_3-3\phi_1$, by examining the phase space for several values of $\alpha > 0$ (For $\alpha < 0$ a similar analysis holds, see [2]).
We start with $\alpha=\frac{1}{10}$. The behaviour of the solutions of (34)-(37) in the (r_1, r_3, ψ) phase space is given in Fig. 4.
For a point on the line $(0, \sqrt{16/3}, \psi)$ we have $\dot{r}_1=\dot{r}_3=0$ and $\dot{\psi} > 0$, which means we have a singular line with movement along it. The line is stable in r_1, r_3-direction and therefore is a stable periodic solution. We see that the saddle point in the plane $\psi=2\pi$ starts moving downwards and the saddle

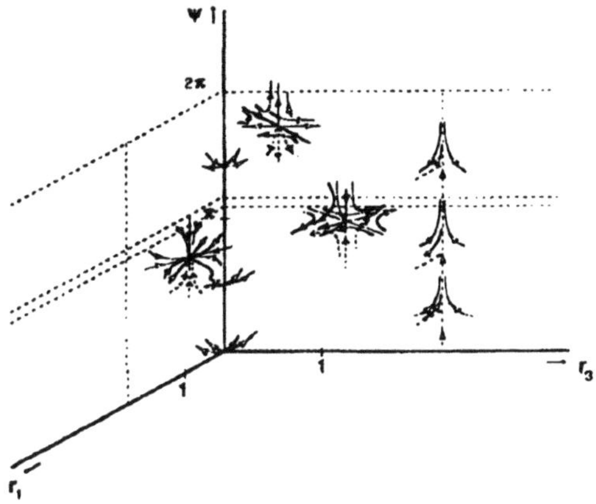

Figure 4. phase space for $p^2=9+\epsilon\alpha$ with $\alpha=\frac{1}{10}$

point in the plane $\psi=\pi$ upwards. The stable point changes from a *3d*-node to a spiral-node for $\alpha > 0.1$. With increasing α we see that the two saddles move closer towards each other. The stable point moves towards the r_1-axis. The line $(0, \sqrt{16/3}, \psi)$ is still a stable periodic solution. The line $(0,0,\psi)$ remains a line of singular points. At some point the two saddle-points coincide and then vanish. This happens for $\alpha\approx0.5649$. In [2] we find the behaviour of the solutions of (34)-(37) in the (r_1, r_3, ψ) phase space as given in Fig. 5.

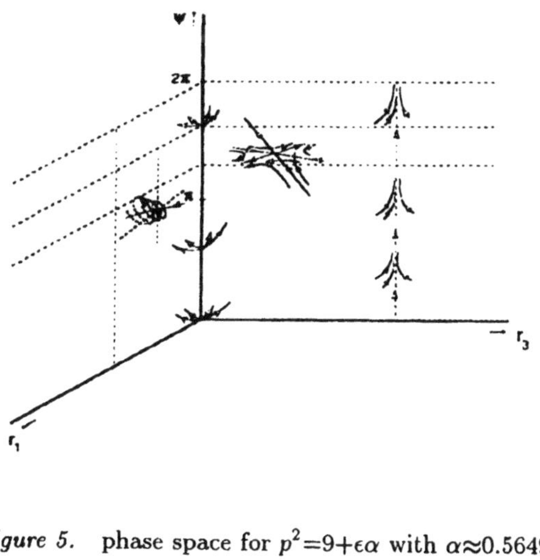

Figure 5. phase space for $p^2=9+\epsilon\alpha$ with $\alpha\approx0.5649$

For $\alpha > 0.5649$ a separatrix region remains. The stable point moves further towards the r_1-axis, and the line $(0, \sqrt{16/3}, \psi)$ remains a stable periodic solution. For α=1 we find the behaviour of the solutions of (34)-(37) in the (r_1, r_3, ψ) phase space as given in Fig. 6.

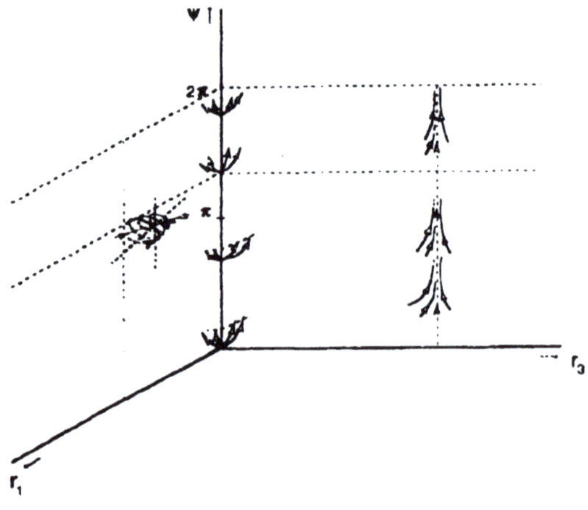

Figure 6. phase space for $p^2=9+\epsilon\alpha$ with $\alpha=1$

3. Conclusions

In this paper we considered an initial-boundary value problem for a weakly nonlinear beam equation. We have constructed formal approximations of order ϵ and considered the interaction between different oscillation modes. We showed that for most p^2-values the behaviour of solutions of the Rayleigh beam equation is similar to that of solutions of the Rayleigh wave equation as presented in [5]. In that case mode interactions occur only between modes with non-zero initial energy (up to $\mathcal{O}(\epsilon)$). We then say the coupling between the modes is of $\mathcal{O}(\epsilon)$, and truncation is allowed, restricted to those modes that have non-zero initial energy. We showed that for large times the system will oscillate in only one mode. However, for some p^2-values extra mode interactions occur, which cause difficult internal resonances. For $0 < p^2 < 10$ these values are p^2=9 and $p^2=\frac{693}{152}$. In a similar way special values can be found for $p^2 \geq 10$. For p^2=9 the equations have been studied in detail. Energy transfer occurs between modes 1 and 3 even if mode 3 initially has zero energy. We call this coupling of $\mathcal{O}(1)$. Truncation to one mode will give loss of information, and approximations will not be valid. Examining the behaviour of the oscillations, we showed that a new equilibrium state arises, namely a combined oscillation of modes 1 and 3. The analysis performed in this paper can be extended to other values of p^2.

References

1. Abdel-Rohman, M. (1992) Galloping of Tall Prismatic Structures: A Two-Dimensional Analysis, *J. Sound and Vibration*, **Vol. no. 153**, pp. 97–111
2. Boertjens, G.J., Horssen, W.T. van (1996) On Mode Interactions for a Weakly Nonlinear Beam Equation, *TU Delft, Faculty of Techn. Math. and Inf.*, **Report no. 96-148**
3. Horssen, W.T van (1988) An Asymptotic Theory for a Class of Initial-Boundary Value Problems for Weakly Nonlinear Wave Equations with an Application to a Model of the Galopping Oscillations of Overhead Transmission Lines, *SIAM J. Appl. Math.*, **Vol. no. 48**, pp. 1227–1243
4. Horssen, W.T van (1992) Asymptotics for a Class of Semilinear Hyperbolic Equations with an Application to a Problem with Quadratic Nonlinearity, *Nonlinear Analysis*, **Vol. no. 19**, pp. 501–530
5. Keller, J.B., Kogelman, S. (1970) Asymptotic Solutions of Initial Value Problems for Nonlinear Partial Differential Equations, *SIAM J. Appl. Math.*, **Vol. no. 18**, pp. 748–758
6. Lazer, A.C., McKenna, P.J. (1990) Large-Amplitude Peridic Oscillations in Suspension Bridges: Some New Connections With Nonlinear Analysis, *SIAM Review*, **Vol. no. 32**, pp. 537–578
7. Nayfeh, A.H., Abdel-Rohman, M. (1991) Analysis of Galloping Responses in Cantilever Beams, *J. Sound and Vibration*, **Vol. no. 144**, pp. 87–93
8. Wang, L.,Bosley, D.L., Kevorkian, J. (1995) Asymptotic Analysis of a Class of Three-Degree-of-Freedom Hamiltonian Systems Near Stable Equilibria, *Physica D*, **Vol. no. 88**, pp. 87–115

GENUS-TWO SOLUTIONS TO THE KADOMTSEV - PETVIASHVILI EQUATION

E. CAHYONO
FKIP Haluoleo University
Kendari, Indonesia

E. VAN GROESEN
Department of Applied Mathematics, University of Twente
7500 AE Enschede, Netherlands

E. SOEWONO
Department of Mathematics & Center of Mathematics
Institut Teknologi Bandung
Ganesha 10, Bandung 40132

AND

S. SUBARINAH
FKIP University of Mataram
NTB, Indonesia

Abstract. In this paper we consider dynamical aspects of multi-directional waves described by the Kadomtsev-Petviashvili (KP) equation. We investigate some analytically known solutions: the two-soliton interacting waves and their periodic equivalents. It is shown that the behaviour of the interaction of two-solitons can be classified by a parameter $A \geq 0$ (depending on the amplitudes of pure one-solitons and the angles of interactions). In the limiting case when $A = 0$, it is found that the two-soliton reduces to a three-branch soliton.

1. Introduction

The Kadomtsev-Petviashvili (KP) equation

$$(f_t + 6ff_x + f_{xxx})_x + 3f_{yy} = 0, \tag{1}$$

which was introduced in 1970 (see [1]), is of both physical and mathematical interest. It is the generalization of the following well known Korteweg-de Vries (KdV) equation

$$f_t + 6ff_x + f_{xxx} = 0. \tag{2}$$

The KP equation has been known as a model for dispersive, weakly nonlinear waves which are essentially uni-directional with weak transverse effects. In this paper we discuss exact solutions of KP equation. Analytic solutions of KP equation are available in the form of theta function (see for example in [4], [6]), but it is difficult to understand. Genus-two solutions are discussed briefly in section 2, mainly to show that in limiting cases some genus-two solutions can be seen as a superposition of two corresponding cnoidal waves. Some investigations on the effect of the interaction are also included. The effect of the interaction is discussed analytically in section 3 for the simpler case; i.e. the interaction of two solitons.

2. Genus-Two Solutions

KP-equation is a nonlinear equation that has very special properties: it is completely integrable. We will use only one facet of this integrability, namely the fact that special solutions can be written down explicitly. We will show these formulae below. Usually with nonlinear equations, it is not possible to write down solutions in an analytic way (not even a few, let alone whole classes of solutions as we will see below). On the other hand, the interpretation of these analytic expressions is not easy, and we will investigate some aspects below.

In the first subsection we consider periodic wave trains, in particular the interaction of two colliding trains. The explicit expression with Riemann theta function is shown in a limiting case to consist of a linear superposition of two wave trains. In the next subsection we focus on the interaction process that is in fact highly nonlinear. Then it is no restriction to consider the two-soliton waves that exist as an interaction of two solitons (instead of wave trains).

For the following (both the periodic and the soliton case), the basic observation is that the exact solutions are easiest obtained after the transformation:

$$f = 2\partial_x^2 \, ln \, \theta. \tag{3}$$

2.1. GENUS-TWO, LIMITING CONDITION AND SUPERPOSITION OF CNOIDALS

When looking for periodic waves, the simplest exact solution is the one parameter family of KdV cnoidal wave. For KP this amounts to the plane wave

$$f(x,y,t) = cn(\phi;m). \tag{4}$$

Here cn is the cnoidal function, with elliptic modulus m, and the phase variable is given by $\phi = \mu x + \nu y - \omega t + \delta$. The wave vector (μ, ν) determines the propagation direction (perpendicular to the plane wave). The propagation speed of the wave is a rather complicated expression depending on the wave vector and the modulus m (see [4]). An alternative formulation of the solution (4) in terms of a series, with the transformations (3), is given by the Riemann theta function of genus-one:

$$\theta = \sum_{m=-\infty}^{\infty} e^{\frac{b}{2}m^2+im\phi}. \tag{5}$$

Here the parameter $b < 0$ determines the amplitude and the speed of the wave. In a good approximation, the amplitude a as a function of b is obtained from(5) when $\phi = 0$ as follows

$$\sum_{m=-\infty}^{\infty} e^{\frac{1}{2}bm^2}. \tag{6}$$

The larger $-b$, the lower the amplitude, and for $b \mapsto -\infty$ the solution f approaches the sinusoidal waves of infinitesimal amplitude (the solutions of the linearized equation).
More interesting solutions, the so called genus-2 solutions, are given as follows. For two phase factors ϕ_1, ϕ_2 that determine two directions of propagation :

$$\phi_j = \mu_j x + \nu_j y - \omega_j t + \delta_j, j = 1,2 \tag{7}$$

the Riemann-theta function of genus two

$$\theta(\phi_1,\phi_2) = \Sigma_{m_1\ m_2}\ e^{\frac{1}{2}bm_1^2}\ e^{im_1\phi_1}\ e^{\frac{1}{2}(d+b\lambda^2)m_2^2}\ e^{im_2\phi_2}\ e^{b\lambda m_1 m_2} \tag{8}$$

defines the class of solutions

$$f(x,y,t) = 2\ \partial_x^2\ ln\ \theta(\phi_1,\phi_2). \tag{9}$$

The parameters are $b < 0, d < 0$ and λ, which should satisfy $0 < \lambda^2 \leq \frac{1}{2}$, and $d + b\lambda^2 \leq b$. The interpretation of the formula (8) becomes somewhat clearer by taking the limit $b\lambda \mapsto 0$. Then, formally, the double series (8) reduces to a product

$$\theta = \theta_1 \times \theta_2, \tag{10}$$

and the solution f becomes a superposition of two non interacting waves of genus-one, determined by the parameters b and d

$$f = f_1 + f_2, \tag{11}$$

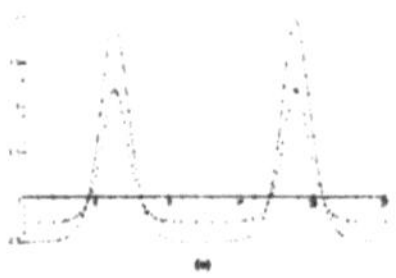

(a)

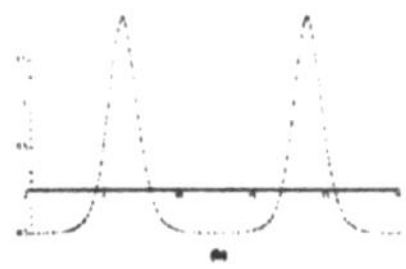

(b)

Figure 1. The difference between a genus 2 solution where $\mu_1 = \mu_2 = 0.5$, $\nu_1 = -\nu_2 = 0.2$ $b = d = -2$ (+++), with the superposition of two corresponding cnoidal waves (—) (a) $\lambda = 0.7$, (b) $\lambda = 0.001$

with $f_k = 2\partial_x^2 \, ln \, \theta_k$. For $b\lambda$ not too small there is a clear nonlinear interaction which is difficult to analyze from the formula. An investigation is presented in the next subsections.

2.2. AN INVESTIGATION ON THE EFFECT OF THE INTERACTION

Since $\lambda = 0$ is not allowed and $b = 0$ is impossible, we are interested in observing how the genus-2 solutions behave for small λ. For certain value of b with λ varies the result is that the effect of interaction is small (large) for λ small (large). At a fixed point y, this can be seen in fig. 1. On 2-D surface the difference between linear superposition of two wave trains and genus-2 solution for λ large is recognized easily in [5], but for small λ it is hardly distinguishable, see fig. 2.

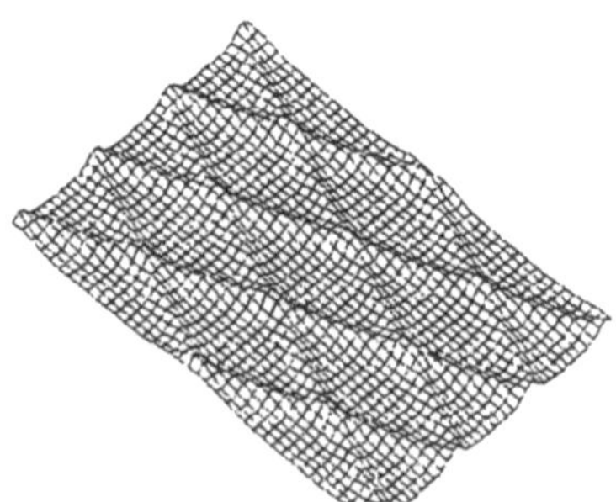

(a)

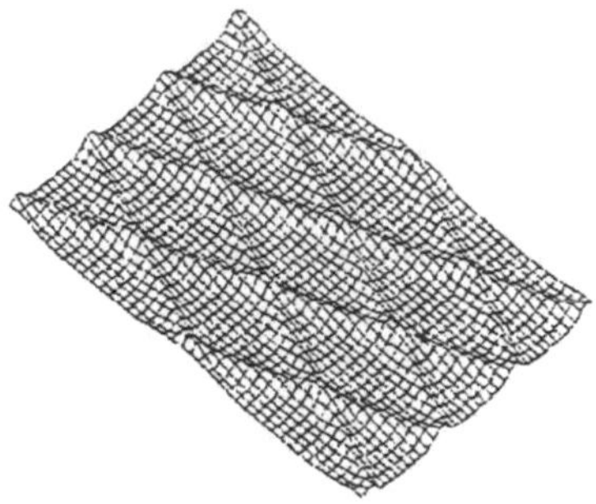

(b)

Figure 2. (a) Genus 2 solution with $\mu_1 = \mu_2 = 0.5$, $\nu_1 = -\nu_2 = 0.2$, $b = d = -2$, $\lambda = 0.001$ (b) Superposition of two corresponding cnoidal waves at (a).

3. Interaction of two-solitons

In this section we will study the nonlinear interaction process in somewhat more detail. As is quite common in the literature (see [3] for obliquely interacting wave, and [2] for uni directional waves), we will simplify the analysis by replacing the periodic interaction by the interaction of two solitary waves. In the first subsection 3.1 we will present the solitary equivalent of the periodic genus-2 solution. Restricting to an investigation of the spatial pattern, we will analyze the pattern at a fixed time by treating the y-variable as a timelike quantity.

3.1. TWO-SOLITONS, SUPERPOSITION REPRESENTATION

A solitary wave in the direction of a wave vector $\mathbf{k} = (\mu, \nu)^T$ is a single hump wave form that decays rapidly to zero in the propagation direction. For KP equation, the governing formula of the solitary wave reads

$$f = 2\partial_x^2 \ ln \ \theta \tag{12}$$

with the phase function $\phi = \mathbf{k}.\mathbf{x} - ct$, where $k = \mid \mathbf{k} \mid$ and

$$\theta = 1 + e^{2\phi}. \tag{13}$$

A simple calculation produces the well known result

$$f = \mu\partial_x \ tanh \ \phi = \mu^2 \ sech^2\phi. \tag{14}$$

Multi-direction waves, known as genus-two solutions, can be written down in the non-periodic case in an easier way than in the periodic case. We look for solutions depending on two phases in the form

$$f = u(x + \rho_1 y - c_1 t, \quad x + \rho_2 y - c_2 t). \tag{15}$$

This solution is a steady state in a moving coordinate frame of reference:

$$x \longmapsto x - \nu_1 t, \quad y \longmapsto y - \nu_2 t$$

where $\nu_1 = \frac{\rho_1 c_2 - \rho_2 c_1}{\rho_1 - \rho_2}$, $\nu_2 = \frac{c_1 - c_2}{\rho_1 - \rho_2}$. Then we have $f = u(x + \rho_1 y, \ x + \rho_2 y)$. Furthermore, the directions with respect to the y-axis, determined by $\rho_{1,2}$ can be made symmetrical ($\rho_1 = -\rho_2 = r$) by another transformation

$$x \longmapsto x + py, \quad y \longmapsto y + qy$$

where

$$p = \frac{r(\rho_1 + \rho_2 + 2\rho_1\rho_2)}{\rho_2 - \rho_1}, \quad q = \frac{r(2 + \rho_1 + \rho_2)}{\rho_2 - \rho_1}. \tag{16}$$

From now on the variable t is dropped from the representation of genus-two solutions. Then a genus-2 solution (two-soliton solution) can be written down in two different equivalent ways.
A first way, based on the transformation (3), is written as (see [7] and [4])

$$\theta = 1 + e^{2\phi_1} + e^{2\phi_2} + Ae^{2\phi_1 + 2\phi_2}, \tag{17}$$

where A is given as

$$A = \frac{4(\mu_1 - \mu_2)^2 - (\rho_1 - \rho_2)^2}{4(\mu_1 + \mu_2)^2 - (\rho_1 - \rho_2)^2}. \tag{18}$$

In the absence of ρ, A reduces to the corresponding A for KdV. Note that, unlike in KdV where $0 < A < 1$, here A can take any value.
Another way, following the approach in [2] describing the two-soliton solutions for KdV, is to write the solution directly as a linear superposition:

$$f = u_1 + u_2 \tag{19}$$

where (for symmetrical interaction)

$$\begin{aligned}
u_j &= 2\mu_j \partial_x \, tanh \; g_j \\
g_1 &= \phi_1 + \tfrac{1}{2} \, ln\{\tfrac{1+Ae^{2\phi_2}}{1+e^{2\phi_2}}\}, \quad g_2 = \phi_2 + \tfrac{1}{2} \, ln\{\tfrac{1+Ae^{2\phi_1}}{1+e^{2\phi_1}}\} \\
\phi_1 &= \mu_1(x + ry) + \delta_1, \quad \phi_2 = \mu_2(x - ry) + \delta_2 \\
A &= \frac{(\mu_1 - \mu_2)^2 - r^2}{(\mu_1 + \mu_2)^2 - r^2}.
\end{aligned}$$

We assume from now on that $\mu_1 > \mu_2$, and consider y as time-like variable. As in KdV, conservation of mass for each individual soliton u_1 and u_2 is immediately shown as follows (assuming that $A > 0$, which implies that the range of g_j is $(-\infty, \infty)$)

$$\int_{-\infty}^{\infty} u_j dx = \int_{-\infty}^{\infty} 2\mu_j \; sech^2(v)d(v) = 4\mu_j, \quad j = 1, 2. \tag{20}$$

Define the center of masses $x_1(y)$ and $x_2(y)$ such that $\int_{-\infty}^{x_j(y)} u_j dx = \int_{x_j(y)}^{\infty} u_j dx = 2\mu_j, \; j = 1, 2$. Then we have

$$\int_{-\infty}^{g_j} sech^2(v)d(v) = \int_{g_j}^{-\infty} sech^2(v)d(v) = 1 \tag{21}$$

where g_j are evaluated at $x = x_j(y), \; j = 1, 2$ respectively. We have from (21) that the path of each soliton is given by

$$g_1 \equiv \phi_1 + \tfrac{1}{2}\; ln\{\tfrac{1+Ae^{2\phi_2}}{1+e^{2\phi_2}}\} = 0, \quad g_2 \equiv \phi_2 + \tfrac{1}{2}\; ln\{\tfrac{1+Ae^{2\phi_1}}{1+e^{2\phi_1}}\} = 0, \tag{22}$$

respectively. The phase shift of each soliton is $-\frac{1}{2}\; lnA$. This follows from (22) by fixing one of the phase variable and letting the second tend to $\pm\infty$. The intersection point which can be interpreted as the peak of the interaction, is obtained by intersecting (22)

$$\phi_1 + \tfrac{1}{2}\; ln\{\tfrac{1+Ae^{2\phi_2}}{1+e^{2\phi_2}}\} = 0 = \phi_2 + \tfrac{1}{2}\; ln\{\tfrac{1+Ae^{2\phi_1}}{1+e^{2\phi_1}}\}, \tag{23}$$

which gives

$$\phi_1 = \phi_2 = -\tfrac{1}{4}\; lnA. \tag{24}$$

When the constants δ_i that appear in the definitions of ϕ_i are assigned the values $\delta_i = -\frac{1}{4}\ln A, i = 1, 2$, the point of interaction of the paths as given by (24) becomes the origin $(0, 0)$ in the (x, y) plane. Interacting paths which differ from KdV are shown in figure 3.

3.2. LENGTH OF THE INTERACTION

As in [2], we define the beginning of the interaction by y_0 and the end of interaction by y_1. Let Δ be a small positive number which is used to indicate the portion of the overlapping solitons. Then y_0 and y_1, are defined implicitly as follows

$$\int_{A_0}^{x_0} u_1 dx = 2\mu_1(1 - \Delta), \quad \int_{x_0}^{B_0} u_2 dx = 2\mu_2(1 - \Delta) \tag{25}$$

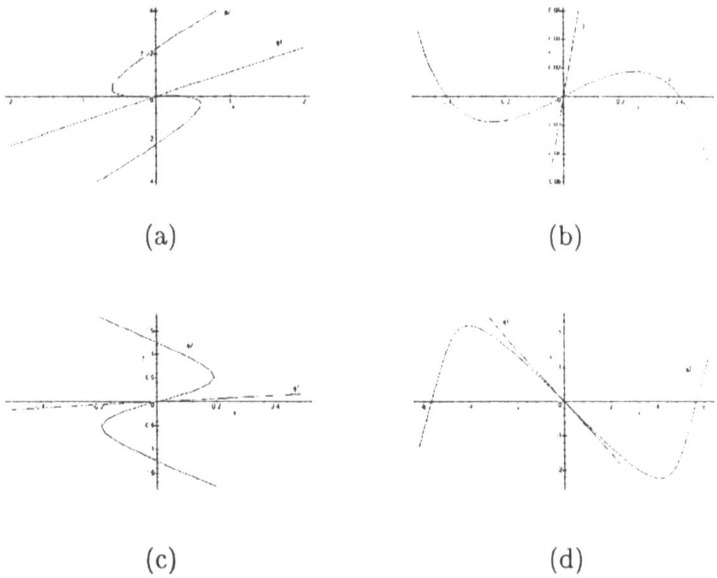

Figure 3. (a) The path of solitons for $\mu_1 = 0.9,\ \mu_2 = 0.35,\ \rho_1 = -0.81,\ \rho_2 = -0.5$, (b) blowup of (a), (c) The path of solitons for $\mu_1 = 0.9,\ \mu_2 = 0.1,\ \rho_1 = -3,\ \rho_2 = 0.36$, (d) blowup of (c)

$$\int_{B_1}^{x_1} u_2 dx = 2\mu_2(1-\Delta), \quad \int_{x_1}^{A_1} u_1 dx = 2\mu_1(1-\Delta). \tag{26}$$

Here $A_0,\ A_1$ and $B_0,\ B_1$ are the center of masses of u_1 and u_2 at the point y_0 and y_1 respectively and $x_0,\ \ x_1$ represent the points between the two solitons where the "Δ overlap" occurs corresponding to before and after interaction respectively. Since the path of each soliton is symmetrical with respect to $(0,0)$, we have $x_1 = -x_1$ and $y_1 = -y_0$ and the length of interaction $L = 2y_1$. Then direct calculation of integrals in (25) and (26) gives

$$L = \frac{\mu_1 + \mu_2}{-2\mu_1\mu_2 r}\, ln\{\frac{(1-\Delta)A^{\frac{1}{2}} + \sqrt{(1-\Delta)^2 A + (2-\Delta)\Delta}}{\Delta}\}. \tag{27}$$

We will now summarize in the following theorem.

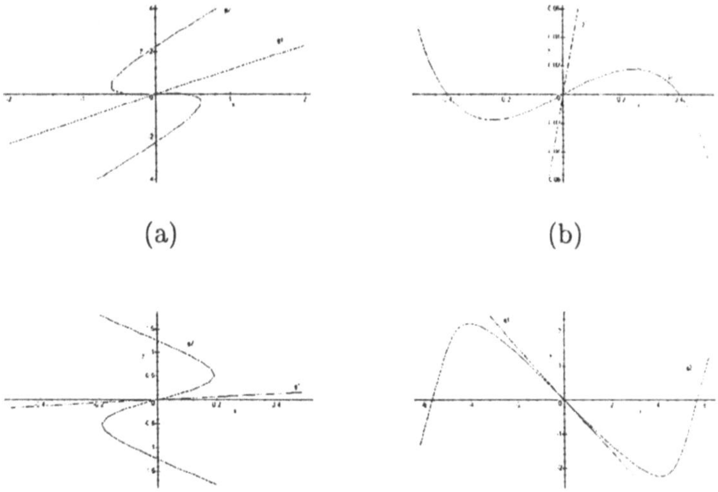

(a) (b)

(c) (d)

Figure 4. Interacting solitons (a) for $0 < A < 1$, (b) for $A = 0$, (c) for $A > 1$ and (d) for $A < 0$

Proposition 1

(i) *If $r < \mu_1 - \mu_2$, then $0 < A < 1$, and each soliton undergoes negative shift ($= \frac{1}{2}\ ln(A)$), mass conservation, exact interaction and splitting. The length of the interaction approaches ∞ as $r \longrightarrow 0$.*

(ii) *If $r > \mu_1 + \mu_2$, then $A > 1$, and each soliton undergoes positive shift ($= \frac{1}{2}\ ln(A)$), mass conservation, exact interaction and splitting. The length of interaction approaches 0 as $r \longrightarrow \infty$.*

(iii) *If $\mu_1 - \mu_2 < r \leq \mu_1 + \mu_2$, then singularity (at finite time) occurs in the form of unbounded solutions.*

(iv) *If $r = \mu_1 - \mu_2$, then the solution u forms a three-branch soliton along ϕ_1 = constant on the upperhalf plane, along ϕ_2 = constant on the lowerhalf plane, and along $\phi_1 - \phi_2$ = constant on the righthalf plane.*

Proof:
The mass conservation in (20) is valid when g ranges from $-\infty$ to ∞, i.e. when $A > 0$. The phase shifts ($\frac{1}{2}ln(A)$) is positive when $A > 1$ and negative when $0 < A < 1$. The conclusion about the length of interaction follows immediately from (27). This completes the proof of parts (i) and (ii).
For the case (iii), we have $A < 0$ and therefore g_1, g_2 are no longer defined

in the whole R^2 and singularities occur when $1 + Ae^{2\phi_i} = 0$, $i = 1$ or 2. When $A = 0$, then using (12) and (17) u reduces to $u = 2\partial_x^2 ln(\theta)$, where $\theta = 1 + e^{2\phi_1} + e^{2\phi_2}$. The first and the second solitons reduce to
$u_1 = 2\mu_1 \frac{\mu_1+(\mu_1-\mu_2)e^{2\phi_2}}{1+e^{2\phi_2}}\ sech^2(\phi_1 - \frac{1}{2}\ ln(1+e^{2\phi_2}))$, and
$u_2 = 2\mu_2 \frac{\mu_2+(\mu_1-\mu_2)e^{2\phi_1}}{1+e^{2\phi_1}}\ sech^2(\phi_1 - \frac{1}{2}\ ln(1+e^{2\phi_1}))$ respectively.
We will show that there are exactly three nonvanishing phase directions $\phi_1 = constant, \phi_2 = constant$ and $\phi_1 - \phi_2 = constant$.
Along $\phi_1 = constant$, as $y \to +\infty$, we have $\phi_2 = \frac{\mu_2}{\mu_1}\phi_1 - 2\mu_2 ry \to -\infty$, and therefore $\theta \to 1 + e^{2\phi_1}$. Hence along $\phi_1 = constant$, as $y \to +\infty$, $u \to 2\partial_x^2 ln(1+e^{2\phi_1})$, i.e. the pure first soliton.
Along $\phi_2 = constant$, as $y \to -\infty$, we have $\phi_1 = \frac{\mu_1}{\mu_2}\phi_2 + 2\mu_1 ry \to -\infty$, and therefore $\theta \to 1 + e^{2\phi_2}$. Hence along $\phi_2 = constant$, as $y \to -\infty$, $u \to 2\partial_x^2 ln(1+e^{2\phi_2})$, i.e. the pure second soliton.
Along $\Phi = \phi_1 - \phi_2 = constant$, as $y \to -\infty$, we have
$\phi_2 = \frac{\mu_2}{\mu_1-\mu_2}\Phi - \frac{2\mu_1\mu_2 r}{\mu_1\mu_2}y \to +\infty$. Rewrite θ as $\theta = e^{2\phi_2}[1 + e^{2(\phi_1-\phi_2)} + e^{-2\phi_2}]$. The first exponential factor vanishes under the operator $2\partial_x^2 ln$ and the second factor $\to 1 + e^{2(\phi_1-\phi_2)}$ as $y \to -\infty$. Hence, as $y \to -\infty$ we have $u \to 2\partial_x^2 ln(1 + e^{2(\phi_1-\phi_2)})$, i.e. the pure one-soliton in the phase direction $\phi_1 - \phi_2$. Finally it will be shown that there is no other nonvanishing phase direction. Suppose there is a nonvanishing phase direction $\xi = \phi_1 + \alpha\phi_2 = constant$ on the upperhalf plane. If ξ for large y lies above $\phi = 0$ ($\alpha > 0$ or $\alpha < -\frac{\mu_1}{\mu_2}$), then along $\xi = constant$, $\phi_1 \to \infty$ as $y \to \infty$. By rewriting $\theta = e^{2\phi_2}[1 + e^{-2\phi_1} + e^{\frac{2}{\alpha}\xi - \frac{2(1+\alpha)}{\alpha}\phi_1}]$, then the second factor $\to 1$ as $y \to \infty$. If ξ for large y lies below $\phi_1 = 0$, then along $\xi = constant$, as $y \to \infty$ we have $\phi_1, \phi_2 \to -\infty$. This implies that $\theta \to 1$ as $y \to \infty$ (along $\xi = constant$). This contradicts the fact that ξ is a nonvanishing phase direction. Similar argument applies in the lower half plane.

4. CONCLUSION

A classification of genus-two solutions of KP equation has been obtained in which for limiting cases some genus-2 solutions can be seen as a superposition of two genus-1 solutions. In getting indepth view we considered the simpler case, i.e. interaction of two-solitons. During the interaction we recognize different types of behavior (parameterized by A): positive phase shift with the peak larger than their amplitudes, negative phase shift with the peak smaller than the largest amplitude, mass conservation, period of interaction and singularities. In the limiting case when $A = 0$, the two-soliton reduces to a three-branch soliton of amplitudes $2\mu_1^2, 2\mu_2^2$ and $2(\mu_1 - \mu_2)^2$. This can be viewed as an interaction of two solitons which results in one other soliton of different amplitude.

Acknowledgement Part of the research is sponsored by (1) Commision of the European Communities, Directorate General XII-B, Joint Project CI1*-CT93-0018 between the Department of Mathematics, Institut Teknologi Bandung, Indonesia, and the Faculty of Applied Mathematics, University of Twente, the Netherlands, (2) Hibah Tim URGE Project No. 007/HTPP/-URGE/1995, DIKTI, and (3) Royal Netherlands Academy of Arts and Sciences in the programme for scientific cooperation Netherlands-Indonesia.

References

1. Kadomtsev, B.B. & Petviashvili, V.I. : *On the stability of solitary waves in weakly dispersive media*, Soviet Phys. Dokl. 15(6) (1970), 539-541
2. Hodnett, P.F. & Moloney, T.P. : *On the structure during interaction of the two-soliton of the Korteweg-de Vries equation*, SIAM, J. Appl. Math., Vol. 49, 1989.
3. Miles, J.W. : *Obliquely interacting solitary waves*, J. Fluid Mech. Vol. 79 part 1 pp. 157-169, 1977.
4. Segur, H. & Finkel, A. : *An Analytical model of periodic waves in shallow water*, Studies in applied Mathematics 73, 1985.
5. Hammack, J.; Scheffner, N., Segur, H. : *Two-dimensional periodic waves in shallow water*, J. Fluid Mechanics, Vol. 209 pp. 567-589, 1989.
6. Dubrovin, B.A.: Russian Math. Surveys 36:11-92, 1981
7. Whitham, G.B. : Linear and Nonlinear Waves J.Wiley, 1974

ON THE FORMATION OF HOTSPOT IN MICROWAVE HEATING

DANIEL CHANDRA AND TOTO NUSANTARA
Department of Mathematics Education,
FPMIPA IKIP MALANG
Jl. Surabaya 6, Malang

MOCH. ARUMAN IMRON
Department of Mathematics, FMIPA UNIBRAW MALANG
Jl. MT. Haryono 169, Malang

AND

BASUKI
Department of Mathematics, FMIPA UAD YOGYAKARTA
Jl. Kapas 9, Yogyakarta

Abstract. We consider a simple model for microwave heating. The effect of non-uniform diffusivity as well as non-uniform electric fields on the formation of hotspots is investigated. A fundamental mode approximation is used in the analysis. It is shown that non-uniform diffusifity and electric fields contribute to the occurrence of hotspots. To illustrate, numerical results are carried for a unit slab geometry.

1. Introduction

The use of microwave heating in industries such as melting and drying as well as in domestic environments now is a common technique. This technique proves to have advantages over the use of a conventional oven, as it quickly provides heating throughout the material.

In general the problem involves a reaction-diffusion equation for the temperature with a source term which is proportional to the square of the intensity of the electric fields. This heating source also depends nonlinearly on the temperature as a result of the auto-catalytic chemical reaction of the heated medium.

In this paper we would like to investigate the influence of non-uniform diffusivity of material as well as non-uniform intensity of electric fields on the formation of a hotspot which is a small region in the geometry considered that has much higher temperature than the other part. We assume that the magnetic and electric properties of the material are constant. In this case the intensity of the electric fields is decaying exponentially in space.

As described in [2], the governing equation for the model can be written in the form

$$\frac{\partial \theta}{\partial t} = \nabla.(k(\theta)\nabla\theta) + \delta E(x) f(\theta). \tag{1}$$

Here, θ is the temperature; $k(\theta)$ is diffusivity with the properties $k(\theta) > 0, k'(\theta) > 0$ and in this study is taken to be $k(\theta) = \mu e^{\gamma\theta}$, for some positive parameter γ and μ; δ is a positive parameter related to the intensity of electric fields; $E(x)$ is the heating source due to electric fields; and $f(\theta)$ is the auto-catalytic chemical heating source with properties $f(\theta) > 0, f'(\theta) > 0$. As observed by Smyth [3] it is realistic to take $f(\theta)$ to be Arrhenius type, that is $f(\theta) = e^{\frac{\alpha\theta}{\alpha+\theta}}$ for some $\alpha > 0$.

We consider equation (1) in open bounded domain D subject to the initial and boundary condition

$$\theta(x,0) = h(x) \geq 0 \;, \theta(x,t) = 0 \;\; on \;\; \partial D. \tag{2}$$

In this paper we will employ the fundamental mode approximation described in [1]. In the next section we describe some previous results. The analysis of the formation of hotspots due to the effect of diffusivity as well as intensity of electric fields is given in section 3. Numerical result for a unit slab geometry is given in the section 4. Further discussion on a more complicated problem and the conclusion are given in the section 5 and section 6 respectively.

2. Previous Results

Let

$$v = \int_0^\theta k(s)ds. \tag{3}$$

Using this transformation, equation (1) becomes

$$\frac{\partial v}{\partial t} = K(v)\{\nabla^2 v + \delta E(x)F(v)\}, \tag{4}$$

where $K(v) = k(\theta(v))$ and $F(v) = f(\theta(v))$. Since $v(\theta)$ is monotonically increasing, we observe that both $K(v)$ and $F(v)$ have the same features as $k(\theta)$ and $f(\theta)$ respectively. The initial and the boundary condition (2) become

$$v(x,0) = H(x), \tag{5}$$

and

$$v(x,t) = 0 \ on \ \partial D, \tag{6}$$

where $H(x) = \int_0^{h(x)} k(s)ds$.

Then we use time scaling : τ be such that $\frac{d\tau}{dt} = K(v(x,t))$ and $u(x,\tau) = v(x,t)$. Since $K(v(x,t))$ is positive then $\tau(t)$ is monotonically increasing and so $u(x,\tau)$ has the same feature as $v(x,t)$. Using this time scaling, we have :

$$\begin{aligned} \frac{\partial u}{\partial \tau}\frac{d\tau}{dt} &= \frac{\partial v}{\partial t} \quad (*) \\ \frac{\partial u}{\partial \tau}K(v(x,t)) &= K(v)[\nabla^2 v + \delta E(x)F(v)] \end{aligned}$$

$$\frac{\partial u}{\partial \tau} = \nabla^2 u + \delta E(x)F(u). \tag{7}$$

Note from (*) we have this relation :

$$\frac{\partial u}{\partial \tau} = 0 \leftrightarrow \frac{\partial v}{\partial t} = 0$$

This means that the above time scaling is suitable. Also note that the steady state solutions of equation (7) are exactly the same as those of equation (4), because the equation (7) is obtained from (4) by introducing time scaling and the steady state solutions is time independent.

3. Hotspot Formation

3.1. THE EFFECT OF DIFFUSIVITY

In this section we investigate the effect of diffusivity on the formation of hot spots.

From the transformation (3), we obtain

$$\theta = \frac{1}{\gamma}\ln(1 + \frac{\gamma v}{\mu}). \tag{8}$$

As the steady state solutions of (4) are also the steady state solutions of (7), we expect that after sufficiently long the temperature θ is approximated by

$$\theta = \frac{1}{\gamma}\ln(1 + \frac{\gamma u}{\mu}). \tag{9}$$

We use fundamental mode approximation to obtain steady state solution u of the equation (7). Let $u(x,t) \approx A(t)\psi_1(x)$, where $A(t)$ satisfies

$$\begin{aligned} \frac{dA}{dt} &= -\lambda_1^2 A + \delta \int_0^1 E(x)F(A\psi_1(x))\psi_1(x)dx, \\ A(0) &= C1, \end{aligned} \tag{10}$$

and $\psi_1(x)$ is the eigenfunction corresponding to the smallest eigenvalue λ_1^2 of the eigenvalue problem

$$\begin{aligned} \nabla^2\varphi &= -\lambda^2\varphi \\ \varphi &= 0 \text{ on } \partial D. \end{aligned} \tag{11}$$

Let $I(A) = \int_0^1 E(x)F(A\psi_1(x))\psi_1(x)dx$, where

$$F(A\psi_1(x)) = \exp\left(\frac{\frac{\alpha}{\gamma}\ln(1+\frac{\gamma A\psi_1}{\mu})}{\alpha + \frac{1}{\gamma}\ln(1+\frac{\gamma A\psi_1}{\mu})}\right) \tag{12}$$

The steady state solutions A of (10) can be obtained graphically from the intersection of the line $\frac{A\lambda_1^2}{\delta}$ *vs* A and the curve $I(A)$ *vs* A, see Figure 1.

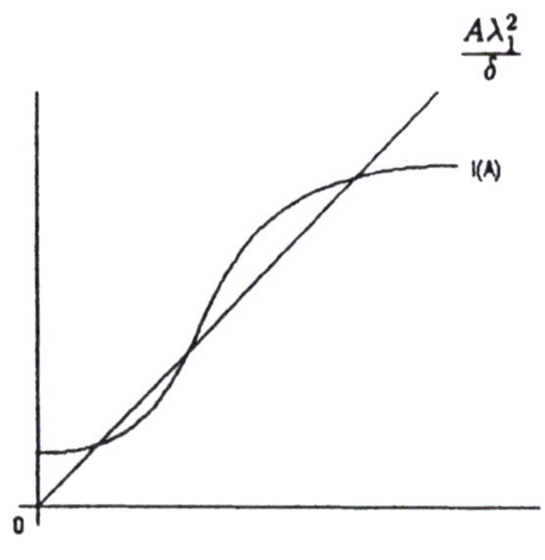

Figure 1. Intersection of the line $\frac{A\lambda_1^2}{\delta}$ vs A and the curve $I(A)$ vs A

From (9), (10), and (12), if we consider two separate domains with diffusivity $k_1 = \mu_1 e^{\gamma\theta}$ and $k_2 = \mu_2 e^{\gamma\theta}$ respectively, where $\mu_1 > \mu_2$, we expect that the domain with diffusivity k_2 will achieve a higher temperature than that of the one with diffusivity k_1.

3.2. THE EFFECT OF INTENSITY OF ELECTRIC FIELDS

In the previous section we have investigated the influence of diffusivity of material on the formation of hotspots. In this section we will study the effect of intensity of electric fields this hotspot formation.

From (9) we obtain that the temperature θ at the steady state condition is a function of u where u is a steady state solution of (7). Using fundamental mode approximation for u, we write $u \approx A(t)\psi_1(x)$, where A satisfies (10) and $\psi_1(x)$ is the eigenfunction corresponding to smallest eigenvalue λ_1^2 of the eigenvalue problem (12). The steady state solution A can be obtained graphically from the intersection of the line $\frac{A\lambda_1^2}{\delta}$ vs A and the curve $I(A)$ vs A as shown in details in Figure (2.a.). The bifurcation diagram of A as a function of parameter δ, which corresponds to the intensity of electric fields, is given in Figure (2b). In Figure (3), we present the bifurcation diagram of the temperature θ as a function of δ.

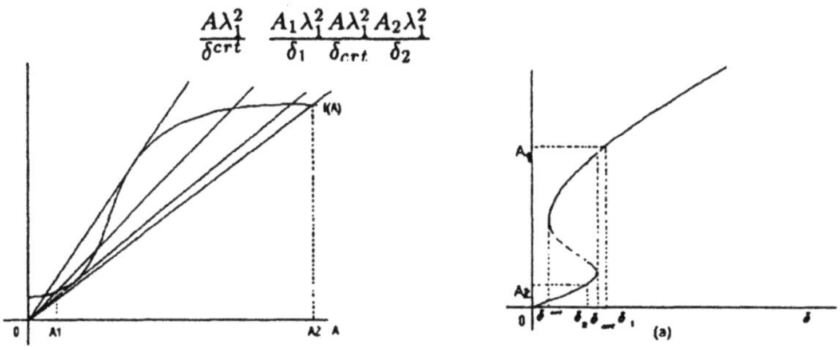

Figure 2. (a) Intersection of the line $\frac{A\lambda_1^2}{\delta}$ vs A and the curve $I(A)$ vs A for some δ, (b) bifurcation diagram of A as a function of δ.

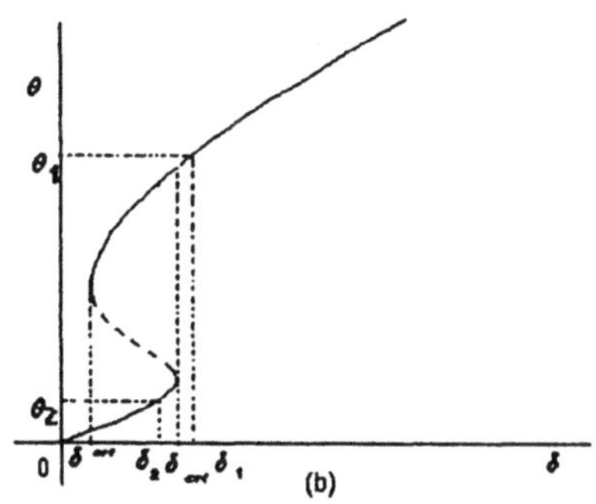

Figure 3. Bifurcation diagram of temperature θ as a function of intensity of electric fields δ

Here δ_{crt} and δ^{crt} are critical values of δ when the line $\frac{A\lambda_1^2}{\delta}$ vs A tangent to the curve $I(A)$ vs A. For $\delta < \delta^{crt}$ the equation (10) admits only one steady state solution A which is of $O(1)$, while for $\delta > \delta_{crt}$, the steady state solution A is of $O(e^{\alpha})$. This means that there is a maximum intensity of electric fields below which it only produces a low temperature and there is a minimum intensity of electric field above which it only admits a very high steady state temperature. This fenomena can be observed from the bifurcation diagram in Figure (3).

Let $E_1 = \delta_1 e^{-x}$ and $E_2 = \delta_2 e^{-x}$, where $\delta_1 > \delta_2$, be the intensity of electric fields entering two different separated domains. Let the initial temperature θ be zero. From the bifurcation diagram of the approximated steady state temperature θ vs δ in Figure (3), it is clear that the temperature θ_1 corresponding to E_1 is greater than θ_2 corresponding to E_2. In particular if $\delta_1 > \delta_{crt}$ and $\delta_2 < \delta_{crt}$, then θ_1 is substantially higher than θ_2. It is interesting to see that E_1 and E_2 need to differ very little to produce a substantial difference in the steady state temperature.

4. Numerical Results

4.1. NON-HOMOGENEOUS DIFFUSIVITY

In 1-D problem, as an example, we consider a unit slab with non homogeneous diffusivity, see Figure (4).

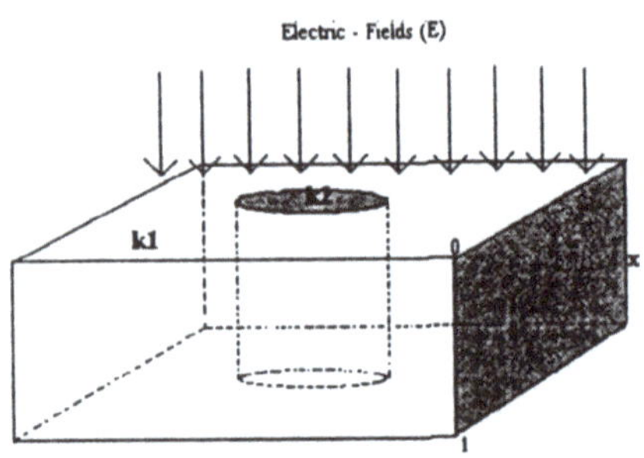

Figure 4. A unit slab with non-homogeneous diffusivity, $k_1(\theta) = \mu_1 e^{\gamma\theta}$ and $k_2(\theta) = \mu_2 e^{\gamma\theta}$.

Let the diffusivity $k_1(\theta) = \mu_1 e^{\gamma\theta}$ at a certain location of the slab, and $k_2(\theta) = \mu_2 e^{\gamma\theta}$ for the rest, with $\mu_1 > \mu_2$. Then from (9), temperature θ can be calculated, where u in this expression is approximated by $A(t)\psi_1(x)$. Note that for this unit slab the eigenvalues and eigenfunctions of (12) are $\{n^2\pi^2\}$ and $\{\sqrt{2}\sin(n\pi x)\}$ respectively. Steady states A in this approximation are obtained from the steady state solutions of (10).

We carry out computations for $\alpha = 10, \gamma = 0.1, \mu = 1$ and several values of δ in Table (1) and we give some examples of these steady states A_{std}.

From numerical result below, we can see that, the part of the material with lower diffusivity has higher temperature than the other part, and so forming hotspots.

δ	A_{std1}	A_{std2}	A_{std3}
0.1	0.00570		
0.5	0.02925		
1	0.06053		
1.31991	0.08175	50.13980	
2	0.13051	12.94107	237.97319
3	0.21368	7.19111	499.86165
4	0.31667	4.86877	785.24760
5	0.45340	3.48129	1008.70087
6	0.66423	2.44446	1406.33609
6.5	0.84955	1.93896	1596.61829
6.8690		1.29000	1691.82067
7.2			1802.58391
8			2074.43088

TABLE 1. The steady states A_{std} corresponding to some values of α and δ for a unit slab. A_{std1}, A_{std2}, and A_{std3} are the first, the second, and the third steady state respectively.

$$\begin{aligned}
\mu &= 1, & A &= 0.45340, & \theta &= 0.62148 \\
\mu &= 0.8, & A &= 0.59506, & \theta &= 1.00019 \\
\mu &= 0.5, & A &= 1388.09414, & \theta &= 59.75366 \\
\mu &= 0.1, & A &= 1949.35925, & \theta &= 79.22192 \\
\mu &= 0.01, & A &= 2547.69125, & \theta &= 104.92129
\end{aligned}$$

For the calculations above, we use $\alpha = 10, \gamma = 0.1, \delta = 5, x = 0.5$, and initial value of $A = 0$

4.2. NON-HOMOGENEOUS INTENSITY OF ELECTRIC FIELDS

In 1-D problem, as example, we consider a unit slab with non homogeneous electric field, see Figure (5) below.

Let the intensity of electric fields E_1 be larger than E_2, so we have $\delta_1 > \delta_2$. From the analysis we expect that the part of material with the intensity of electric fields E_1 must be hotter than that of with the intensity

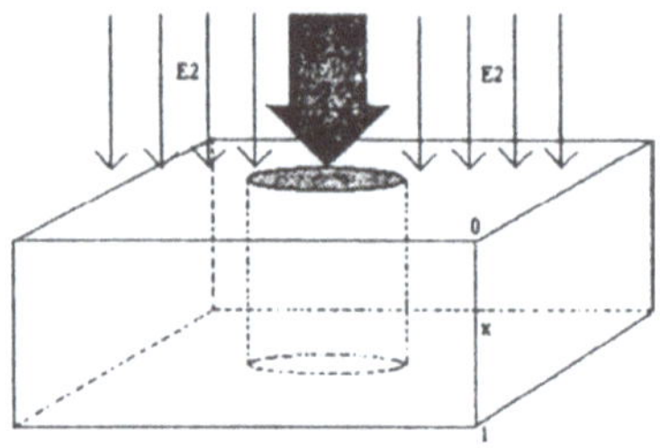

Figure 5. A unit slab with non-homogeneous electric fields, where $E_1 = \delta_1 e^{-x}$ and $E_2 = \delta_2 e^{-x}$, $\delta_1 \neq \delta_2$

of electric fields E_2. In particular when $\delta_1 > \delta_{crt}$ and $\delta_2 < \delta_{crt}$, then the part with intensity of electric fields E_1 has a substantially higher temperature than that of with E_2.

We carry out computations for several values of α in Table (2) and summarize these critical values δ.

α	δ_{crt}	A	δ^{crt}	A
10	6.870	1.290	1.3199	0.08175
12	6.696	1.211	-	-
14	6.581	1.162	-	-
16	6.499	1.128	-	-
18	6.437	1.104	-	-
20	6.389	1.085	-	-

TABLE 2. Critical values δ corresponding to some values of α for a unit slab. A is the smallest steady state corresponding to the critical value δ.

The steady state temperature θ as a function of δ at $\alpha = 10$, $\mu = 10^{-5}$, $\gamma = 10^{-1}$ and $x = 0.5$ is given in Table (3).

From the Table (3), the corresponding bifurcation diagram for temperature θ as a function of δ is given in Figure (6).

From the above bifurcation diagram if we start with initial temperature zero, and the intensity of electric fields E_1 and E_2 lower than the critical intensity of electric fields, then temperature difference between the part with the intensity E_1 and E_2 is not very high. However, if the intensity of electric fields E_1, E_2 are larger and smaller than the critical intensity

δ	θ		
	θ_1	θ_2	θ_3
0.1	0.00806	-	-
0.5	0.04128	-	-
1	0.08523	-	-
1.31991*	0.11494	20.90732	-
2	0.18289	10.40328	35.45427
3	0.29771	7.01599	42.72367
4	0.43810	5.23869	47.18952
5	0.62148	4.00338	49.67393
6	0.89782	2.96913	52.97747
6.5	1.13458	2.42326	54.240506
6.8690*	-	1.67575	54.81720
7.2	-	-	55.44880
8	-	-	56.84834

TABLE 3. Bifurcation table of the steady state temperature θ as a function of δ. θ_1, θ_2, and θ_3 are the first, the second, and the third steady state of temperature respectively. The upper star and the lower star indicate the values of δ^{crt} and δ_{crt} respectively.

respectively, then the temperature of the part with the intensity of electric fields E_1 is substantially higher than that of with E_2. It can be concluded that the non-homogeneous intensity of electric fields contributes to the hotspot formation. As an example, we take $\delta_1 = 6.5$ which is slightly smaller than $\delta_{crt} = 6.8690$ as shown in Table 3, and $\delta_2 = 7.2$ which is slightly larger than δ_{crt}, we will have the values of the temperature corresponding to δ_1 and δ_2 are 1.13458 and 55.44880 respectively. We note that a small difference of δ as in the above case can give a substantial difference in the temperature.

5. Further Discussion

In the previous section we have assumed that the diffusivity of the material is constant throughout the domain. In general the diffusivity depends on the space variable, hence we need to treat the problem differently. To illustrate

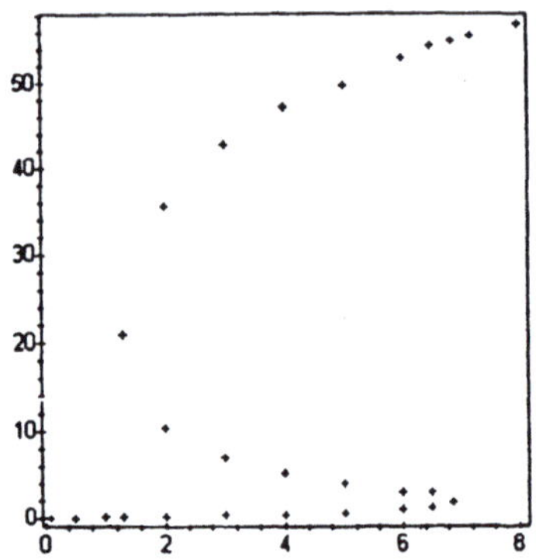

Figure 6. Bifurcation diagram of temperature θ as a function of δ

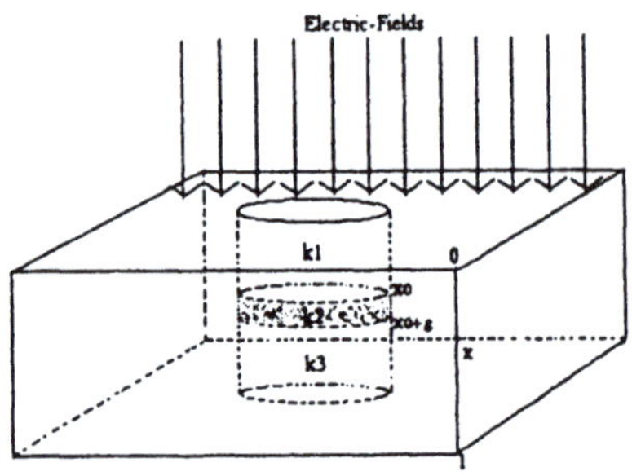

Figure 7. A small region in unit slab with different diffusivity $k1, k2, k3$

the problem consider a unit slab with non-homogeneous diffusivity as it is given in the figure below.

In this case we formulate the governing equation (1) as follows :

$$\frac{\partial \theta}{\partial t} = \nabla.(k(\theta)\nabla(\theta)) + \delta E(x) f(\theta)$$

where the diffusivity $k(\theta)$ is taken to be

$$k(\theta) = \begin{cases} \mu_1 e^{\gamma\theta} & \text{if } 0 \leq x < x_0 \\ \mu_2 e^{\gamma\theta} & \text{if } x_0 \leq x \leq x_0 + \epsilon \\ \mu_1 e^{\gamma\theta} & \text{if } x_0 + \epsilon < x \leq 1 \end{cases}$$

For simplification, we may investigate the occurrence of hotspots at the steady state. At this state the temperature at $x = x_0$ and at $x = x_0 + \epsilon$, say $\theta = \theta_1$ and $\theta = \theta_2$ respectively, no longer depend on time. These value have to be determined and are part of the problem.

6. Concluding Remarks

We have used a fundamental mode approximation to analyse the formation of hotspots in microwave heating. For a unit slab geometry, it is shown that non-uniform diffusivity as well as electric fields contribute to the occurence of hotspots. Further it is also shown that a slight difference in the intensity of electric fields entering different part of the slab may produce a substancial difference in the temperature of those parts. Thus there is a critical the intensity of electric fields below which it admits a low temperature and above which it admits a high temperature.

Acknowledgments

We would like to thank RWS organizers who have given us the opportunity to do this research. Our knowledge on mathematical research has been expanded after attending this RWS. We especially wish to acknowledge two persons : our supervisor Andonowati who has given her best guidance so that our research can be done well and Edy Soewono who has given many valuable suggestions.

Finally, we would like to express many thanks to other RWS participants for their cooperation throughout this research workshop.

References

1. Andonowati. *A study of some problems arising from combustion and the microwave heating*. Ph.D. thesis, McGill University, Montreal, Canada (1995).
2. J. M. Hill and A. H. Pincombe. *Some similarity temperature profiles for the microwave heating of a half-space*. J. Austral. Math. Soc. B. 33:290-320 (1992).
3. N. F. Smyth. *The effect of conductivity on hotspots*. J. Austral. Math. Soc. B. 33:403-413 (1992).

MODEL MANIFOLDS FOR ANALYSING DATA OF WAVE SPLITTING

E. VAN GROESEN † AND W.L. IJZERMAN
University of Twente, Department of Applied Mathematics
P.O. Box 217, 7500 AE Enschede, The Netherlands

Abstract. When large amounts of data are available, such as from numerical calculations of a physical-technical process, insight in the underlying process can be gained if the data are used to specify adjustable parameters in a preliminary model of the phenomenon. Information from the signal can thus be obtained that would otherwise be difficult to detect. A general approach to the modelling of data is presented in this paper, and is illustrated for numerical data of the splitting of a wave due to bottom variations.

1. Introduction and motivation

In analysing a spatial-temporal signal, the *method of principal components* , also called POD: proper orthogonal decomposition, is often used nowadays (Aubry *et al*, 1991; Berkooz *et al*, 1993; Sirovich *et al*, 1989). This method has proven to be useful in various cases when the signal is generated by a system with a (low-dimensional) attractor; the spatial components determine the geometry of the attractor, while the temporal components describe the evolution on the attractor.

This method uses the signal as its only input: no knowledge of the underlying system is required. If some (qualitative) knowledge of the system is known, it can be expected that this knowledge can be used profitable in the analysis of the signal. In particular the signal can be used in the "calibration" of parameters that appear in an adjustable preliminary model:

†Part of the research is sponsored by the Commission of the European Communities, Directorate General XII-B, Joint Research Project CI1*-CT93-0018, and by the Royal Netherlands Academy of Arts and Sciences in the programma for scinetific cooperation Netherlands-Indonesia.

the signal will determine the evolution of the parameters.

To be more specific, we start to review briefly the method of principal components.
Let the signal be given by $s(x,t)$ where x is the spatial variable and t the time; it is no restriction to assume that s is a scalar quantity. Then N spatial components (topos) $\phi_k(x), k = 1 \dots N$ and temporal components (chronos) $a_k(t), k = 1 \dots N$ are sought from the minimization problem

$$\text{Min} \left\{ \int\int [s(x,t) - \Sigma a_k(t)\phi_k(x)]^2 dxdt \mid a_k(t), \phi_k(x), k = 1 \dots N \right\}.$$

The spatial components are normalized, and additional conditions like boundary conditions may be imposed.
The solution of this minimization problem, say $\Sigma a_k^*(t)\phi_k^*(x)$, is then the ($N$-dimensional) representation of the signal.
Remarks

1. The components can also be found in another way: the spatial (temporal) components are the eigenfunctions of the integral operator with the temporal (spatial) correlation function as kernel. For instance, the first spatial component ϕ_1 is the solution of the maximization problem

$$\text{Max} \left\{ \int\int C(x,y)\phi(x)\phi(y)dxdy \mid \int \phi^2(x)dx = 1 \right\} \tag{1.1}$$

   with C the temporal correlation function

$$C(x,y) = \int s(x,t)s(y,t)dt \tag{1.2}$$

   The maximizing characterization gives the interpretation of $\phi_1^*(x)$ as being the time-averaged main spatial structure that is present in the signal. The corresponding temporal component $a_1^*(t)$ is the projection of the signal along $\phi_1^*: \quad a_1^*(t) = \int s(x,t)\phi_1^*(x)dx$.
2. The other spatial components can be found in a successive way by maximizing over the orthogonal complement. These components characterize in decreasing order of importance the remaining spatial structures in the signal.
3. The number N that is required to represent the signal in a sufficient degree of accuracy (for instance to capture 95% of the energy) is an indication of the complexity of the signal and, under some ergodicity properties, of the complexity of the attractor of the underlying system; the spatial components then represent the attractor.

For the following it is useful to interpret the method of principal components as the search for spatial components ϕ_k that are such that the signal $s(x,t)$ can be represented as good as possible by a trajectory $t \mapsto u(t)$ in their linear span

$$M = \{u = \sum_{k=1}^{N} a_k \phi_k(x) \mid a_k \in \mathcal{R} \}.$$

The measure to determine the optimal trajectory is the spatial-temporal L_2-norm (denoted by $|||\ldots|||$) of the difference of the signal from the trajectory:

$$\text{Min } \{ \; |||s - u(t)||| \;\mid\; u(t) \in M \;\}.$$

This interpretation allows for a simple modification when some information can be extracted a priori from the underlying system, or, more generally, when some qualitative information is known about the signal. In fact, suppose that a manifold $\mathcal{M}$ can be defined consisting of (spatial) structures that are expected to be present in the signal dominantly; then a trajectory in $\mathcal{M}$ will be sought to represent the signal by minimizing some distance function between the trajectory and the signal.

Both the choice of the manifold $\mathcal{M}$ and the choice of distance function are essential in the resulting representation of the signal. The manifold $\mathcal{M}$ will be called the *model manifold*; the choice of the distance function is equivalent to specifying a projection onto this manifold.

In the next section we present a simple example to motivate the necessity of looking for modifications of the principal component method described above. The example is a single wave travelling at constant speed; we propose a modification that determines both the profile and the speed. In section 3 we consider the much more difficult case of a signal that describes the splitting of one wave into two waves with different amplitudes and speeds. In the final section we make some remarks and formulate conclusions.

2. Reconstruction of a travelling wave

Consider a signal of a travelling wave of permanent wave profile f that is translated with a fixed speed c:

$$s(x,t) = f(x - ct). \tag{2.3}$$

Analytic consideration

First assume that the signal is known for all x and t; then the temporal correlation function becomes a function of the difference $C(x,y) = C_0(x - y)$, and the corresponding integral operator is a convolution; the eigenvalue problem for the spatial principal components becomes

$$\int C_0(x-y)\phi(y)dy = \mu\phi(x),$$

or after Fourier transformation (denoted by hats)

$$\hat{C}_0(k)\hat{\phi}(k) = \mu\hat{\phi}(k).$$

This leads to a continuous spectrum with harmonic functions as principal components:

$$\phi(x) = e^{ikx} \text{ with } \mu = \hat{C}_0(k) = |\hat{f}(k)|^2 \tag{2.4}$$

The conclusion is that the direct application of the method of principal components does not lead to a simple reconstruction of the signal.

Practical reconstruction
Practically speaking, the data of the wave are not available for all x and t and $C(x,y) \neq C_0(x-y)$. When the above method is then applied to the signal, the result will depend in an essential way on the length of these intervals.

By way of demonstration consider a wave profile of the form:

$$f(x) = \exp(-x^2/2)$$

with x in a spatial interval which is large enough to have no influences of the boundaries and $t \in [-T,T]$. For $T = 10, 20$ and 30 the first principal component ϕ_1 is shown in Fig 1.a. Its support increases with T; for $T \to 0$, $\phi_1 = f$ and for $T \to \infty$ we get (2.4).
In Fig 1.b the number of modes necessary to cover a fraction of the "energy" (L^2-norm) is shown for $T = 30$. The 95 % criterion indicates that the signal is rather complex, which is certainly not the case.

The conclusion is that a direct application of the method of principal components is *not suited* to detect the wave profile or velocity from a signal consisting of a translating wave at constant speed.

Modification
The performance of the POD method can be improved using the qualitative information that the signal represents some travelling wave. Considering the wave in a moving frame of reference separates the translation symmetry from the other (trivial) dynamics.

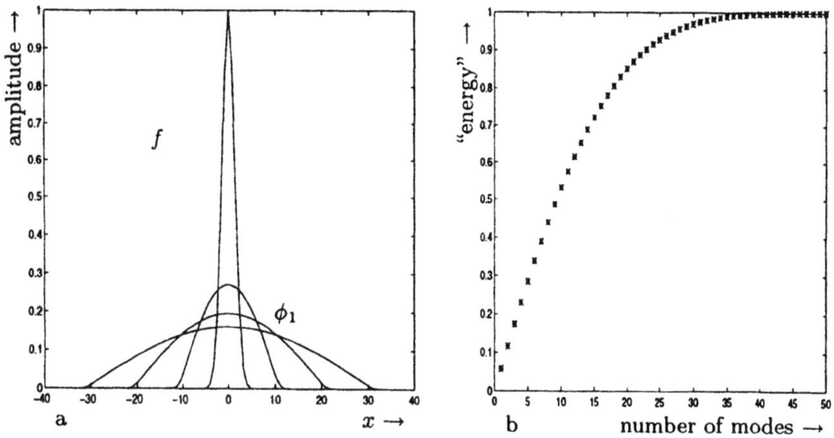

Figure 1. In a. we see the original signal f, and with an increasing support the principal topos ϕ_1 calculated from data for which T=10,20 and 30 respectively. In b. we see the number of modes necessary to cover a certain fraction of the "energy" of the signal when $T = 30$.

In a coordinate system moving with velocity λ the spatial autocorrelation function of the wave (2.3) reads:

$$C(x, y; \lambda) = \int f(x - (c - \lambda)t) f(y - (c - \lambda)t) dt \tag{2.5}$$

For $\lambda = c$ this reduces to $C(x, y; c) = \mathcal{T} f(x) f(y)$ and the corresponding principal components read:

$$a_1 = \mathcal{T}\|f\|, \phi_1 = f/\|f\|, \quad \text{and} \quad a_i = 0 \qquad \text{for } i \geq 2,$$

where $\mathcal{T}$ is the length of the time interval. For $\lambda \neq c$ we deal with a travelling wave with velocity $c - \lambda \neq 0$ and the POD method gives the same complications as described above.
A priori the velocity is not known and has to be determined form the signal too. To motivate the procedure, suppose the wave is known for all x and $t \in [-T, T]$ with $T < \infty$ and that it has compact support. There holds

$$\frac{1}{2T} \int_{-T}^{T} \left[\int u(x + \lambda t, t) \phi(x) dx \right]^2 dt \leq \int f^2(x) dx$$

for all λ and functions ϕ with $\int \phi^2 dx = 1$. Equality holds when $\lambda = c$ and $\phi = f(x)/\|f\|$, for $\lambda \neq c$ strict inequality holds. This shows that the velocity and the first principal component are characterized by:

$$\max_{\lambda, \phi} \left\{ \int \left[\int u(x + \lambda t, t) \phi(x) dx \right]^2 dt \mid \int \phi^2(x) dx = 1 \right\} \tag{2.6}$$

which is a generalization of (1.1) with (1.2) replaced by (2.5).
For a numerical treatment of (2.6) we have to discretise the signal; for example $u(x,t)$ becomes $u_{ij} = u(x_i, t_j)$ and ϕ becomes $\phi_i = \phi(x_i)$. The finite dimensional equivalent of (2.6) then reads

$$\max_{\lambda, \underline{\phi}} \left\{ \underline{\phi}^* A(\lambda) \underline{\phi} \mid \underline{\phi}^* B \underline{\phi} \right\},$$

where $A(\lambda)$ is a matrix which components depend on λ and u_{ij}.
Solving (2.6) corresponds to the maximization of the eigenvalue $\mu(\lambda)$ of the generalized eigenvalue problem:

$$A(\lambda)\underline{\phi} = \mu(\lambda) B \underline{\phi}.$$

This largest eigenvalue can be obtained without calculation of $\underline{\phi}$ and smaller eigenvalues by use of the bisection or Given's method (Wilkinson *et al*, 1971).
Implementation of the above method has been done using piecewise linear basefunctions (IJzerman, 1996). The velocity and wave profile could be obtained up to machine precision when the grid size of the data is small enough.

3. Analysis of wave splitting

In this section we consider a substantially more difficult problem. The signal consists of numerical data of the following characteristic phenomenon in wave dynamics.
On a layer of water of constant depth, the theory of free surface waves predicts the existence of solitary waves: a single hump wave that is practically confined (exponentially decaying) and translates at a constant speed. Actually, a whole family of such waves exists: for every amplitude the solitary wave has a specific width and velocity:

$$u(x,t) = S(h, \lambda t, a)(x)$$

where S denotes the spatial profile of amplitude a on a layer of depth h and with its top at λt; the speed λ depends on a. In a certain approximation (KdV) the profile function is given by

$$S(h, \phi, a) = a \operatorname{sech}^2[b(x - \phi)] \tag{3.7}$$

where the width b is related to a by (Van Groesen, 1996)

$$3a = 4b^2 h^3 \tag{3.8}$$

Starting with an initial solitary wave above a flat bottom, the wave will deform when it runs into a region of different depth. In particular, when the depth decreases, nonlinear effects will become more dominant than dispersive effects, and the wave may undergo changes that resemble the splitting into several distinct waves. In an analytic approximation, it has been predicted that under certain circumstances a second wave will be formed at the back of the first wave (Van Groesen, 1996; Van Groesen and Pudjaprasetya, 1993).
In this section we will verify this splitting process by analysing a signal that consists of numerical data. These data are obtained by calculating the evolution of a solitary wave during run up from the exact water wave equations with a boundary element method, see (Van Daalen *et al*, 1996). In Fig 2 the deformation of the wave is shown graphically; the appearance of a second wave is clearly visible.

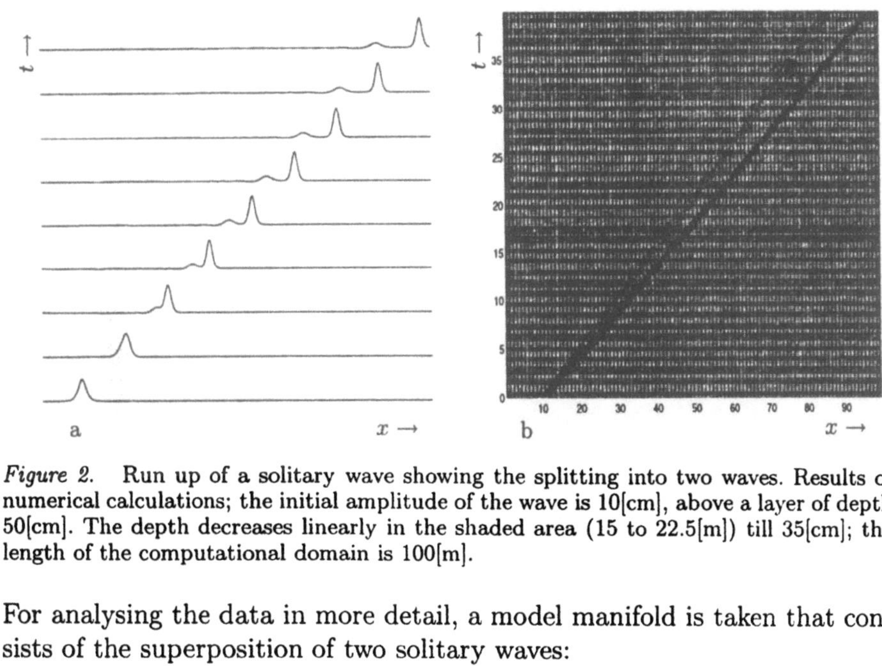

Figure 2. Run up of a solitary wave showing the splitting into two waves. Results of numerical calculations; the initial amplitude of the wave is 10[cm], above a layer of depth 50[cm]. The depth decreases linearly in the shaded area (15 to 22.5[m]) till 35[cm]; the length of the computational domain is 100[m].

For analysing the data in more detail, a model manifold is taken that consists of the superposition of two solitary waves:

$$\mathcal{M} = \{ \; u(a_{+,-}, b_{+,-}, \phi_{+,-}) = a_+ \operatorname{sech}^2[b_+(x - \phi_+)] + a_- \operatorname{sech}^2[b_-(x - \phi_-)] \;\mid\; a_{+,-}, b_{+,-}, \phi_{+,-} \; \} \tag{3.9}$$

The parameters a, b, ϕ determine the amplitude, width and position of each solitary wave respectively. Note that we do not prescribe the amplitude and width to be related by (3.8); see the remarks in the next section. The numerical data are now used to determine the evolution of these parameters:

$$t \mapsto a_{+,-}(t), b_{+,-}(t), \phi_{+,-}(t).$$

The choice of a projection of the data on the manifold $\mathcal{M}$ will determine this evolution, and hence the representation of the phenomenon. We use the simplest choice for this projection: at each instant, the L_2-norm of the difference of the data with the manifold $\mathcal{M}$ is minimized.

The results for the parameter evolution are shown in Fig 3; in Fig 4 the

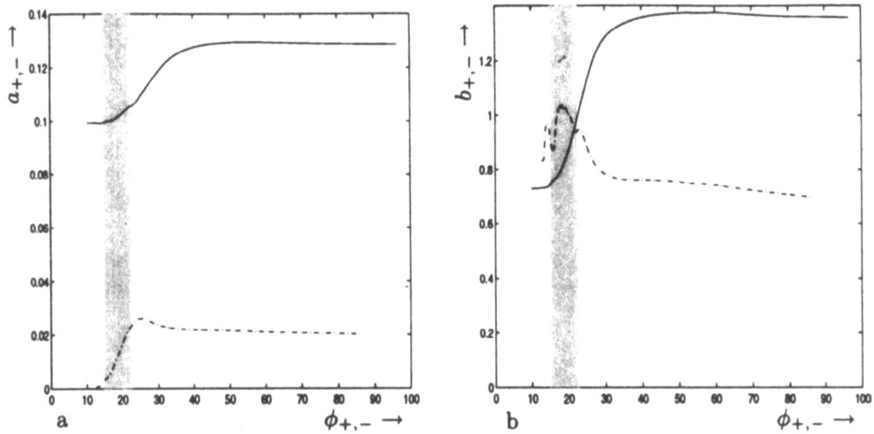

Figure 3. In a. we see the amplitude a_+, (—) and a_-, (– · –·) of the approximations within $\mathcal{M}$ and in b. the width b_+ (—) and b_- (– · –·) as a function of $\phi_{+,-}$.

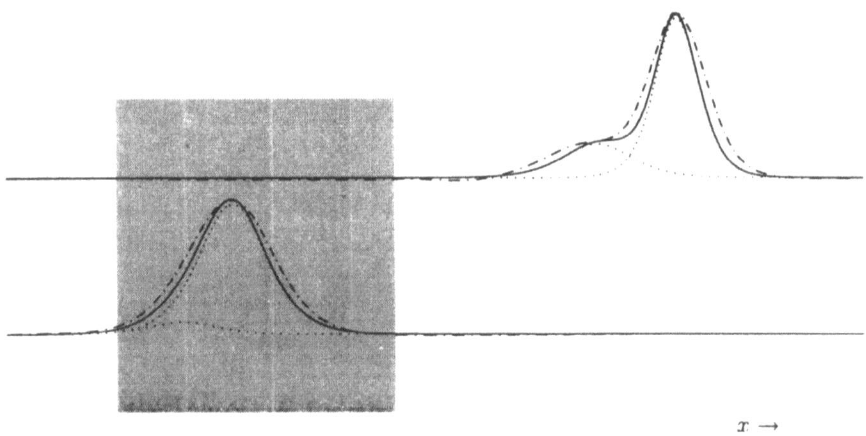

Figure 4. The approximation in the model manifold (—) at an instant during the splitting (at the left) and at an instant when the wave has passed the depth increase compared to the original data (– · –·) . The dotted lines are the constituent waves of the model manifold.

wave profile from the data is compared to the reconstructed element from $\mathcal{M}$ at two characteristic times.

From these results we can draw the conclusion that the reconstruction using the manifold $\mathcal{M}$ is quite successful: the splitting process is quantified in the

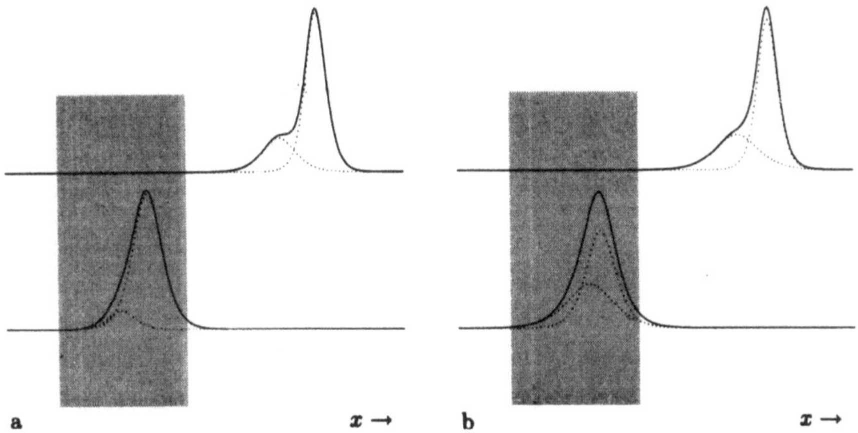

Figure 5. The approximation in the model manifold (—) at an instant during the splitting (at the left) and at an instant when the wave has passed the depth increase. In a. we see the result for the manifold $\mathcal{M}$ and in b. for $\mathcal{M}^*$; the dotted lines are the constituent waves.

evolution of the parameters of the manifold, and the reconstructed profile matches the profile of the data quite well. The evolution of width b_- as shown in Fig 3 is initially rather insecure; this can be expected since then this wave is still of small amplitude; the effect on the total wave profile is negligible. Furthermore, during the splitting process the assumption of the total profile as being a superposition of two waves is rather questionable; see also the next section.

4. Conclusions and remarks

In the previous section, the amplitude and width of the waves that determine the model manifold were taken as independent variables, and were determined independently in the minimization process. When the full free surface wave problem is approximated by a KdV-model, the solitary waves are of this form with amplitude and width related by (3.8). It is tempting to prescribe this relation in the definition of a smaller model manifold:

$$\mathcal{M}^* = \{\, u(a_{+,-}, b_{+,-}, \phi_{+,-}) \in \mathcal{M} \mid a_{+,-} \text{ and } b_{+,-} \text{ related by (3.8)} \,\} \subset \mathcal{M}.$$

This implies that now only $a_{+,-}$ and the positions $\phi_{+,-}$ are the parameters to be adjusted from the signal. The projection of the signal on $\mathcal{M}^*$ leads at first sight to different results: see Fig 6 where $b_{+,-}$ is calculated from $a_{+,-}$ according to relation (3.8). Observe that now the amplitude of the initial wave decreases drastically, while that of the new wave has much larger amplitude than previously. Remarkably enough, although the constituent waves that determine the profile by superposition are rather different for

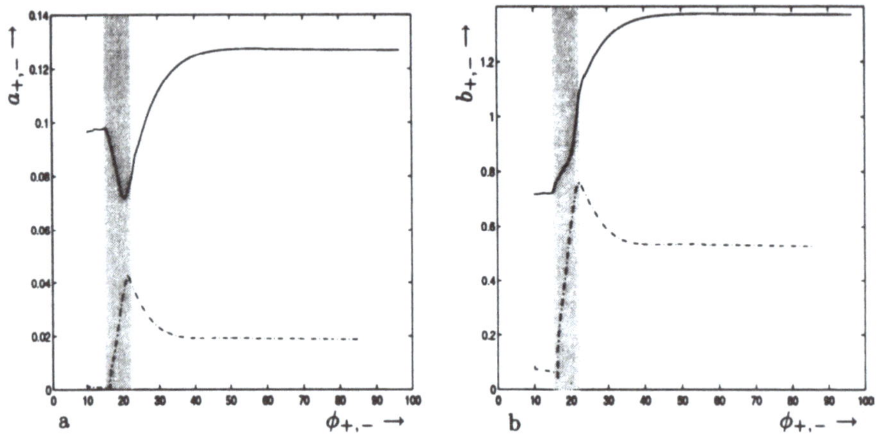

Figure 6. In a. we see the amplitude a_+, (—) line and a_-, (– · –·) line of the approximations within $\mathcal{M}^*$ and in b. the width b_+ and b_- as a function of $\phi_{+,-}$.

the model manifold $\mathcal{M}$ and $\mathcal{M}^*$, the actual superpositions agree quite well and both predict the profile of the signal quite well.

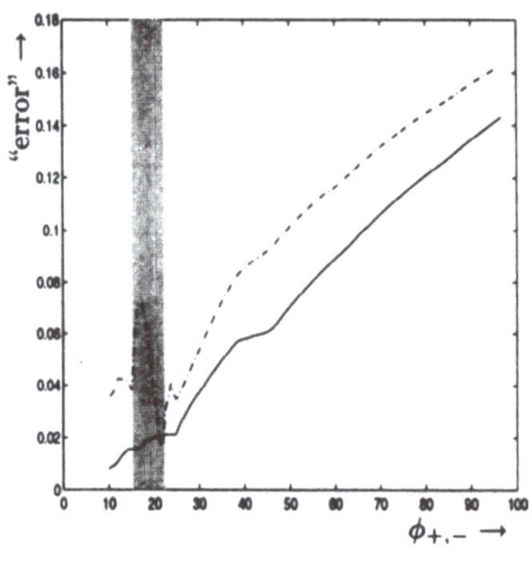

Figure 7. In this figure we see the relative "error" between the data and the approximation within $\mathcal{M}$, (—) and within $\mathcal{M}^*$, (– · –·) .

Fig 5 shows this complication that may arise in other, more complicated, cases as well. This picture shows that one and the same profile can be represented (in the order of approximation that is used) by two elements from the manifold $\mathcal{M}$ that have two different constituent waves but give the same superposition:

$$u(a^1_{+,-}, b^1_{+,-}, \phi^1_{+,-}) = u(a^2_{+,-}, b^2_{+,-}, \phi^2_{+,-}).$$

The difference of the reconstruction in $\mathcal{M}$ and that in $\mathcal{M}^*$ should be a consequence of deviation of an actual solitary wave with the representation (3.7) where the parameters a, b are related to (3.8); freeing the parameters from this relation gives more flexibility ($\mathcal{M}^* \subset \mathcal{M}$) and hence a better approximation [1].

More quantitatively, the validity of the reconstruction can be investigated from Fig 7. The information in this picture should be interpreted with care. First the unexpected (almost linear) increase of the relative "error" has been found to be caused by errors in the numerical calculation of the waves, in particular by an incorrect treatment of the boundary conditions in the numerical grid which causes a non-zero level of the undisturbed water level. The remaining "error" in the reconstruction with the manifold $\mathcal{M}$ can be attributed partly to an inaccurate description of the main waves with a sech^2-profile, and partly to "radiation" that is known to emerge in a splitting process as considered here.

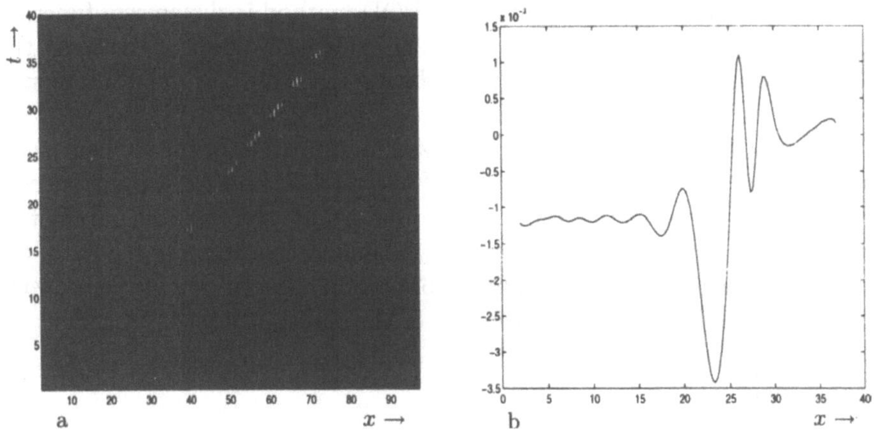

Figure 8. In a. we see the gray scale plot of the residual and in b. the first topos obtained in a moving frame of reference.

This "radiation" shows itself at the downstream side of the two solitary wave: there is a third wave with a maximum amplitude of ≈ 4[mm], ($a_+ \approx$ 13[cm] and $a_- \approx 2$ [cm]). Since this wave is small it follows the linear wave equation and its velocity should be $c = \sqrt{g\,h} = 1.85\ldots$[m/s].
To analyze this wave by use of the modified POD method we considered the residue: the signal minus the approximation in $\mathcal{M}$, see Fig 8.a. The white line from the lower left to the upper right is the third wave, the two

[1]This is verified in a different way by analysing the data of numerical calculations of the wave problem in the approximation for which the solitary waves are exactly given by (3.7); then the reconstruction in $\mathcal{M}$ and that in $\mathcal{M}^*$ agree better.

dark lines on the right are the residues of the solitary waves.
Also reflection is visible in this figure: when the solitary wave passes the uneven bottom, small waves are reflected to the left which are reflected once again at the boundary $x = 0$ of the computational domain. In the gray scale plot these waves are seen as the dark bands on the left of the third wave.
The first (normed) topos is shown in Fig 8.b; the oscillations on the right

Table 1. The estimated velocity from the POD method, for several temporal intervals: $[T, 40]$[s].

T [s]	30	25	20	15	10	5
$\hat{c}$ [m/s]	1.829	1.828	1.828	1.829	1.831	1.833

are the residues of the the larger solitary waves; the waves on the left are probably caused by dispersion.
The velocity obtained by the modified POD method is shown as a function of the temporal interval in Tab. 1. We see that the difference between the obtained and theoretical predicted velocity is within 2%, for all intervals used.
Concluding we can say that the proposed modification of the POD method is quite successful: the construction of the model manifold for the specific example of wave splitting allows one to determine the relevant parameters from the data; subtracting this primary phenomenon from the the data, the residue can be analyzed to obtain a more refined description of the contents of the data and a more refined model of the phenomenon.

References

Aubry, N., Guyonnet, R. and Lima R. (1991) Spatiotemporal analysis of complex signals: Theory and applications. *Journal of Statistical Physics*, 64(3/4).

Berkooz, G., Holmes, P. and Lumley, J.L. (1993) Th proper orthogonal decomposition in the analysis of turbulent flows, *Ann. Rev. Fluid Mech.*, volume 25.

Van Daalen, E.F.G., Groesen, E., Pudjapraserya, S.R. (1996) BEM-numerics and KdV-model analysis for solitary wave split-up. to appear in Comp. Mech.

Van Groesen, E. (1996) A phenomenological description of soliton splitting during run up, to appear in Contemp. Math.

Van Groesen, E. and Pudjaprasetya, S. R. (1993) Uni-directional Waves over slowly varying bottom. Part I: Derivation of a KdV-type of equation. *Wave Motion*, 18, 345-370.

Sirovich, L. and Kirby, M. (1989) An eigenfunction approach to large scale transitional structures in jet flow. *Physical Fluids A*, 2(2):127-135.

Wilkinson, J.H., and Reinsch, C. (1971) *Linear Algebra*, volume 2. Springer Verlag, New York.

IJzerman, W.L. (1996) Model Manifolds and the reconstruction of waves, *Master Thesis*, University of Twente.

MODELLING OF SEDIMENT TRANSPORT AND BATHYMETRIC CHANGE IN LEMAH ABANG, JEPARA COASTAL WATERS

S. HADI AND S. PERMONO
Department of Geophysics and Meteorology
Institut Teknologi Bandung
Ganesha 10, Bandung 40132, Indonesia

Abstract. A real time sediment transport and bathymetric change model is developed to simulate short time beach evolution in Lemah Abang, Jepara, Central Java. The inputs for the model are wave condition at the offshore and current field derived from a nearshore current model.

The model equations is solved by finite difference method using staggered mesh scheme. The simulation results of beach change for the period of two weeks and one month are compared with the field data and the results of the model developed by NEWJEC, Japan.

1. Introduction

Utilization of coastal areas of northern coast of Java has steadily been increasing for human activities such as coastal real estate, industry, transportation, aquaculture, and recreation areas. As a result coastal environment has inevitably been changing mainly due to these activities and some time has frequently brought about severe beach erosion.

Significant environmental problems will persist during the accelerated future development of northern coastal areas of Java. Therefore men's impacts on the natural environment due to development of coastal areas should be evaluated and careful long-term assessment to ensure adequate protection should be prepared.

The capability to predict beach change due to natural causes or human activities will facilitate the effort in preventing coastal areas from erosion and siltation. With the progress in computer software and hardware, numerical models have been extensively used in beach change prediction.

The present paper describes numerical model of sediment transport and bathymetric change. The model is based on the predictive formulation developed by Watanabe, et.al. (1986) and then applied in Lemah Abang, Jepara, Central Java, coastal waters.

2. Model Equations

Numerical model for beach change prediction which is described in this paper simulates sediment transport and bottom topography changes caused by local changes in waves and current. The model consists of two sub models i.e. sub model for wave-induced currents, and sub model for sediment transport and bathymetric change. This paper mainly focuses on the second sub model namely sediment transport and bathymetric change model. Detail explanation about wave-induced current model can be found in Watanabe (1982, 1988).

The change in local bottom elevation, $\mathbf{Z}_b$, or water depth, h, is determined by solving the conservation equation for sediment mass (Watanabe et.al. 1986) :

$$\frac{\partial Z_b}{\partial t} = -\frac{\partial h}{\partial t} = -\frac{\partial q_x}{\partial x} - \frac{\partial q_y}{\partial y} \tag{1}$$

where qx and qy are the component of sediment transport rate per unit width in the x and y direction, respectively. x coordinate is perpendicular to the shore line while y coordinate parallel to the shoreline.

Sediment transport is produced both by waves and currents. Watanabe et. al. (1986) employed sediment transport formula as :

$$(q_x, \ q_y) = (q_{cx}, \ q_{cy}) + (q_{wx}, \ q_{wy}) \tag{2}$$

where :

(q_{cx}, q_{cy})	is the transport rate due to mean current
(q_{wx}, q_{wy})	is the transport rate due to wave

The transport rate due to mean current is determined by

$$q_{cx} = Q_c U \text{ and } q_{cy} = Q_c V$$

$$Q_c = \frac{A_c(\tau - \tau_{cr})}{\rho g} \tag{3}$$

where :

A_c is a non dimensional coefficient (of order 0.1 - 1)

τ is the maximum value of the bottom shear stress in a wave and current coexistent field

τ_{cr} is the critical shear stress for the onset of sediment movement

ρ is the density of water

g is the gravitational acceleration

Q_c is the volume of sediment set in motion per unit area

If $\tau \leq \tau_{cr}$, Q_c is zero

U and V are mean current velocity in x and y directions, respectivelly. They are derived from a nearshore current model which is based on:
Equation of motion:

$$\frac{\partial U}{\partial t} + U\frac{\partial U}{\partial x} + V\frac{\partial U}{\partial y} + F_x + M_x + R_x + g\frac{\partial \bar{\zeta}}{\partial x} = 0 \tag{4}$$

$$\frac{\partial V}{\partial t} + U\frac{\partial V}{\partial x} + V\frac{\partial V}{\partial y} + F_y + M_y + R_y + g\frac{\partial \bar{\zeta}}{\partial y} = 0 \tag{5}$$

and Continuity equation:

$$\frac{\partial \bar{\zeta}}{\partial t} + \frac{\partial (h + \bar{\zeta})U}{\partial x} + \frac{\partial (h + \bar{\zeta})V}{\partial x} = 0 \tag{6}$$

where :

$\bar{\zeta}$ is water elevation

Fx and Fy are bottom friction in x and y directions, respectivelly

Mx and My are lateral mixing in x and y directions, respectivelly

Rx and Ry are radiation stress in x and y directions, respectivelly

The expression for estimating the rate of sediment transport due wave motion is given by

$$q_{wx} = F_d \; Q_w \; \hat{u}_b \cos\alpha$$

$$q_{wy} = F_d \; Q_w \; \hat{u}_b \sin\alpha$$

$$Q_w = A_w(\tau - \tau_{cr})/\rho g \tag{7}$$

where :

$\hat{u}_b$ is the amlitude of near-botton wave orbital velovity

F_b is a transport direction function

$$F_d = \tanh(k_d \frac{\Pi_c - \Pi}{\Pi_c})$$

$\prod$ is criteria for the local cross-shore sediment transport direction

$$\prod = \frac{\hat{u}_b}{sgd}\frac{h}{L_o} < \prod_c \ : \ \text{onshore transport}$$
$$> \prod_c : \ \text{offshore transport}$$

d is the diameter of sediment particle

s is defined as : $s = \frac{\rho_s - \rho}{\rho}$; ρ_s is the density of sediment

Lo is the deep water wave length

$\prod_c$ is a critical value of $\prod$ at the null-point where the net rate of transport is zero. The value of $\prod_c$ is determined empirically through trial computation

k_d is the coefficient (of order 1) that controls the degree of change in the cross-shore transport rate around the null-point.

A_w is the non didimensional coefficient (of order 0.2-0.9)

To incorporate the bottom slope, Watanabe et.al. (1986) modified equation (1) into

$$\frac{\partial z_b}{\partial t} = -\frac{\partial}{\partial x}(q_x - \varepsilon_s |q_x| \frac{\partial z_b}{\partial_x}) - \frac{\partial}{\partial y}(q_y - \varepsilon_s |q_y| \frac{\partial z_b}{\partial_y}) \tag{8}$$

where ε is a positive constant; its value is determined empirically.
By defining:

$q'_x = (q_x - \varepsilon_s |q_x| \frac{\partial z_b}{\partial x})$ and

$q'_y = (q_y - \varepsilon_s |q_y| \frac{\partial z_b}{\partial y})$

equation (8) can be written as

$$\frac{\partial z_b}{\partial t} = -\frac{\partial}{\partial x}(q'_x) - \frac{\partial}{\partial y}(q'_y) \tag{9}$$

which is similar to equation (1).

Equation (8) is solved numerically using finite difference method. A staggered meshed scheme is used, in which the bottom elevation, Z_b, and sediment transport rates (q_x, q_y) are defined at staggered points (see Figure 1).

The difference equation of elevation (5) is given

$$z'_{bi,j} = z_{bi,j} - \frac{\Delta t}{\Delta s}[(q_{xi+1,j} - q_{xi,j})$$
$$-\varepsilon_s \left\{ |q_{xi+1,j}| \frac{z_{bi+1,j} - z_{bi,j}}{\Delta s} - |q_x, j| \ \frac{z_{bi,j} - z_{bi-1,j}}{\Delta s} \right\} + (q_{yi,j+1} - q_{yi,j})$$

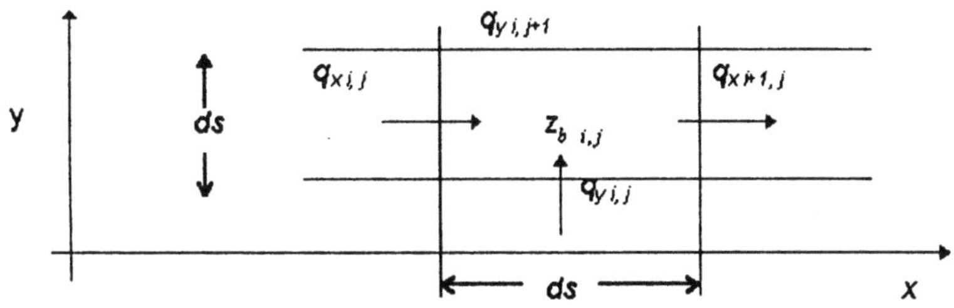

Figure 1. Staggered Mesh Scheme

$$-\varepsilon_s \left\{ |q_{yi,j+1}| \frac{z_{bi,j+1} - z_{bi,j}}{\Delta s} - |q_{yi,j}| \frac{z_{bi,j} - z_{bi,j-1}}{\Delta s} \right\}] \tag{10}$$

Stability condition that should be satisfied is

$$\Delta t \leq \left\{ \frac{1}{2} \frac{(\Delta s)^2}{\varepsilon_s |q_x|_{\max}}, \frac{1}{2} \frac{(\Delta s)^2}{\varepsilon_s |q_y|_{\max}} \right\} \tag{11}$$

where $|q_x|_{\max}$ and $|q_y|_{\max}$ are the maximum values of $|q_x|$ and $|q_y|$, respectively.

3. Model Application

Lemah Abang coastal waters is chosen as the area for the model application. It is located at $110, 47.45^o$ E and $6, 25.37^o$ S in the northern coast of Central Java, namely in Jepara (see Figure 2.a.). It is a shallow area with the bottom contours relatively parallel to the shoreline (see Figure 2.b.).

The model covers the area of 120 m times 90 m. The mesh size, Δs, is taken to be 2.0 meters. The time step is 60 seconds. The wave comes from the northwest ($\alpha = 29^o$) with the hight of 0.58 meters and the period of 6.0 seconds. The sediment in Lemah Abang are sand with density of 2640 kg.m^{-3} and particle diameter 0.2 mm. The current field is derived from the nearshore current model developed by Hadi, et.al. (1994) based on equations (4) - (6).

4. Result and Discussion

The model is run to simulate bathymetric change of Lemah Abang coastal waters caused by waves and currents. The periods of simulation are taken to be two weeks and one month. The model simulations are then compared to the field observation and the results of the numerical model developed by NEWJEC. The comparison of the model simulations and the field observation for the period of two weeks and one month are shown in

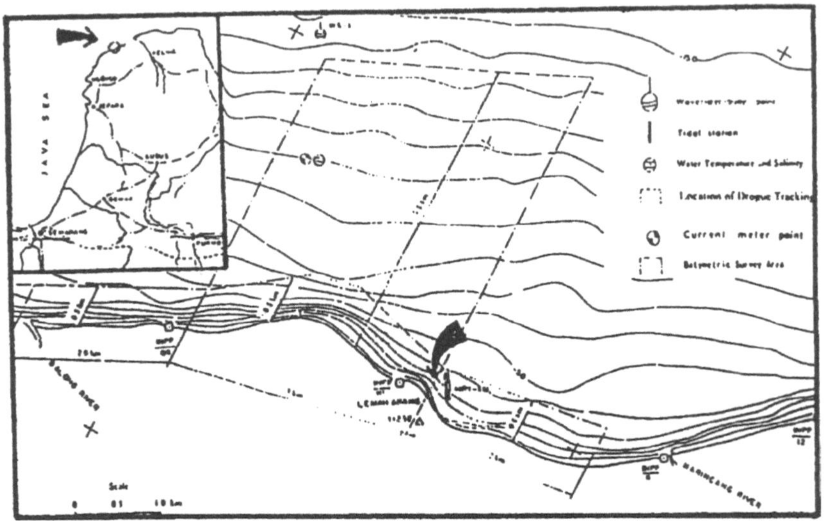

Figure 2.a. Model area and location of field obsevation performed by NEWJEC in january 1994

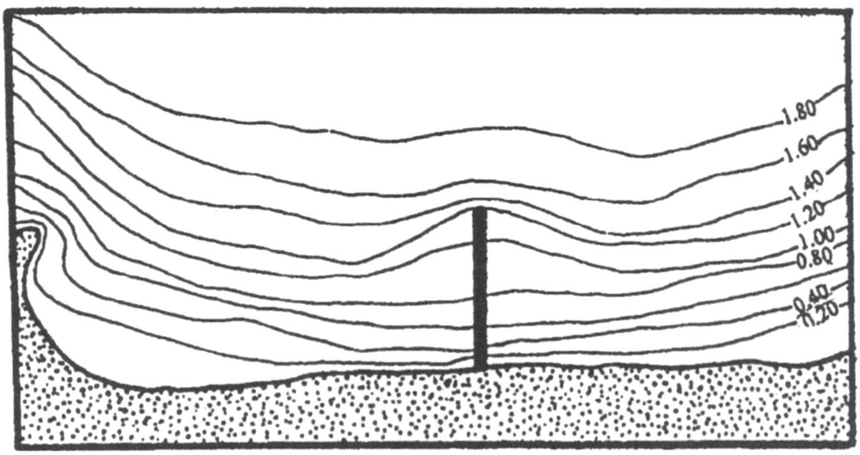

Figure 2.b. Bathymetry of Lemah Abang coastal water around a temporary groin

Figures 3.a., 3.b., 3.c., and 3.d. The simulation result of the NEWJEC's model for the period of one month is given in Figure 3.e. From Figures 3.a. through Figure 3.e. one can see that, in general, the model developed in this work and the NEWJEC's model shows a good agreement with the observed data. The waves that come from the northwest direction generate longshore current which moves the sediment toward the east as shown in Figure 4. The eastward movement of the sediment is hindered by a groin located in the middle of the area. As a result, siltation occurs at the left foot of the groin while erosion occurs at its right foot. These sedimentation and erosion processes are nicely simulated by the model. By comparing Figures 3.b. and 3.d. one can see the outward movement of the shoreline at the left foot of the groin and the retreat at the right foot.

The pattern of the bottom topography simulated by the model is not quite agree with field data at the western part of the area. This might be due to the eddy that is generated at the western side of the area (see Figure 4). The effect of this eddy to sediment transport and bathymetric change at the western side is not really shown by the field observation.

The pattern of bathymetric change outside of the surfzone simulated by the model also does not quite agree with the observation. This might be due to tidal current which does not taken into account in the model.

The model is able to simulate the features of the bathymetric change in Lemah Abang coastal waters especially in the surfzone and at the areas around the groin.

5. Conclusion

In general, the model of sediment transport and bathymetric change developed in this work is able to simulate short time beach change due to the effects of waves and currents in Lemah Abang coastal waters. It also shows a good agreement with the result of the NEWJEC's model.

The model, however, needs further development by including the terms which represent the effects of wind and tidal current to the sediment transport in order to give better prediction of bathymetric change outside of the surfzone.

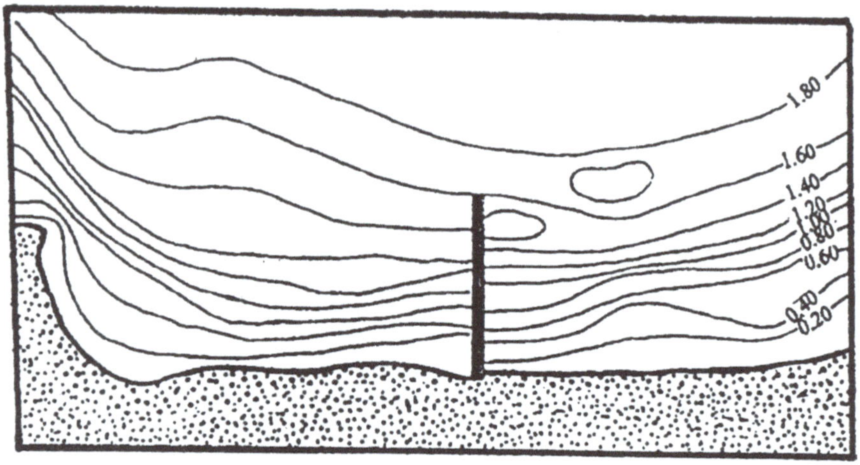

Figure 3.a. Field data of bathymetric change after 2 weeks observation

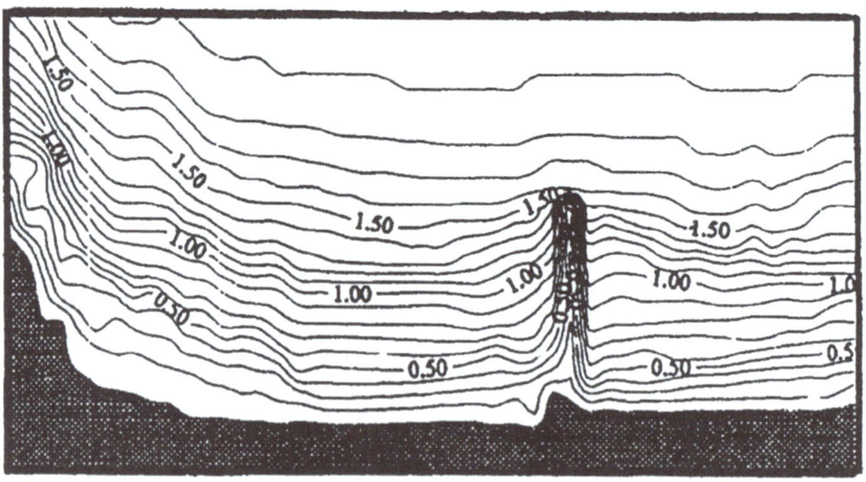

Figure 3.b. Model simulation bathymetric change for 2 weeks

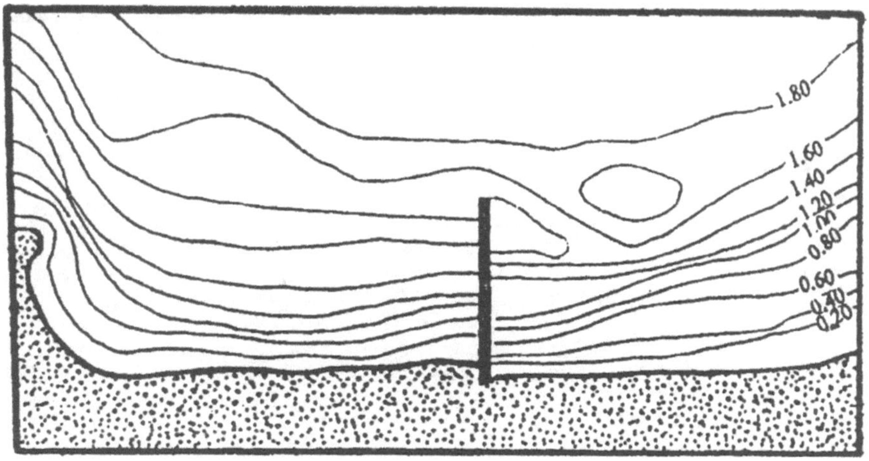

Figure 3.c. Field data of bathymetric change after one month

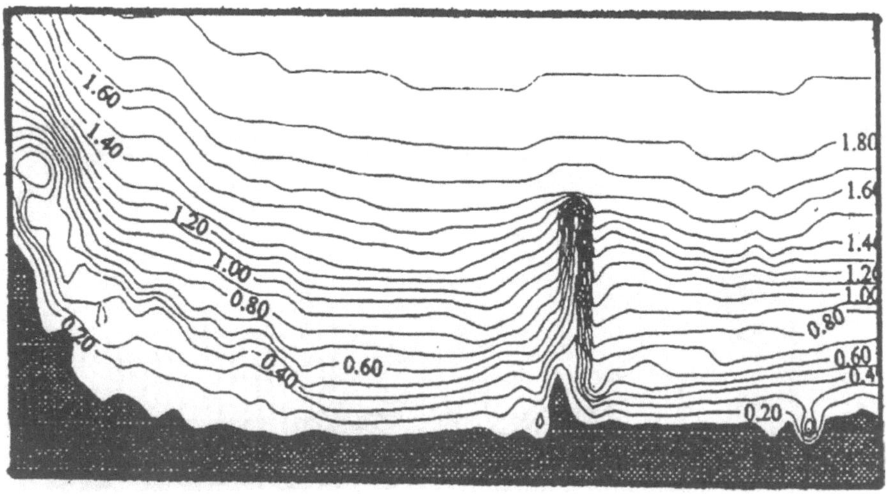

Figure 3.d. Model simulation of bathymetric change for one month

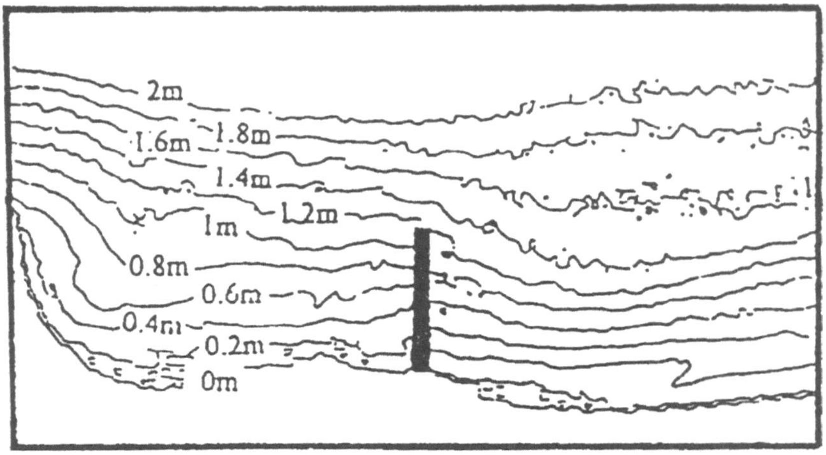

Figure 3.e. Result of NEWJEC's model simulation of bathymetric change for one month

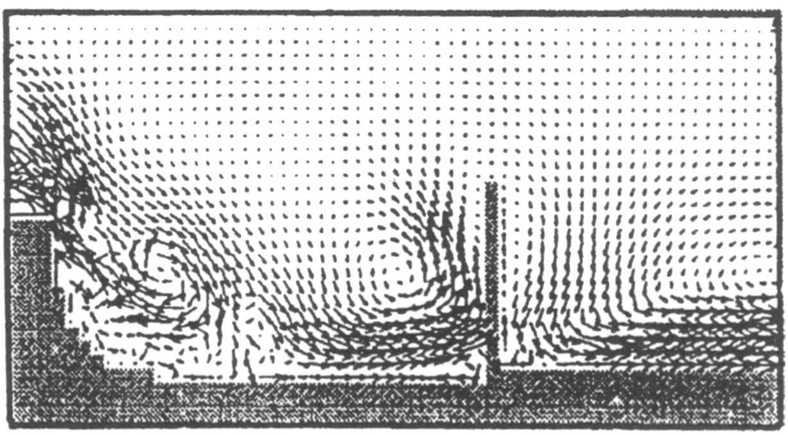

Figure 4. Simulation of current patern around a temporary groin

Acknowledgements

The authors wish to thank the Directorate General of Higher Education, Department of Education and Culture, for providing the financial support to this work.

References

1. Hadi, S. et.al., *"Modelling of Wave-induced Current; Case Study Anyer-Carita Coastal waters"*, Technical Report of University Competitive Research II/2, 1994/95 (In Indonesian).
2. Watanabe, A., *"Numerical Models of Nearshore Currents and Beach Deformation"*, Coastal Eng. in Japan, Vol 25, pp. 147-161, 1982.
3. Watanabe A. in Horikawa, K., *"Nearshore Dynamics and Coastal Processes, Theory, Measurement, and Predictive Models"*, University of Tokyo Press, 1988.
4. Watanabe, A. and K. Maruyama, T. Shimizu, and T. Sakakiyama, *"Numerical Prediction Model of Three Dimensional Beach Deformation around a Structure"*, Coastal Eng. in Japan, Vo. 29, pp.179-194, 1986.

A DOUBLE PENDULUM IN A WIND FIELD

The non-resonant case

M.J. HUISKES AND A.H.P VAN DER BURGH
Department of Technical Mathematics and Informatics,
Delft University of Technology, P.O. Box 5031, 2600 GA Delft,
The Netherlands.

Abstract. In this paper an aeroelastic oscillator of a double pendulum type is studied by using a quasi-steady approach to model the flow-induced vibrations. The attention is focused to the analysis of the qualitive changes in the structure's dynamics for increasing flow velocity. By making extensive use of computer algebra techniques new results have been obtained for the description of the so-called jump phenomenon in this system with two degrees of freedom.

1. Introduction

We consider the galloping behaviour of a two degrees of freedom aeroelastic oscillator of which one degree of freedom interacts directly with a steady flow. The oscillator consists of a so-called seesaw oscillator with an extra

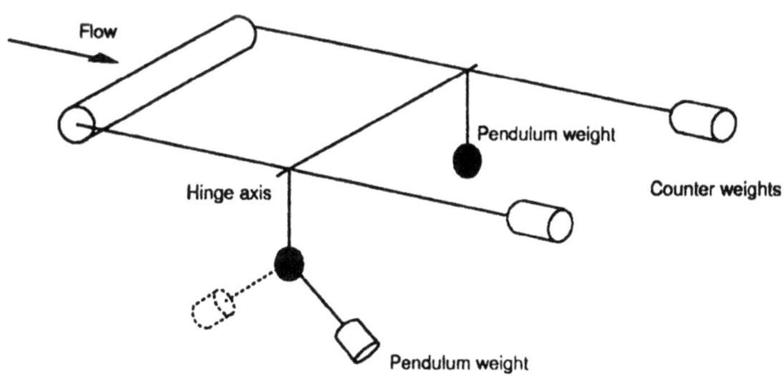

Figure 1. Seesaw-pendulum oscillator.

pendulum attached to its pendulum weight. We refer to the entire structure, which behaves as a double pendulum in the absence of fluid forces, as a seesaw-pendulum oscillator. See Fig. 1. The seesaw oscillator consists of a cylinder fixed between the ends of two parallel rigid bars. On the other end of the bars two counterweights balance the cylinder with respect to a hinge axis through the middle of the bars. Two bars with pendulum weights providing a restoring force are fixed perpendicular to the main bars. The fluid forces exerted on the cylinder will depend on the shape of the cylinder's cross-section.

The dynamical behaviour of the seesaw oscillator (i.e. without the additional pendulum) has been investigated for several cross-section shapes in a windtunnel by Van Oudheusden ([12]).

Quasi-steady theory is rather well established for cylinders that perform a pure translational motion with respect to a steady flow ([13]), but this is not the case for cylinders that may rotate about their own axis. The seesaw oscillator model can be regarded as an intermediate step between one degree of freedom translational galloping and two degrees of freedom galloping with uncoupled translation and rotation: its motion combines both translation and rotation while the flow-induced forces can still be modelled using quasi-steady theory ([11]).

Another important characteristic of the seesaw oscillator is that it allows very low natural frequencies ([4, 11]). This is important in preventing the simultaneous occurence of vortex-induced vibrations ([2]).

A detailed mathematical analysis of the seesaw oscillator based on the quasi-steady theory is given by Haaker and Van der Burgh in [7] and Haaker in [8, 9].

The seesaw-pendulum oscillator with an extra pendulum attached to the pendulum weight of the seesaw oscillator was first described by Van der Burgh in [3]. The extra pendulum which does not interact with the flow is used to study the effect of an additional degree of freedom. We will be especially interested in the effects on jump phenomena which have been observed in a number of experiments with one degree of freedom oscillators ([11]).

2. Equations of motion

In order to obtain equations of motion for the seesaw-pendulum oscillator the well-known Lagrangian method is used, as described in for instance [5]. Fig. 2 shows a cross-section in which generalized coordinates θ and ϕ are indicated. The arm length of the bars from the cylinder to the hinge axis is equal to the arm length of the bars from the counterweights to the hinge axis and is denoted by R. The cylinder's mass is equal to the counterweights' mass and is denoted by M_c.

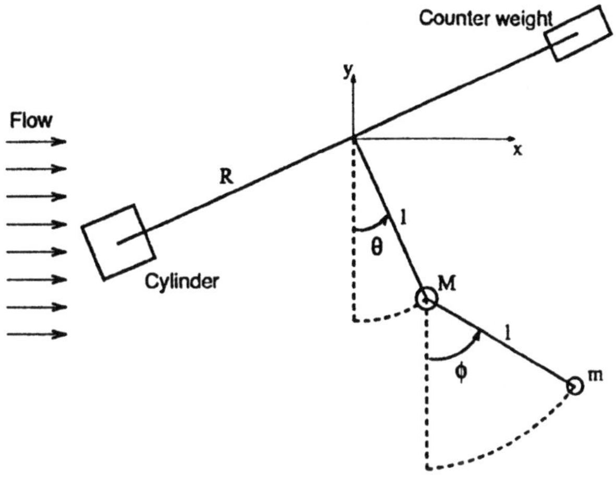

Figure 2. **Generalized coordinates for the seesaw-pendulum oscillator.**

The pendulum mass and bar length of the seesaw oscillator are M and l, respectively. Mass and length of the additional pendulum are denoted by m and l, respectively.

Expressions for kinetic and potential energy in terms of the generalized coordinates are given by

$$T = \frac{1}{2}I\dot{\theta}^2 + \frac{1}{2}ml^2(\dot{\theta}^2 + 2\dot{\theta}\dot{\phi}\cos(\theta - \phi) + \dot{\phi}^2) \quad (1)$$

$$V = Mgl(1 - \cos\theta) + mgl(2 - \cos\theta - \cos\phi) \quad (2)$$

where $I = Ml^2 + 2M_cR^2$ is the moment of inertia with respect to the hinge axis. Using Lagrange's equations of motion we find

$$(I + ml^2)\ddot{\theta} + ml^2\cos(\phi - \theta)\ddot{\phi} - ml^2\dot{\phi}^2\sin(\phi - \theta) + (M + m)gl\sin\theta = Q_\theta \quad (3)$$

$$ml^2\cos(\phi - \theta)\ddot{\theta} + ml^2\ddot{\phi} + ml^2\dot{\theta}^2\sin(\phi - \theta) + mgl\sin\phi = 0 \quad (4)$$

in which Q_θ is a generalized force that will be used to incorporate the force exerted by the flow on the cylinder and a small linear viscous damping force. Damping forces on the additional pendulum are neglected.

The seesaw-pendulum oscillator without flow-induced forces and damping is a conservative system which can adequately be described in the neighbourhood of its neutral equilibrium position by retaining only linear terms in the lefthandsides of (3) and (4). Setting $k_1^2 = \frac{Mgl}{I}$, $k_2^2 = \frac{g}{l}$, $n^2 = \frac{k_1^2}{k_2^2} = \frac{Ml^2}{I}$, $\tau = k_2 t$, and $\mu = \frac{ml^2}{I}$ and solving the two equations for $\ddot{\theta}$

and $\ddot{\phi}$ we find as equations of motion

$$\ddot{\theta} + (n^2 + \mu)\theta = \mu\phi + \frac{Q_\theta}{Ik_2^2} \tag{5}$$

$$\ddot{\phi} + (1 + \mu)\phi = (n^2 + \mu)\theta - \frac{Q_\theta}{Ik_2^2} \tag{6}$$

where a dot now denotes differentiation with respect to τ.

2.1. QUASI-STEADY MODELLING OF FLUID FORCES

Fluid forces on the structure are modelled by using the quasi-steady theory as described for instance in [2, 11]. In order to be able to define a meaningful average relative velocity of the cylinder with respect to the flow, the ratio of the cylinder cross-section diameter to the seesaw armlength R must be small. It is further assumed that the reduced frequencies of the seesaw-pendulum oscillator are sufficiently small to avoid vortex-induced vibrations.

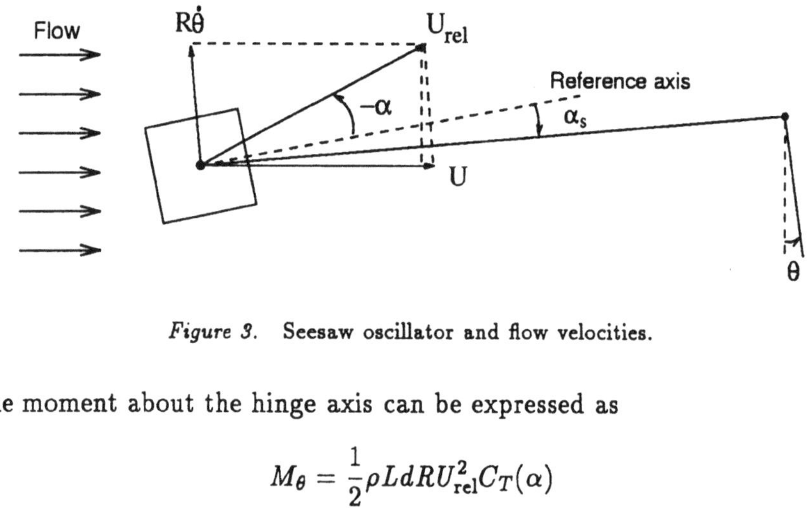

Figure 3. **Seesaw oscillator and flow velocities.**

The moment about the hinge axis can be expressed as

$$M_\theta = \frac{1}{2}\rho L d R U_{\mathrm{rel}}^2 C_T(\alpha) \tag{7}$$

where ρ is the density of the flowing medium, L is the length of the cylinder, d is the cross-section diameter of the cylinder perpendicular to the flow and $C_T(\alpha)$ is a dimensionless aerodynamic coefficient depending on the angle of attack α of the relative flow velocity U_{rel}. For a given reference angle α_s $C_T(\alpha)$ can be expressed in experimentally determined $C_D(\alpha)$ and $C_L(\alpha)$ functions by the relation

$$C_T(\alpha) = C_D(\alpha)\sin(\alpha - \alpha_s) - C_L(\alpha)\cos(\alpha - \alpha_s) \tag{8}$$

where $C_D(\alpha)$ and $C_L(\alpha)$ are the well-known drag- and lift coefficients curves. Fig.3 shows the oscillator in a uniform homogeneous flow with velocity U at an angle θ moving in positive θ direction with velocity $R\dot{\theta}$. This motion induces a flow relative to the cylinder in opposite direction and with the same velocity. The vector sum of this velocity and the uniform flow velocity is the relative flow velocity U_{rel}. From Fig. 3 we can conclude

$$U_{\text{rel}}^2 = (Rk_2\dot{\theta}\cos\theta)^2 + (U - Rk_2\dot{\theta}\sin\theta)^2 \tag{9}$$

and

$$\alpha - \alpha_s = \theta - \arctan\left(\frac{Rk_2\dot{\theta}\cos\theta}{U - Rk_2\dot{\theta}\sin\theta}\right) \tag{10}$$

where a dot denotes differentiation with respect to $\tau = k_2 t$. We approximate $C_T(\alpha)$ locally by

$$C_T(\alpha) = c_1(\alpha - \alpha_s) + c_3(\alpha - \alpha_s)^3 + c_5(\alpha - \alpha_s)^5 + c_7(\alpha - \alpha_s)^7 \tag{11}$$

By using an approximation to the seventh degree we are able to model jump phenomena in the amplitude of the oscillator as we shall show in what follows.

We substitute (9), (11) and (10) in (7) and approximate M_θ by its Taylor series expansion with respect to θ and $\dot{\theta}$ up to terms of seventh degree, giving

$$\frac{M_\theta}{Ik_2^2} = \epsilon f(\theta, \dot{\theta}) \tag{12}$$

in which

$$\begin{aligned} f(\theta,\dot{\theta}) = {} & a_1\,\theta + a_2\,\dot{\theta} + a_3\,\theta^3 + a_4\,\theta^2\,\dot{\theta} + a_5\,\theta\,\dot{\theta}^2 + a_6\,\dot{\theta}^3 \\ & + a_7\,\theta^5 + a_8\,\theta^4\,\dot{\theta} + a_9\,\theta^3\,\dot{\theta}^2 + a_{10}\,\theta^2\,\dot{\theta}^3 + a_{11}\,\theta\,\dot{\theta}^4 + a_{12}\,\dot{\theta}^5 \\ & + a_{13}\theta^7 + a_{14}\theta^6\,\dot{\theta} + a_{15}\theta^5\,\dot{\theta}^2 + a_{16}\theta^4\,\dot{\theta}^3 + a_{17}\theta^3\,\dot{\theta}^4 \\ & + a_{18}\theta^2\,\dot{\theta}^5 + a_{19}\theta\,\dot{\theta}^6 + a_{20}\,\dot{\theta}^7 \end{aligned} \tag{13}$$

and where

$$u_d = \frac{U}{Rk_2} \tag{14}$$

and

$$\epsilon = \frac{\rho d L R^3}{2I} \tag{15}$$

and expressions for $a_2, \ldots, a_{20}$ are given in Table 1.
If we now take

$$Q_\theta = \varepsilon f(\theta,\dot{\theta}) - 2\beta\varepsilon\dot{\theta} \tag{16}$$

a_2	$-c_1 u_d$	a_4	$(-\frac{3}{2}c_1 - 3c_3)u_d$
a_6	$(-\frac{2}{3}c_1 - c_3)\frac{1}{u_d}$	a_8	$(\frac{7}{24}c_1 - \frac{1}{2}c_3 - 5c_5)u_d$
a_{10}	$(c_1 - \frac{1}{2}c_3 - 10c_5)\frac{1}{u_d}$	a_{12}	$(\frac{2}{15}c_1 - c_5)\frac{1}{u_d^3}$
a_{14}	$(-\frac{11}{720}c_1 + \frac{5}{24}c_3 + \frac{1}{2}c_5 - 7c_7)u_d$	a_{16}	$(-\frac{7}{12}c_1 + \frac{17}{8}c_3 + \frac{35}{3}c_5 - 35c_7)\frac{1}{u_d}$
a_{18}	$(-c_1 + \frac{7}{5}c_3 + \frac{25}{2}c_5 - 21c_7)\frac{1}{u_d^3}$	a_{20}	$(-\frac{2}{35}c_1 + \frac{1}{15}c_3 + \frac{2}{3}c_5 - c_7)\frac{1}{u_d^5}$

TABLE 1. Expressions for $a_2, \ldots, a_{20}$

where a small linear damping term proportional to $\dot{\theta}$ with damping coefficient $\beta > 0$ is added to represent viscous damping forces on the cylinder, counter weight and pendulum weight, the equations of motion (5) and (6) for the seesaw-pendulum oscillator can be written as

$$\begin{pmatrix} \ddot{\theta} \\ \ddot{\phi} \end{pmatrix} + \begin{pmatrix} n^2 + \mu & -\mu \\ -n^2 - \mu & 1 + \mu \end{pmatrix} \begin{pmatrix} \theta \\ \phi \end{pmatrix} = \epsilon \begin{pmatrix} f(\theta, \dot{\theta}) - 2\beta\dot{\theta} \\ -f(\theta, \dot{\theta}) + 2\beta\dot{\theta} \end{pmatrix} \qquad (17)$$

By using the transformation

$$\begin{pmatrix} \theta \\ \phi \end{pmatrix} = \begin{pmatrix} 1 & 1 \\ \frac{n^2+\mu-\sigma_1^2}{\mu} & \frac{n^2+\mu-\sigma_2^2}{\mu} \end{pmatrix} \begin{pmatrix} \psi_1 \\ \psi_2 \end{pmatrix} \qquad (18)$$

the linear parts of these equations can be decoupled giving

$$\begin{pmatrix} \ddot{\psi}_1 \\ \ddot{\psi}_2 \end{pmatrix} + \begin{pmatrix} \sigma_1^2 & 0 \\ 0 & \sigma_2^2 \end{pmatrix} \begin{pmatrix} \psi_1 \\ \psi_2 \end{pmatrix} = \epsilon \begin{pmatrix} \kappa_2 \tilde{f} \\ -\kappa_1 \tilde{f} \end{pmatrix} \qquad (19)$$

where $\tilde{f} = f(\psi_1 + \psi_2, \dot{\psi}_1 + \dot{\psi}_2) - 2\beta(\dot{\psi}_1 + \dot{\psi}_2)$ and σ_1 and σ_2 are the natural frequencies of the unforced seesaw-pendulum oscillator, given by

$$\sigma_{1,2}^2 = \frac{1}{2}(n^2 + 2\mu + 1 \pm \sqrt{(n^2 + 2\mu - 1)^2 + 4\mu}) \qquad (20)$$

and

$$\kappa_i = \frac{n^2 + 2\mu - \sigma_i^2}{\sigma_1^2 - \sigma_2^2}, \quad i = 1, 2 \qquad (21)$$

In Fig. 4 the eigenmodes corresponding to σ_1 and σ_2 of the seesaw-pendulum oscillator are sketched.

Figure 4. Eigenmodes of the seesaw-pendulum oscillator.

2.2. AVERAGING

Solutions of (19) for $\epsilon = 0$, i.e. for the reduced problem, can be written as

$$\begin{pmatrix} \psi_1 \\ \dot{\psi}_1 \\ \psi_2 \\ \dot{\psi}_2 \end{pmatrix} = \begin{pmatrix} r_1 \cos(\sigma_1 \tau + p_1) \\ -r_1 \sigma_1 \sin(\sigma_1 \tau + p_1) \\ r_2 \cos(\sigma_2 \tau + p_2) \\ -r_2 \sigma_2 \sin(\sigma_2 \tau + p_2) \end{pmatrix} \tag{22}$$

where r_i and p_i for $i = 1, 2$ are real constants. These solutions can be used as a coordinate transformation for the non-reduced problem from coordinates $(\psi_1, \dot{\psi}_1, \psi_2, \dot{\psi}_2)$ to (r_1, p_1, r_2, p_2), where now $r_i = r_i(\tau)$ and $p_i = p_i(\tau)$ for $i = 1, 2$. This yields

$$\frac{d}{d\tau} \begin{pmatrix} r_1 \\ p_1 \\ r_2 \\ p_2 \end{pmatrix} = \epsilon \begin{pmatrix} \frac{-\kappa_2}{\sigma_1} \tilde{f}(\psi_1 + \psi_2, \dot{\psi}_1 + \dot{\psi}_2) \sin(\sigma_1 \tau + p_1) \\ \frac{-\kappa_2}{\sigma_1 r_1} \tilde{f}(\psi_1 + \psi_2, \dot{\psi}_1 + \dot{\psi}_2) \cos(\sigma_1 \tau + p_1) \\ \frac{\kappa_1}{\sigma_2} \tilde{f}(\psi_1 + \psi_2, \dot{\psi}_1 + \dot{\psi}_2) \sin(\sigma_2 \tau + p_2) \\ \frac{\kappa_1}{\sigma_2 r_2} \tilde{f}(\psi_1 + \psi_2, \dot{\psi}_1 + \dot{\psi}_2) \cos(\sigma_2 \tau + p_2) \end{pmatrix} \tag{23}$$

Since the righthandside of (23) can be expressed as a finite sum of periodic functions, system (23) is in so-called quasi-periodic standard form for averaging. The method of averaging, as described for instance in [14], yields $\mathcal{O}(\epsilon)$ approximations on the timescale $\mathcal{O}(\frac{1}{\epsilon})$ for the normal mode amplitudes. Setting

$$x = r_1^2 \qquad \text{and} \qquad y = r_2^2 \tag{24}$$

we find for the case of no resonance

$$\begin{aligned} \dot{x} &= \epsilon \kappa_2 x((a_2 - 2\beta) + A_1 x + 2A_2 y + A_3 x^2 + A_4 xy \\ &\quad + 3A_5 y^2 + A_6 x^3 + 2A_7 x^2 y + 3A_8 xy^2 + 4A_9 y^3) \end{aligned} \tag{25}$$

$$\begin{aligned}\dot{y} &= -\varepsilon\kappa_1 y((a_2 - 2\beta) + 2A_1x + A_2y + 3A_3x^2 + A_4xy \\ &\quad + A_5y^2 + 4A_6x^3 + 3A_7x^2y + 2A_8xy^2 + A_9y^3)\end{aligned} \tag{26}$$

where expressions for $A_1, \ldots, A_9$ are given in Table 2.

A_1	$\frac{1}{4}a_4 + \frac{3}{4}\sigma_1^2 a_6$
A_2	$\frac{1}{4}a_4 + \frac{3}{4}\sigma_2^2 a_6$
A_3	$\frac{1}{8}a_8 + \frac{1}{8}\sigma_1^2 a_{10} + \frac{5}{8}\sigma_1^4 a_{12}$
A_4	$\frac{3}{4}a_8 + \frac{3}{8}(\sigma_1^2 + \sigma_2^2)a_{10} + \frac{15}{4}\sigma_1^2\sigma_2^2 a_{12}$
A_5	$\frac{1}{8}a_8 + \frac{1}{8}\sigma_2^2 a_{10} + \frac{5}{8}\sigma_2^4 a_{12}$
A_6	$\frac{5}{64}a_{14} + \frac{3}{64}\sigma_1^2 a_{16} + \frac{5}{64}\sigma_1^4 a_{18} + \frac{35}{64}\sigma_1^6 a_{20}$
A_7	$\frac{15}{32}a_{14} + \frac{3}{32}(2\sigma_1^2 + \sigma_2^2)a_{16} + \frac{5}{32}(\sigma_1^4 + 2\sigma_1^2\sigma_2^2)a_{18} + \frac{105}{32}\sigma_1^4\sigma_2^2 a_{20}$
A_8	$\frac{15}{32}a_{14} + \frac{3}{32}(\sigma_1^2 + 2\sigma_2^2)a_{16} + \frac{5}{32}(\sigma_2^4 + 2\sigma_1^2\sigma_2^2)a_{18} + \frac{105}{32}\sigma_1^2\sigma_2^4 a_{20}$
A_9	$\frac{5}{64}a_{14} + \frac{3}{64}\sigma_2^2 a_{16} + \frac{5}{64}\sigma_2^4 a_{18} + \frac{35}{64}\sigma_2^6 a_{20}$

TABLE 2. Expressions for $A_1, \ldots, A_9$

3. Analysis

System (25), (26) describing the normal mode amplitude behaviour is a so-called quartic Kolmogorov system. We first investigate the influence of the uniform flow velocity and damping coefficient β on the galloping amplitude for systems in one of their normal modes which results in a list of possible bifurcation diagrams. Then a Gröbner basis approach is used to investigate mixed mode galloping. In the following subsection realistic values for various physical parameters are listed which are obtained from experiments performed by Van Oudheusden.

3.1. EXPERIMENTAL CONFIGURATIONS

As mentioned before, experiments with a seesaw oscillator were performed in a windtunnel by Van Oudheusden ([11, 12]). Two of his configurations are used, denoted by configuration I and configuration II. The cylinder of configuration I has a 40 × 40 mm square cross-section. The orientation of the cylinder is such that $\alpha_s = 0$, so from (11) we have $C_T(\alpha) = c_1\alpha + c_3\alpha^3 + c_5\alpha^5 + c_7\alpha^7$. Notice that $C_T(\alpha)$ may be used despite the fact that in Van Oudheusden's experiments no additional pendulum was attached to the pendulum weight of the seesaw oscillator. The values obtained for $c_1, \ldots, c_7$

and values for the various other quantities describing the properties of the oscillator configuration are listed in Table 3. Values for k_2, n and ε are obtained from the other listed quantities. The cylinder of configuration II

Quantity	Symbol	I	II	Unit
Cylinder arm length	R	0.30	0.30	m
Cylinder length	L	0.35	0.35	m
Cylinder heigth	d	0.040	0.025	m
Pendulum mass	M	1.72	1.74	kg
Pendulum length	l	0.15	0.15	m
Moment of inertia	I	0.203	0.170	kg m^2
Constant of gravity	g	9.8	9.8	m s^{-2}
Fluid density (air)	ρ	1.2	1.2	kg m^{-3}
Characteristic frequency	k_2	8.1	8.1	s^{-1}
Natural frequency seesaw	n	0.43	0.48	
Perturbation parameter	ε	0.0011	$8.3\ 10^{-4}$	

Parameters aerodynamic moments :

I: $c_1 = -4.82$, $c_3 = 297$, $c_5 = -9104$, $c_7 = 85981$

II: $c_1 = -3.0$, $c_3 = 330$, $c_5 = -34500$, $c_7 = 850000$

TABLE 3. Parameter values configurations I and II.

has a rectangular cylinder cross-section of 25 × 40 mm, with the 25 mm side facing the flow.

3.2. NORMAL MODE BEHAVIOUR

Behaviour of the oscillator in one of its (invariant) normal modes is governed by the equation

$$\dot{x} = \varepsilon\kappa_g x\big((-c_1 u_d - 2\beta_g) + \Omega_1(\sigma_g)\, x + \Omega_2(\sigma_g)\, x^2 + \Omega_3(\sigma_g)\, x^3\big), \quad x \geq 0 \quad (27)$$

in which the natural frequency σ_g and factor κ_g may be chosen as needed to obtain the equation for either the first or second mode. It can also be shown that the dynamics of an oscillator without the additional pendulum is governed by this equation for the choices $\sigma_g = n$ and $\kappa_g = 1$. Expressions for $\Omega_i(\sigma_g)$, $i = 1, 2, 3$ are listed in Table 4.

Ω_1	$\frac{1}{4}a_4 + \frac{3}{4}\sigma_g^2 a_6$
Ω_2	$\frac{1}{8}a_8 + \frac{1}{8}\sigma_g^2 a_{10} + \frac{5}{8}\sigma_g^4 a_{12}$
Ω_3	$\frac{5}{64}a_{14} + \frac{3}{64}\sigma_g^2 a_{16} + \frac{5}{64}\sigma_g^4 a_{18} + \frac{35}{64}\sigma_g^6 a_{20}$

TABLE 4. Expressions for $\Omega_1, \ldots, \Omega_3$.

This equation will be analysed for cylinder cross-sections similar to those of Van Oudheusden in the sense that we assume that the aerodynamic force parameters satisfy

- $c_1 < 0$, $c_3 > 0$, $c_5 < 0$ and $c_7 > 0$
- $\|c_1\| < \|c_3\| < \|c_5\| < \|c_7\|$

If these conditions are met we have the following sign information: $a_2 > 0$ and $a_4, a_6 < 0$ and $a_8, a_{10}, a_{12} > 0$ and $a_{14}, a_{16}, a_{18}, a_{20} < 0$. From this we can conclude that $\Omega_1 < 0$, $\Omega_2 > 0$ and $\Omega_3 < 0$. Scaling time by $\bar{\tau} = \varepsilon\kappa_g u_d \tau$, (27) becomes

$$\dot{x} = x(\,(-c_1 - \frac{2\beta_g}{u_d}) + \frac{\Omega_1}{u_d}x + \frac{\Omega_2}{u_d}x^2 + \frac{\Omega_3}{u_d}x^3), \quad x \geq 0 \tag{28}$$

where a dot now denotes differentiation with respect to $\bar{\tau}$ and we further assume that ε, κ_g and σ_g are fixed. We define

$$f(x, u_d, \beta_g) = x(\,(-c_1 - \frac{2\beta_g}{u_d}) + \frac{\Omega_1}{u_d}x + \frac{\Omega_2}{u_d}x^2 + \frac{\Omega_3}{u_d}x^3\,) \equiv x\,F(x, u_d, \beta_g) \tag{29}$$

so (28) becomes

$$\dot{x} = f(x, u_d, \beta_g) = x\,F(x, u_d, \beta_g), \quad x \geq 0 \tag{30}$$

Onset of galloping

The origin is a critical point of (30) for any choice of parameters. It corresponds to an equilibrium of the oscillator in its neutral position. Its stability is governed by $\frac{\partial f}{\partial x}(0, u_d, \beta_g) = -c_1 - \frac{2\beta_g}{u_d}$. For fixed β_g, (30) undergoes a transcritical bifurcation at $u_d = \frac{-2\beta_g}{c_1} \equiv \bar{u}_d$. For velocities smaller than $\bar{u}_d$ the origin is asymptotically stable. At the critical velocity $u_d = \bar{u}_d$ the origin loses its stability and galloping becomes possible. For varying β_g and fixed u_d a similar critical damping value is found at $\beta_g = \frac{-c_1 u_d}{2}$.

Amplitude vs. flow velocity

We now investigate the effect of an increase of the uniform flow velocity on the galloping amplitude. It will be shown that depending on the shape of the cylinder's cross-section, which is represented by the aerodynamic force coefficients $c_1, \ldots, c_7$, five different amplitude characteristics can be distinguished. In two of these characteristics the so-called jump phenomenon can be observed. Amplitude jumps may occur for different values of the flow velocity depending on whether the velocity is being increased or decreased and can be characterized as a form of hysteresis. Another interesting effect is hard-galloping: for a certain flow velocity the equilibrium position is stable, but for the same flow velocity galloping amplitudes are possible for large enough initial displacements.
Equation (30) depends on u_d in a rather complicated way. Notice that the Ω_i/u_d, $i = 1, 2, 3$ in (28) consist of a constant part and a part of the order $1/u_d^2$. It is assumed that the part of order $1/u_d^2$ may be neglected. Of course, this approach may not be valid for u_d small. In the next subsection we have an interesting way to check how this approach works out for the concrete configurations I and II. The part of Ω_i/u_d that does not vary with u_d is denoted by $\bar{\Omega}_i$. Expressions for $\bar{\Omega}_1$, $\bar{\Omega}_2$ and $\bar{\Omega}_3$ are given in Table 5.

$\bar{\Omega}_1$	$\frac{1}{4}(-\frac{3}{2}c_1 - 3c_3)$
$\bar{\Omega}_2$	$\frac{1}{8}(\frac{7}{24}c_1 - \frac{1}{2}c_3 - 5c_5)$
$\bar{\Omega}_3$	$\frac{5}{64}(-\frac{11}{720}c_1 + \frac{5}{24}c_3 + \frac{1}{2}c_5 - 7c_7)$

TABLE 5. Expressions for $\bar{\Omega}_1, \ldots, \bar{\Omega}_3$.

Taking

$$\bar{F}(x, u_d, \beta_g) = (-c_1 - \frac{2\beta_g}{u_d}) + \bar{\Omega}_1 x + \bar{\Omega}_2 x^2 + \bar{\Omega}_3 x^3 \tag{31}$$

we consider

$$\dot{x} = x\bar{F}(x, u_d, \beta_g) \tag{32}$$

Notice that the function

$$\tilde{F} = -c_1 + \bar{\Omega}_1 x + \bar{\Omega}_2 x^2 + \bar{\Omega}_3 x^3 \tag{33}$$

is independent of u_d. $\bar{F}(x, u_d, \beta_g)$ is obtained from $\tilde{F}(x)$ by a translation of $\frac{2\beta_g}{u_d}$. So to investigate critical points and possible bifurcations for (32) it is sufficient to study all possible shapes of functions $\tilde{F}(x)$. This shape is completely determined by the aerodynamic force coefficients. Although the analysis depends on translations by $\frac{2\beta_g}{u_d}$, we do not consider variations in β_g here. These are considered in the following subsection without making use of (32) instead of (30), so here we fix $\beta_g > 0$. $\bar{F}(x, u_d)$ may have more than one root for some values of u_d only if $\tilde{F}$ has two extreme values. The number of extreme values of $\tilde{F}$ depends on the sign of

$$D = \bar{\Omega}_2^2 - 3\bar{\Omega}_1\bar{\Omega}_3 \tag{34}$$

First consider the case $D > 0$. Extreme values of $\tilde{F}$ occur for

$$x_{\min} = \frac{-\bar{\Omega}_2 + \sqrt{D}}{3\bar{\Omega}_3} \quad \text{and} \quad x_{\max} = \frac{-\bar{\Omega}_2 - \sqrt{D}}{3\bar{\Omega}_3} \tag{35}$$

It is easy to show (using $\bar{\Omega}_1 < 0$, $\bar{\Omega}_2 > 0$ and $\bar{\Omega}_3 < 0$) that we have $0 < x_{\min} < x_{\max}$. Define

$$\tilde{F}_{\min} = \tilde{F}(x_{\min}) \quad \text{and} \quad \tilde{F}_{\max} = \tilde{F}(x_{\max}) \tag{36}$$

The five possible shapes of $\tilde{F}$, denoted by Shape A through E, with respect to the relative positions of $-c_1$, $\tilde{F}_{\max}$, $\tilde{F}_{\min}$ and zero are listed in Fig. 5.

If $\tilde{F}_{\max} > 0$ then for $\tilde{u}_d = \frac{2\beta_g}{\tilde{F}_{\max}}$ we have $\bar{F}(x_{\max}, \tilde{u}_d) = 0$ and it can be shown that (32) undergoes a saddle-node bifurcation at $(x_{\max}, \tilde{u}_d)$. For $\tilde{F}_{\min} > 0$ a similar analysis can be performed. Now a saddle-node bifurcation occurs for $(x_{\min}, \frac{2\beta_g}{\tilde{F}_{\min}})$ at which two critical points disappear for increasing u_d.

In Fig. 5 bifurcation diagrams of equilibrium amplitude $r = \sqrt{x}$ vs. velocity u_d are depicted for each of the five possible shapes of $\tilde{F}$. For shapes A and B the jump phenomenon occurs. From now on we will refer to the interval of u_d values between the two saddle-node bifurcations as the hysteresis interval. Hard-galloping is exhibited in configurations corresponding to shapes B and D.

If $D \leq 0$, (32) has one real root which is positive for $u_d > \frac{-2\beta_g}{c_1}$. The corresponding bifurcation diagram is qualitively the same as that for shape E.

Amplitude vs. damping

Amplitude changes due to the variation of the damping coefficient β_g can be analysed analogously to the amplitude changes due to the variation of

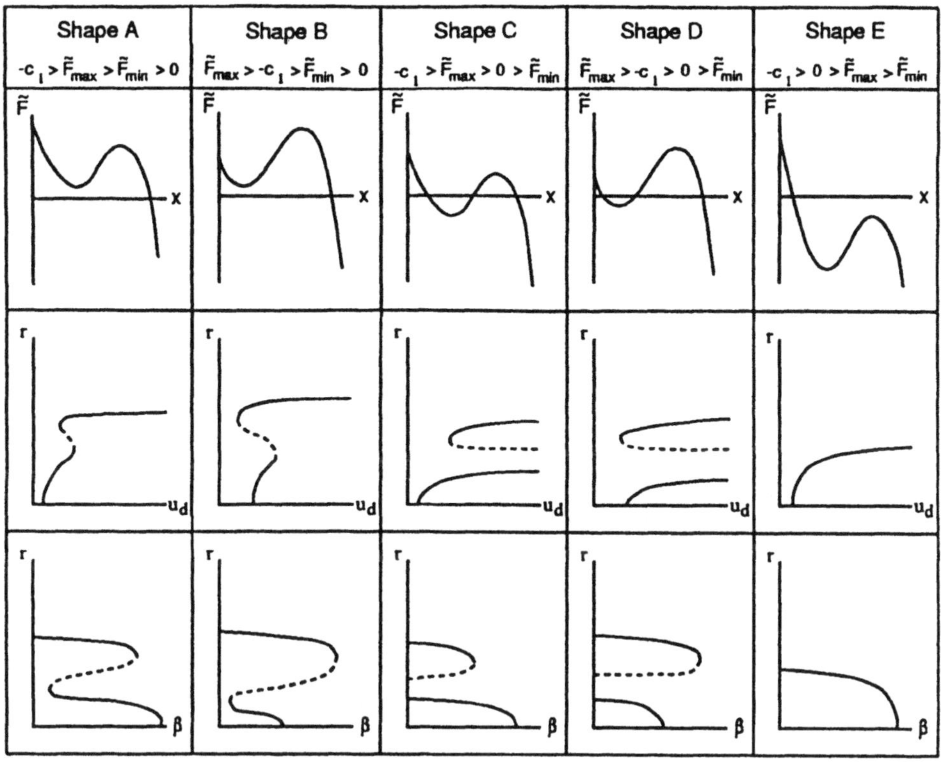

Figure 5. **Bifurcation diagrams (velocity and damping).**

u_d. However since Ω_1, Ω_2 and Ω_3 do not depend on β_g we do not need the approximations $\bar{\Omega}_i, i = 1,2,3$. Analysing (30), i.e. $\dot{x} = xF(x, u_d, \beta_g)$ instead of (32), bifurcation diagrams are obtained by the same method as in the previous paragraph, i.e. by considering possible shapes of

$$\tilde{F} = -c_1 + \frac{\Omega_1}{u_d}x + \frac{\Omega_2}{u_d}x^2 + \frac{\Omega_3}{u_d}x^3 \tag{37}$$

Notice this function now does not only depend on the aerodynamic force coefficients but also on u_d. We define the same expressions as in the previous paragraph but with $\bar{\Omega}_{1,2,3}$ replaced by $\Omega_{1,2,3}$: $D = \Omega_2^2 - 3\Omega_1\Omega_3$ and for $D > 0$

$$x_{\min} = \frac{-\Omega_2 + \sqrt{D}}{3\Omega_3} \quad \text{and} \quad x_{\max} = \frac{-\Omega_2 - \sqrt{D}}{3\Omega_3}$$

and $\tilde{F}_{\min} = \tilde{F}(x_{\min})$ and $\tilde{F}_{\max} = \tilde{F}(x_{\max})$. Saddle-node bifurcations now occur for $\beta_g = u_d\tilde{F}_{\min}/2$ (if $\tilde{F}_{\min} > 0$) and for $\beta_g = u_d\tilde{F}_{\max}/2$ (if $\tilde{F}_{\max} > 0$). See again Fig. 5 for the bifurcation diagrams.

We now introduce an u_d vs. β_g diagram that can be computed for any aerodynamic force coefficients tuple $c_1, \ldots, c_7$, and that provides useful information on the galloping behaviour of the corresponding cylinder's cross-section. It consists of the solid curves $\beta_g = u_d\tilde{F}_{\min}(u_d)/2$ and $\beta_g = u_d\tilde{F}_{\max}(u_d)/2$ corresponding to the values (u_d, β_g) for which the saddle-node bifurcations occur and the dashed line $\beta_g = -c_1u_d/2$ of points for which the transcritical bifurcations occur.

From this diagram immediately follows the order of the respective bifurcations both for fixed u_d and varying β_g and for fixed β_g and varying u_d. It also provides us with a way to check the validity of the approximations of the previous subsection. Using these approximations the points at which saddle-node bifurcations occur are on the lines $\beta_g = \tilde{F}_{\min}u_d/2$ and $\beta_g = \tilde{F}_{\max}u_d/2$ in which $\tilde{F}_{\min}$ and $\tilde{F}_{\max}$ are here defined by (35) , i.e. using $\tilde{\Omega}_{1,2,3}$ instead of $\Omega_{1,2,3}$. We can conclude from the diagrams by comparing the slopes of the solid curves to $\tilde{F}_{\min}/2$ and $\tilde{F}_{\max}/2$ that the approximations are accurate for most u_d. However for configuration I the discriminant D becomes zero for small values of u_d contrary to what one would expect from the approximations.

We also conclude from the diagrams that the hysteresis interval is considerably smaller for the square cross-section of configuration I than that for the rectangular cross-section of configuration II. Furthermore we observe that the hysteresis interval grows when damping is increased.

3.3. MIXED NORMAL MODES

Important cases of mixed mode galloping are represented by critical points of (25), (26) which are not on the x-axis and y-axis. These critical points are solutions of

$$(a_2 - 2\beta) + A_1x + 2A_2y + A_3x^2 + A_4xy + 3A_5y^2 \\ + A_6x^3 + 2A_7x^2y + 3A_8xy^2 + 4A_9y^3 = 0 \tag{38}$$

$$(a_2 - 2\beta) + 2A_1x + A_2y + 3A_3x^2 + A_4xy + A_5y^2 \\ + 4A_6x^3 + 3A_7x^2y + 2A_8xy^2 + A_9y^3 = 0 \tag{39}$$

Solutions of these equations can be obtained by a computer algebraic approach which allows obtaining all real solutions within any desired accuracy.

This has been realised by computing a Gröbner basis for system (38), (39) in which $a_2, A_1, \ldots, A_9$ are kept as parameters, using the Wagram software system for performing Gröbner basis calculations. For an introduction to Gröbner basis theory we refer to [1]; the Wagram software system is described in [15].

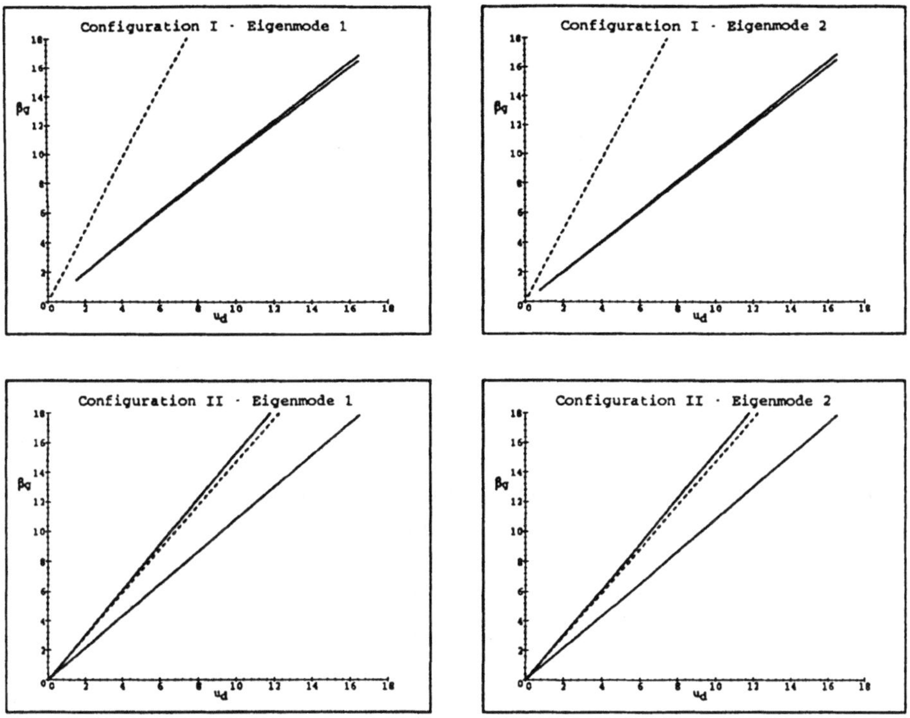

Figure 6. u_d vs. β_g diagrams for configurations I and II

The Gröbner basis contains a polynomial in y of degree nine that does not depend on x. Solutions of this polynomial can be substituted in another Gröbner basis element that is linear in x to obtain possible solutions of (38), (39).

The actual critical points are determined using a program written for the Computer Algebra System Maple. The program, which may be obtained from the authors and contains the Gröbner basis elements, lists the location of all critical points, their topological type and the eigenvectors corresponding to the derivative of the vectorfield at the critical point.

Using this information global phase flow structures can be accurately determined using fourth order Runge-Kutta integration. In our phase plots S denotes a saddle point, sN and uN are stable and unstable nodes, respectively.

For $\beta = 0$ we find that the qualitive structure of the phase flow of (25), (26) is the same for all $u_d > 0$ for both configuration I and II, see Fig. 7. Depending on the initial normal mode amplitudes the system either tends to its first or its second normal mode.

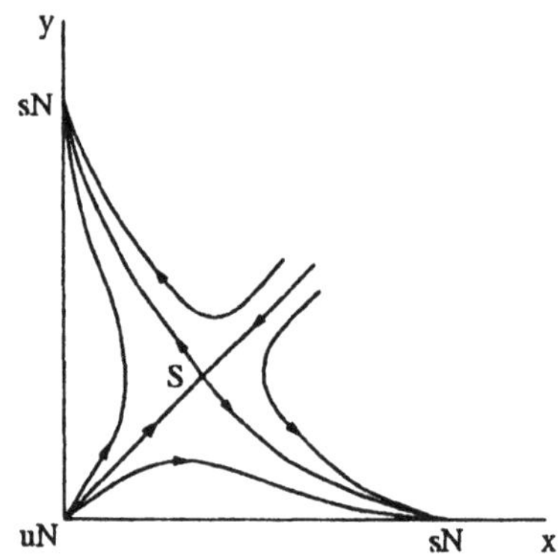

Figure 7. Phase flow for systems without damping.

Because of the invariance of the x- and y-axis and since the only critical point in the region $x > 0$, $y > 0$ is a saddle point (index -1) limit cycles do not exist.

This phase flow structure is particularly important since it will become clear that this structure is reached for all damping levels if u_d is chosen sufficiently large.

We have used our approach to investigate the dynamics of the oscillator corresponding to configuration I for damping coefficients β large enough for the jump phenomenon to be present in both normal modes.

We find that at the critical velocity $\bar{u}_d = \frac{-2\beta_g}{c_1}$ transcritical bifurcations occur for both normal modes and for the same value a saddle point passes from the region $x < 0$, $y < 0$ through the origin into the region $x > 0$, $y > 0$. At the critical velocity the origin becomes unstable and galloping becomes possible, see again Fig. 7.

The bifurcation scenario that follows is the same near both normal mode axes and is depicted in Fig 8.

It consists of a succession of saddle-node and transcritical bifurcations. For instance starting from situation A and increasing the uniform flow velocity, first a saddle-node bifurcation occurs in the normal mode (at a value of u_d which can be found in the u_d vs. β_g diagrams of the normal mode analysis) closely followed by a transcritical bifurcation to arrive at situation B.

For situation C, in which we have two stable nodes corresponding to solutions that are a combination of the two modes, the global phase flow structure has been investigated using numerical integration. We find the phaseportrait depicted in Fig. 9. Notice that although the bifurcation sequences near the normal modes are qualitively symmetrical, the global

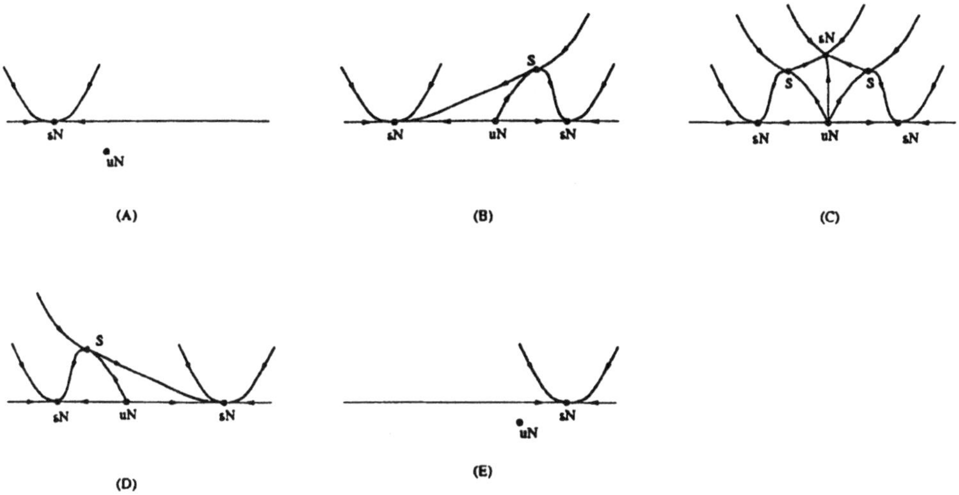

Figure 8. **Bifurcation sequence near an eigenmode.**

phase flow structure is not. The domain of attraction of the stable node near the y-axis is much larger than that of the stable node near the x-axis. We also note that no limit cycles have been observed.

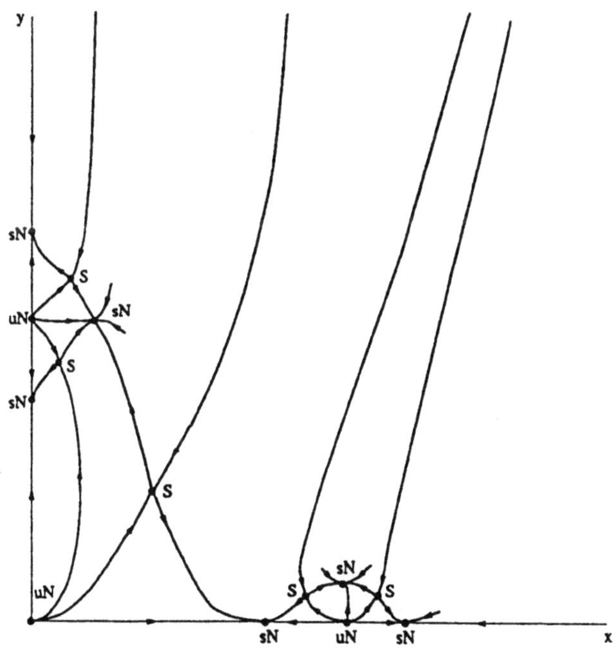

Figure 9. **Phase flow structure during hysteresis interval**

4. Conclusions

In this paper a system of two non-linearly coupled harmonic oscillators is studied with repect to the behaviour of periodic solutions.

The coupled oscillator system (19) describes the non-linear interaction of the two normal modes of a special double pendulum due to aerodynamic forces.

As follows from system (19) the non-linear coupling is of a special type involving polynomials of the seventh degree.

The main qualitative result with respect to the behaviour of unstable solutions can be described as follows. Consider the situation where there is only one mode of oscillation present e.g. look at the critical points on the x-axis in figure 9. One can verify that the interval on the x-axis bounded by two stable critical points involves one unstable critical point. Starting in the unstable point implies that evolution to two stable solutions with different amplitudes (coordinates) is possible. However, the introduction of a second mode of oscillation, i.e. a second degree of freedom, leads to a new stable solution. This solution is characterized by the same amplitude (coordinate) as the one of the unstable critical point on the x-axis and a relatively small amplitude (coordinate) of the second mode.

Apparently small amplitude oscillations of the second mode (second degree of freedom) may stabilize the original unstable solution.

Also a remark should be made on the new quartic Kolmogorov system (25), 26). This system may be regarded as a normal form for the system of coupled oscillators in the non-resonant case. It is believed that this Kolmogorov system has a very rich structure of solutions of which the first results are presented in this paper.

References

1. W. W. Adams and P. Loustaunau, An introduction to Gröbner bases, American Mathematical Society, 1994.
2. R.D. Blevins, Flow induced vibration (2nd edition), Van Nostrand Reinhold, New York, 1990.
3. A.H.P. van der Burgh, Dynamical Systems as Models for Flow-induced Vibrations, *Solitons, Chaos and Fractals*, Vol. 5, No. 9, p. 1563-1577, 1995.
4. A.H.P. van der Burgh, T.I. Haaker and B.W. van Oudheusden, A new aeroelastic oscillator, theory and experiments, *DE-Vol. 84-1, 1995 Design Engineering Technical Conferences, Volume 3-Part A*, ASME 1995.
5. H. Goldstein, Classical Mechanics, 2nd edition, Addison Wesley Publishing Company, Inc., 1980.
6. J. Guckenheimer and P. Holmes, Nonlinear Oscillations, Dynamical Systems, and Bifurcations of Vector Fields, Springer-Verlag, New York, 1990.
7. T.I. Haaker and A.H.P. van der Burgh, On the dynamics of aeroelastic oscillators

with one degree of freedom, *SIAM J. Appl. Math.*, Vol.54 (1994), pp. 1033-1047.

8. T.I. Haaker, Dynamical behaviour of aeroelastic oscillators with one degree of freedom, *Report 94-63, Faculty of Technical Mathematics and Informatics*, Delft University of Technology, 1994.
9. T.I. Haaker, Quasi-steady modelling and asymptotic analysis of aeroelastic oscillators, *Ph.D. Thesis, Department of Technical Mathematics and Computer Science*, Delft University of Technology, Delft, 1996.
10. M.J. Huiskes, An aeroelastic oscillator with two degrees of freedom, *M.Sc. Thesis, Department of Technical Mathematics and Computer Science*, Delft University of Technology, Delft, 1996.
11. B. W. van Oudheusden, On the quasi-steady analysis of one-degree-of-freedom galloping with combined translational and rotational effects, *Nonlinear Dynamics*, Vol.8, no.2, September 1995.
12. B. W. van Oudheusden, Investigation of Large-Amplitude 1-DOF Rotational Galloping, *Report LR-794, Faculty of Aerospace Engineering*, Delft University of Technology, September 1995.
13. G. V. Parkinson and J. D. Smith, The square prism as an aeroelastic non-linear oscillator, *Quart. J. Mech. and Appl. Math.*, Vol.XVII (1964), Pt.2, pp. 225–239.
14. J.A. Sanders and F. Verhulst, Averaging Methods in Nonlinear Dynamical Systems, AMS 59, Springer-Verlag, New-York, 1985.
15. L. A. Timochouk, Wagram: *Yet another Gröbner system*, in Numerical Analysis, Scientific Computing, Computer Science (Proceedings of the 3rd International Congress on Industrial and Applied Mathematics, Hamburg 3-7 July 1995), Ed. by G. Alefeld, *Zeitschrift für Angewandte Mathematik und Mechanik*, Akademie Verlag, Berlin, (1996).

SOLUTION OF THE AFFINE TODA EQUATIONS AS AN INITIAL VALUE PROBLEM

A. A. ISKANDAR
Department of Physics
Institut Teknologi Bandung
Jl. Ganesa 10, Bandung 40132,
INDONESIA

Abstract. Affine Toda field is a multicomponent field in two space-time dimensions, satifying a generalisation of the sinh-Gordon equation $\partial^2\phi + \frac{4\sqrt{2}\mu^2}{\beta}\sinh(\sqrt{2}\beta\phi) = 0$. Solution of the affine Toda field equations is presented as a path ordered exponential integrals of an initial value problem. This is in the same spirit as the work of Mansfield where one evolves the solution at the apex of a backward light-cone from the initial values at some arbitrary points along the legs of this light-cone. These two initial values are then connected by an arbitrary path. Selecting a particular path as the forward light-cone from a point at the infinite past, one obtains the solution to the field equation which reduces to the solution proposed by Olive *et al.* using Kac-Moody algebraic method. This solution as shown by Olive *et al.* yields the soliton solutions provided one chooses the coupling parameter β of the affine Toda fields to be purely imaginary.

1. Introduction

The Toda field theories have been a subject of a vast amount of research for various reasons [1, 2, 3]. Conformal invariance resulted from symmetry under the general coordinate transformation and the integrability of a two dimensional field theory in which exact solution can be constructed has been the starting point for some of those works. A perturbation can be introduced to the conformally invariant Lagrangian of the Toda equations yielding a massive scalar field theory that which is still integrable, the affine Toda field theory [4].

The equation of motion of the affine Toda fields is given as follows

$$\partial^2\phi = -\frac{4\mu^2}{\beta}\sum_{j=0}^{r} m_j \frac{\alpha_j}{\alpha_j^2} e^{\beta\alpha_j\cdot\phi}, \tag{1}$$

where ϕ is an r-dimensional scalar field, μ and β are the mass and coupling parameters, respectively. The vectors α_j are the simple roots of the rank r Lie algebra g underlying the theory with $\alpha_0 = -\psi$, where $\psi = \sum_{j=1}^{r} n_j\alpha_j$ is the highest root of g. The Kac's labels m_j are given by $m_j = n_j\alpha_j^2/\psi^2$; it is customary to take $\psi^2 = 2$ for the simply laced Lie algebras.

Much progress has been achieved in the real coupling regime of the affine Toda field theories [5, 6, 7, 8]. Taking the coupling parameter β to be purely imaginary yields multiple vacua in the Toda potential. These vacua signal the existence of topological soliton solutions which interpolate between them [9]. There is a growing number of results in the understanding of these static soliton solutions as well as the time-dependent soliton solutions (bound state solutions) of the imaginary coupling regime [10, 11]. Recent progress on this front is the calculation of quantum mass correction of these solitons [12] and the construction of soliton S-matrices [13].

The present work concerns with the construction of the solution to the affine Toda equations as an initial value problem. This work follows the work of Mansfield's on the solution of the Toda equations [2]. Using initial values along a backward light-cone, one can construct the solution of the affine Toda equations. When one evalutes these initial values through a particular path connecting them with a point in the infinite past, one obtains the familiar form of the solution to the affine Toda equations as given by Leznov and Saveliev [14] and Olive *et al.* [10].

2. Some Properties of the Kac-Moody Algebra $\hat{g}$

In this section, notion and notation used to represent the generators of an infinite dimensional Lie algebra $\hat{g}$ will be briefly introduced. A more detail introduction can be found in [15].

To a classical simple Lie algebra g of finite dimension according to the Cartan classification, one can construct a *loop algebra* by adding a spectral parameter to the generators of g. The *Kac-Moody* algebra $\hat{g}$ can be viewed as a central extension of this loop algebra with central element k which commutes with all other elements of $\hat{g}$. Further, the elements of the Kac-Moody algebra are *graded* by the *derivation* operator d'. Consider the following elements of $\hat{g}$ in the *modified* Cartan-Weyl basis: $\{H_0^j\}$ which are the affinization of the elements of the Cartan subalgebra of g and the *step* operators $\{E_0^{\pm\alpha_j}, E_{\pm1}^{\alpha_0}\}$. From these operators, one can construct the following alternative operators (see [10])

$$\hat{E}_{\pm 1} = \sum_{j=1}^{r} \sqrt{m_j} E_0^{\pm\alpha_j} + E_{\pm 1}^{\pm\alpha_0}. \tag{2}$$

The commutation relations of the elements of $\hat{g}$ in the *modified* Cartan-Weyl basis can be found in [15]. A few of the commutation relations which will be used are as follows,

$$\begin{aligned} [d', H_0^j] &= 0, \\ [\hat{E}_{+1}, \hat{E}_{-1}] &= k, \\ [d', \hat{E}_{\pm 1}] &= \pm \hat{E}_{\pm 1}. \end{aligned} \tag{3}$$

As in the case of simple Lie algebra, one can construct the root system of the Kac-Moody algebra. The simple roots associated with the operators $\{E_0^{\pm\alpha_j}, E_{\pm 1}^{\alpha_0}\}$ are the weight constructed from the simple roots α_j of g with the addition of two extra components related to the central element k and the grading operator d,

$$\begin{aligned} a_0 &= (-\psi, 0, 1), \\ a_j &= (\alpha_j, 0, 0). \end{aligned}$$

Scalar product of any two weights is defined as,

$$(\alpha, x, y) \cdot (\beta, u, v) = \alpha \cdot \beta + xv + yu.$$

With these affine simple roots, one can write the *step* operators $E_0^{\pm\alpha_j}$ as $E^{\pm a_j}$ and $E_{\pm 1}^{\alpha_0}$ as $E^{\pm a_0}$, thus (2) can also be written as

$$\hat{E}_{\pm 1} = \sum_{j=0}^{r} \sqrt{m_j} E^{\pm a_j}, \tag{4}$$

where $m_0 = 1$.

Furthermore, the fundamental weights of a Kac-Moody algebra are defined by the relation,

$$\frac{2a_j \cdot \Lambda_k}{a_j^2} = \delta_{jk}.$$

With the affine simple roots given as above, the fundamental weights are calculated to be,

$$\begin{aligned} \Lambda_0 &= (0, \frac{1}{2}\psi^2, 0), \\ \Lambda_j &= (\lambda_j, \frac{1}{2} m_j \psi^2, 0), \end{aligned}$$

where λ_j's are the fundamental weights of the corresponding Lie algebra g. The fundamental highest weight representation is built up from these fundamental weight vectors in which the following is fulfilled,

$$E^{+\alpha_j}|\Lambda_k>=0. \tag{5}$$

In particular, the actions of operators H_0^j and k on an arbitrary weight vector $|\Lambda>$ are as follows,

$$\begin{aligned} H_0^j|\Lambda> &= \Lambda^{(j)}|\Lambda>, & (6)\\ k|\Lambda> &= m_\Lambda|\Lambda>, & (7)\end{aligned}$$

where m_Λ is called the *level* of Λ.

3. The Initial Value Problem

Let one introduce the following gauge potentials in the light-cone coordinates $x_\pm = \frac{1}{\sqrt{2}}(t \pm x)$,

$$W_\pm = \pm\frac{\beta}{2}\partial_\pm\Phi \pm \mu e^{\pm\frac{\beta}{2}\mathrm{Ad}\Phi}\hat{E}_{\pm1}, \tag{8}$$

with

$$\Phi = \phi\cdot H_0 + \xi k + \eta d', \tag{9}$$

i.e. one can think of Φ as a superfield comprising of the affine Toda fields ϕ with the auxilary fields ξ and η related to the central element k and the grading operator d' of Kac-Moody algebra $\hat{g}$. The zero curvature condition on this gauge potential, i.e.

$$[\partial_+ + W_+, \partial_- + W_-] = 0, \tag{10}$$

yields the following equation of motions,

$$\begin{aligned} \partial^2\phi &= -\frac{4\mu^2}{\beta}e^{\beta\eta}\sum_{j=0}^{r} m_j\frac{\alpha_j}{\alpha_j^2}e^{\beta\alpha_j\cdot\phi}, & (11)\\ \partial^2\xi &= -\frac{2\mu^2}{\beta}e^{\beta\eta}e^{\beta\alpha_0\cdot\phi}, & (12)\\ \partial^2\eta &= 0. & (13)\end{aligned}$$

Setting $\eta = 0$ (this is called the *conformal gauge*) then (11) as the same with the affine Toda equation (1). Henceforth it is understood that the conformal gauge has been used. Furthermore, one sets the gauge potentials as a pure gauge, i.e. there is a group element g such that

$$W_\pm = g(\partial_\pm g^{-1}). \tag{14}$$

Following Mansfield [2], one takes $|\Lambda>$ to be a highest weight, then one has

$$\partial_+\{g^{-1}e^{-\frac{\beta}{2}\Phi}\}|\Lambda> = \mu g^{-1}e^{\frac{\beta}{2}\Phi}\hat{E}_{+1}e^{-\beta\Phi}|\Lambda> .$$

But one can show that $\Phi|\Lambda> \sim |\Lambda>$, thus since $|\Lambda>$ is a highest weight then using (5) one obtains

$$\partial_+\{g^{-1}e^{-\frac{\beta}{2}\Phi}|\Lambda>\} \; = 0. \qquad (15)$$

Hence, $\{g^{-1}e^{-\frac{\beta}{2}\Phi}\}|\Lambda>$ is independent of the x_+ coordinate. Similarly, one also notes that

$$\partial_-\{<\Lambda|e^{-\frac{\beta}{2}\Phi}g\} = 0, \qquad (16)$$

i.e. $<\Lambda|\{e^{-\frac{\beta}{2}\Phi}g\}$ is independent of the x_- coordinate.

Let (15) and (16) be evaluated at points B and A respectively, then the quantity

$$<\Lambda|g(A)g^{-1}(B)|\Lambda> \; e^{-\frac{\beta}{2}[\Lambda(\Phi_A)+\Lambda(\Phi_B)]} \qquad (17)$$

is independent of the x_- coordinate of point A and of x_+ coordinate of point B. Thus, one can choose the points A and B to lie on the x_+ = constant and x_- = constant world lines of the appropriate backward light-cone (see Fig. 1)

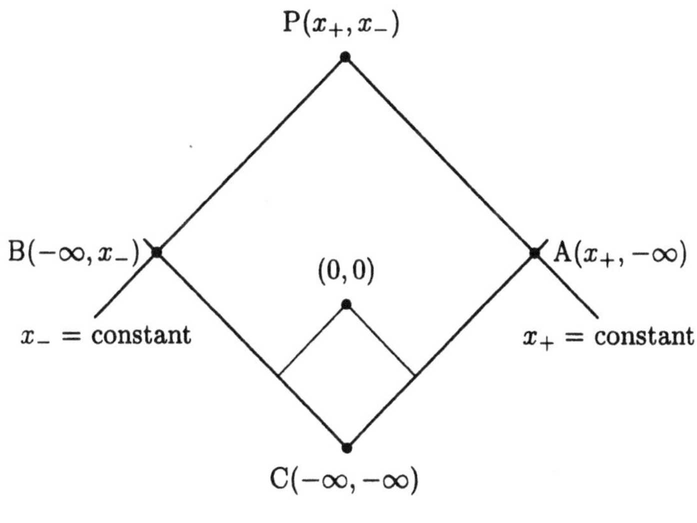

Figure 1. Backward and forward light-cone.

In (17) above one has introduced the following notation,

$$\Lambda(\Phi) = \Phi|\Lambda> = \lambda\cdot\phi + m_\Lambda\xi.$$

Furthermore, following the backward light-cone from points A and B to an arbitrary point $P(x_+, x_-)$, one realises that (17) is equal to

$$e^{-\beta\Lambda(\Phi_P)}. \tag{18}$$

This means that the solution of the affine Toda equations at the arbitrary point $P(x_+, x_-)$, Φ_P is given by the initial values at points A and B,

$$e^{-\beta\Lambda(\Phi_P)} = < \Lambda|g(A)g^{-1}(B)|\Lambda > e^{-\frac{\beta}{2}[\Lambda(\Phi_A)+\Lambda(\Phi_B)]}. \tag{19}$$

To obtain the expression $g(A)g^{-1}(B)$, one has to perform a path ordered integration (putting bigger space argument to the left) along an arbitrary path from point A to B. Let the chosen path be the forward light-cone connecting points A and B to point C (a point in the infinite past).

Since W_- is a pure gauge, i.e. $W_- = -(\partial_- g)g^{-1}$, then integrating it from C to B yields

$$V(x_-) = e^{-\frac{\beta}{2}\Phi(C)}g(C)g^{-1}(B)e^{\frac{\beta}{2}\Phi(B)}, \tag{20}$$

with

$$V(x_-) \equiv \mathrm{P}\exp\left(-\mu\int_C^B e^{-\frac{\beta}{2}\mathrm{Ad}\Phi(\mathrm{y})-\frac{\beta}{2}\mathrm{Ad}\Phi(\mathrm{B})}\hat{E}_{-1}dy\right), \tag{21}$$

where P denotes a path ordered exponential. On the otherhand, one also has

$$\partial_+(e^{\frac{\beta}{2}\Phi}g) = -\mu(e^{\beta\mathrm{Ad}\Phi}\hat{E}_{+1})(e^{\frac{\beta}{2}\Phi}g),$$

which upon integration from A to C yields,

$$e^{\frac{\beta}{2}\Phi(A)}g(A)g^{-1}(C)e^{-\frac{\beta}{2}\Phi(C)} = \mathrm{P}\exp\left(-\mu\int_C^A e^{\beta\mathrm{Ad}\Phi(\mathrm{y})}\hat{E}_{+1}dy\right).$$

Taking conjugate with $e^{-\beta\Phi(C)}$ yields,

$$U(x_+) = e^{-\beta\Phi(C)+\frac{\beta}{2}\Phi(A)}g(A)g^{-1}(C)e^{\frac{\beta}{2}\Phi(C)}, \tag{22}$$

where

$$U(x_+) \equiv \mathrm{P}\exp\left(-\mu\int_C^A e^{\beta\mathrm{Ad}\Phi(\mathrm{y})-\beta\mathrm{Ad}\Phi(\mathrm{C})}\hat{E}_{+1}dy\right). \tag{23}$$

Thus,

$$U(x_+)V(x_-) = e^{-\beta\Phi(C)+\frac{\beta}{2}\Phi(A)}g(A)g^{-1}(B)e^{\frac{\beta}{2}\Phi(B)},$$

and one obtains the solution (19) of the affine Toda equations at an arbitratry point $P(x_+, x_-)$ given as

$$e^{-\beta\Lambda(\Phi_P)} = e^{\beta[\Lambda(\Phi_C)-\Lambda(\Phi_A)-\Lambda(\Phi_B)]} < \Lambda|U(x_+)V(x_-)|\Lambda > . \tag{24}$$

Reasoning that at the infinite past of point C, all the fields are bounded, i.e.

$$\phi_C = 0 \qquad \text{and} \qquad \xi_C = 0,$$

and setting,

$$\begin{aligned} \Lambda(\Phi_A) &= \Lambda(\Phi_0^+) &\quad& \text{right moving free solution at A,} \\ \Lambda(\Phi_B) &= \Lambda(\Phi_0^-) && \text{left moving free solution at B,} \end{aligned}$$

one obtains the familiar solution of Leznov and Saveliev [14] and Olive *et al.* [10]

$$e^{-\beta\Lambda(\Phi_P)} = e^{-\beta\Lambda(\Phi_0)} < \Lambda|U(x_+)V(x_-)|\Lambda >, \tag{25}$$

where

$$\Lambda(\Phi_0) = \Lambda(\Phi_0^+) + \Lambda(\Phi_0^-).$$

Furthermore, the operators $U(x_+)$ and $V(x_-)$ satisfy a differential equation similar to the inverse scattering equation. Differentiating $U(x_+)$ with respect to x_+ and recalling that at point C$(-\infty,-\infty)$ the following gives,

$$\partial_+ \left(\int e^{\beta \mathrm{Ad}\phi} \hat{E}_{+1} dy \right)_C = 0,$$

one can easily show that, because of the path ordered expression, $U(x_+)$ satisfies the following equation

$$\partial_+ U = -\mu \left(e^{\beta \mathrm{Ad}\Phi_0^+} \hat{E}_{+1} \right) U(x_+). \tag{26}$$

In a similar way, one can also show that $V(x_+)$ satisfies the following equation

$$\partial_- V = -\mu V(x_-) \left(e^{-\beta \mathrm{Ad}\Phi_0^-} \hat{E}_{-1} \right). \tag{27}$$

4. Conclusion and Remark

One has seen that the solution of the affine Toda equations can be presented in the form of path ordered exponentials (25) with $V(x_-)$ and $U(x_+)$ given by equations (21) and (23). This solution has been obtained by various groups through various different approaches. The work of Olive *et al.* [10] carries it further by showing that indeed by taking the coupling parameter β to be purely imaginary, one can solve the differential equations (26) and (27). Choosing an approriate basis for $\hat{g}$, the soliton solutions of affine Toda equations can be obtained. This soliton solutions are exactly the same as the soliton solutions obtained by other methods, to name a few: the Hirota method or Bäcklund Transformation [9, 11].

The procedure followed in this work is similar to that of Mansfield's [2], whereas Olive *et al.* [10] have chosen a different gauge for the gauge potential (i.e. $W_{\pm}$ are not pure gauge). Furthermore the path taken in that work is the path avoiding point C (the backward light-cone from the origin in Fig. 1). As a result, the calculations of the present work shows no arbitrary unknown integration constant, in contrast to the result of Olive *et al.* in which, using the zig-zag path, an unknown constant is left in the calculation.

Acknowledgement

I would like to thank the Scientific Committee of the International Conference on Differential Equations held at Institut Teknologi Bandung for the opportunity to present this work. Partial financial support from the Organizing Committee is also acknowledged.

References

1. Arinsthein, A.E., Fateev, V.A. and Zamolodchikov, A.B. (1979) Quantum *S* Matrix of the (1+1)-dimensional Toda Chain, *Phys. Lett.*, **B87**, pp. 389.
   Mikhailov, A.V., Olshanetsky, M.A. and Peremolov, A.M. (1981) Two Dimensional Generalized Toda Lattice, *Comm. Math. Phys.*, **79**, pp. 473.
2. Mansfield, P. (1982) Solution of Toda Systems, *Nucl. Phys.* **B208**, pp. 277.
   Mansfield, P. (1993) Light-cone Quantisation of the Liouville and Toda Field Theories, *Nucl. Phys.* **B222**, pp. 419.
   Mansfield, P. (1985) Solution of the Initial Value Problem for the sine-Gordon Equation using a Kac-Moody Algebra, *Comm. Math. Phys.* **98**, pp. 525.
3. Olive, D.I. and Turok, N. (1983) The Symmetries of Dynkin Diagrams and the Reduction of the Toda Field Equations, *Nucl. Phys.*, **B215**, pp. 470.
   Olive, D.I. and Turok, N. (1983) Algebraic Structures of Toda Systems, *Nucl. Phys.*, **B220**, pp. 491.
   Olive, D.I. and Turok, N. (1985) Local Conserved Densities and Zero Curvature conditions of Toda Lattice Field Theories, *Nucl. Phys.*, **B257**, pp. 277.
   Olive, D.I. and Turok, N. (1985) The Toda Lattice Field Theory Hierarchies and the Zero Curvature Conditions in Kac-Moody Algebras, *Nucl. Phys.*, **B265**, pp. 469.
4. Eguchi, T. and Yang, S-K (1989) Deformations of Conformal Field Theories and Soliton Equations, *Phys. Lett.*, **B224**, pp. 373.
   Hollowood, T. and Mansfield, P. (1989) Rational Conformal Field Theories at, and away from, Criticality as Toda Field Theories, *Phys. Lett.*, **B226**, pp. 73.
5. Braden, H.W., Corrigan, E., Dorey, P.E. and Sasaki, R. (1989) Extended Toda Field Theory and Exact *S* Matrices, *Phys. Lett.*, **B227**, pp. 41.
   Braden, H.W., Corrigan, E., Dorey, P.E. and Sasaki, R. (1990) Affine Toda Field Theory and Exact *S* Matrices, *Nucl. Phys.* , **B338**, pp. 74.
   Braden, H.W., Corrigan, E., Dorey, P.E. and Sasaki, R. (1991) Multiple Poles and Other Features of Affine Toda Field Theory, *Nucl. Phys.*, **B356**, pp. 46.
6. Braden, H.W. and Sasaki, R. (1991) The *S* Matrix Coupling Dependence for A, D and E Affine Toda Field Theory, *Phys. Lett.*, **B255**, pp. 343.
   Braden, H.W. and Sasaki, R. (1992) Affine Toda Perturbation Theory, *Nucl. Phys.*, **B379**, pp. 377.
   Sasaki, R. and Zen, F.P. (1993) The Affine Toda *S* Matrices vs. Perturbation Theory,

Int. J. Mod. Phys., **A8**, pp. 115.

7. Delius, G.W., Grisaru, M.T. and Zanon, D. (1991) Exact S Matrices for the Non Simply Laced Affine Toda Theories $a_{2n-1}^{(2)}$, *Phys. Lett.*, **B277**, pp. 414.
   Delius, G.W., Grisaru, M.T. and Zanon, D. (1992) Exact S Matrices for the Non Simply Laced Affine Toda Theories, *Nucl. Phys.* **B382**, pp. 365.
   Corrigan, E., Dorey, P.E. and Sasaki, R. (1993) On Generalized Bootstrap Principle, *Nucl. Phys.* **B408**, pp. 579.
   Dorey, P. (1993) A Remark on the Coupling Dependence in Affine Toda Field Theories, *Phys. Lett. B312*, pp. 291.
8. Dorey, P.E. (1991) Root Systems and Purely Elastic S Matrices, *Nucl. Phys.* **B358**, pp. 654.
   Freeman, M.D. (1991) On the Mass Spectrum of Affine Toda Field Theory, *Phys. Lett. B261*, pp. 57
   Dorey, P.E. (1992) Root Systems and Purely Elastic S Matrices 2, *Nucl. Phys.* **B374**, pp. 741.
9. Hollowood, T. (1992) Solitons in Affine Toda Field Theories, *Nucl. Phys.* **B384**, pp. 523.
10. Olive, D.I., Turok, N. and Underwood, J.W.R (1993) Solitons and the Energy Momentum Tensor for the Affine Toda Theory, *Nucl. Phys.* **B401**, pp. 663.
11. Olive, D.I., Turok, N. and Underwood, J.W.R (1993) Affine Toda Solitons and Vertex Operators, *Nucl. Phys.* **B409**, pp. 509.
    Hollowood, T. (1993) Quantizing Sl(n) Solitons and the Hecke Algebra, *Int. J. Mod. Phys.* **A8**, pp. 947.
    MacKay, N.J. and McGhee, W.A. (1993) Affine Toda Solitons and Automorphisms of Dynkin Diagrams, *Int. J. Mod. Phys.* **A8**, pp. 2791.
    Liao, H.C., Olive, D.I. and Turok, N. (1993) Topological Solitons in A_r Affine Toda Theory, *Phys. Lett.* **B298**, pp. 95.
    McGhee, W.A. (1994) The Topological Charges of the $a_n^{(1)}$ Affine Toda Solitons, *Int. J. Mod. Phys.* **A9**, pp. 2645.
    Kneipp, M.A.C. and Olive, D.I. (1994) Crossing and Anti-Solitons in Affine Toda Theories, *Nucl. Phys.* **B408**, pp. 565.
    Fring, A., Johnson, P.R., Kneipp, M.A.C. and Olive, D.I. (1994) Vertex Operators and Soliton Time Delays in Affine Toda Field Theory, *Nucl. Phys.* **B430**, pp. 597.
    Harder, U., Iskandar, A.A. and McGhee, W.A. (1995) On the Breathers of $a_n^{(1)}$ Affine Toda Field Theory, *Int. J. Mod. Phys.* **A10**, pp. 1879.
12. Hollowood, T. (1993) Quantum Soliton Mass Correction in Sl(n) Affine Toda Field Theory, *Phys. Lett.* **B300**, pp. 73.
    Watts, G.M.T. (1994) Quantum Mass Corrections for $c_2^{(1)}$ Affine Toda Theory Solitons, *Phys. Lett.* **B338**, pp. 40.
    Delius, G.W. and Grisaru, M.T. (1994) Toda Soliton Mass Corrections and the Particle-Soliton Duality Conjecture, *Nucl. Phys.* **B441**, pp. 259.
    Watts, G.M.T. and MacKay, N.J. (1994) Quantum Mass Corrections for Affine Toda Solitons, *Nucl. Phys.* **B441**, pp. 277.
13. Gandenberger, G.M. (1995) Exact S-matrices for Bound States of $a_2^{(1)}$ Affine Toda Solitons, *Nucl. Phys.* **B449**, pp. 375.
    Gandenberger, G.M. and MacKay, N.J. (1995) Exact S-matrices for $d_{n+1}^{(2)}$ Affine Toda Solitons and Their Bound States, *Nucl. Phys.* **B457**, pp. 240.
    Gandenberger, G.M., MacKay, N.J. and Watts, G.M.T. (1996) Twisted Algebra R-matrices and exact S-matrices for $b_n^{(1)}$ Affine Toda Solitons and Their Bound States, *Nucl. Phys.* **B465**, pp. 329.
    Gandenberger, G.M. and MacKay, N.J. (1996) Remarks on Excited States of Affine Toda Solitons, DAMTP-96-70, hep-th/9608055.
14. Leznov, A.N. and Saveliev, M.V. (1979) *Lett. Math.* **3**, pp. 485.

Leznov, A.N. and Saveliev, M.V. (1992)*Group Theoretical Methods for Integration of Non-linear Dynamical System*, Progress in Physics vol. 15, Birkhäuser-Verlag.
15. Goddard, P. and Olive, D.I. (1986) Kac-Moody and Virasoro Algebras in Relation to Quantum Physics, *Int. J. Mod. Phys.* **A1**, pp. 303.

MATHEMATICAL MODELLING OF CHLORINE DECAY IN A CONTACT TANK

AHMAD IZANI BIN MD. ISMAIL
School of Mathematical Sciences
Universiti Sains Malaysia, 11800 USM, Penang
Malaysia

AND

ROGER A. FALCONER
Department of Civil Engineering
University of Bradford, Bradford, BD7 1DP
England

Abstract. Recently concerns have been expressed regarding the side-effects of some disinfectants used to disinfect drinking water supply. Chlorine is, by far, the most common disinfectant used in the disinfection of drinking water supply. The purpose of a chlorine contact tank is to allow sufficient time for chlorine to achieve full disinfection before the water enters the distribution network. Mathematical models are increasingly being used in the design of efficient chlorine contact tanks. The transport of chlorine in a contact tank is governed by an advective-diffusion partial differential equation with a decay term. In this paper details are given of the application of a finite difference scheme - a 2-D semi-time-centred implicit QUICK scheme - to predict the decay of chlorine as it is being transported from the inlet to the outlet of a scaled serpentine contact tank. The scheme is shown to give stable and reasonable results.

1. Introduction

The aim of the disinfection step in water treatment is to ensure that pathogens which have yet to be destroyed by earlier steps do not enter the water distribution network. Chlorine is the most common disinfectant used since it is effective, inexpensive, persists reasonably well and is easy to apply, measure and control (Culp, 1974; White, 1979). The primary purpose of a chlorine contact tank is to provide sufficient time for the chlorine to achieve full disinfection before the water enters the distribution network. Some recent studies have indicated a higher incidence of carcinoma of the bladder, colon, rectum, thyroid, kidney and liver in those drinking excess chlorinated water (Grizzle, 1984; Bull, 1986; Singer, 1994). Hence, in order to design chlorine contact tanks which uses the optimal amount of chlorine it is important for engineers and scientists to have a full understanding of the disinfection process within such tanks.

The most common chlorine-based disinfectant used in water treatment is free chlorine. Free chlorine is usually produced by the addition of chlorine gas to water. A weaker disinfectant is combined chlorine, which is produced when chlorine is added to ammonia and the combination is added to water in solution. Combined chlorine has the advantage of being more persistent and not producing the harmful by-products associated with free chlorine (McGhee, 1991). Bacterial deactivation by chlorine is strongly dependent upon the chlorine concentration and the contact time. An important parameter in disinfection is the CT value where C = chlorine concentration in milligrammes per litre and T = contact time in minutes (Singer, 1994; Clark et al., 1994). Typically, filtrated water requires 30 minutes of contact with 1 milligramme per litre free chlorine, giving a CT value of 30, to ensure satisfactory disinfection. CT values assume the chlorine concentration remains constant within the tank. However, chlorine will kill bacteria and viruses at a certain rate and will therefore decay in water. It is important to be able to simulate this decay within a contact tank so as to obtain a better understanding of the disinfection process and this decay rate is linked to the distribution of residual bugs (Chick, 1908).

The aim of this paper is to investigate the application of a finite difference scheme to predict the decay of chlorine as it is being transported from the inlet to the outlet of a rectangular scaled serpentine chlorine contact tank (Figures 1,2 and 3).

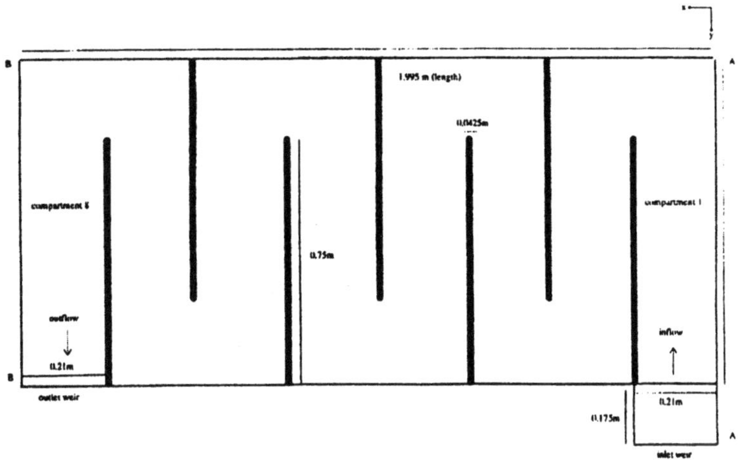

Figure 1. Plan view of contact tank

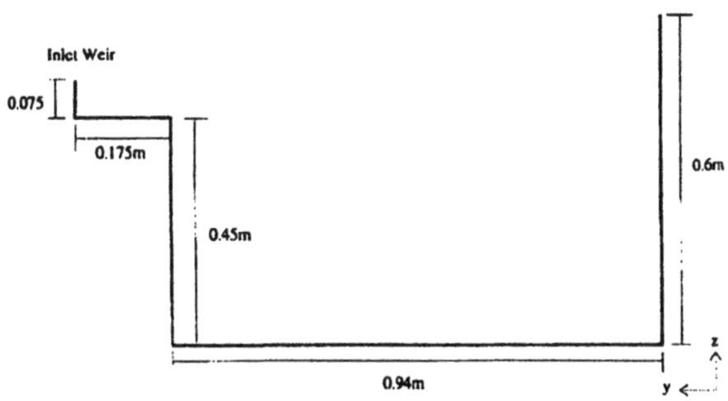

Figure 2. View A-A

Figure 3. View B-B

2. Governing equations

The depth integrated conservation of momentum equations governing the flow of water in a contact tank can be written for the x and y directions respectively as (Falconer, 1993):

$$\begin{aligned} &\frac{\partial UH}{\partial t} + \beta\frac{\partial}{\partial x}(U^2H) + \beta\frac{\partial(UVH)}{\partial y} \\ &= -gH\frac{\partial \eta}{\partial x} - \frac{gU\sqrt{U^2+V^2}}{C^2} + \bar{e}H\left(\frac{\partial^2 U}{\partial x^2} + \frac{\partial^2 U}{\partial y^2} \right) \end{aligned} \tag{1}$$

and

$$\begin{aligned} &\frac{\partial VH}{\partial t} + \beta\frac{\partial}{\partial y}(V^2H) + \beta\frac{\partial(UVH)}{\partial x} \\ &= -gH\frac{\partial \eta}{\partial y} - \frac{gV\sqrt{U^2+V^2}}{C^2} + \bar{e}H\left(\frac{\partial^2 V}{\partial y^2} + \frac{\partial^2 V}{\partial x^2} \right) \end{aligned} \tag{2}$$

where η = water surface elevation above datum; H = total water depth; t = time; U, V= depth mean velocity components in the x, y directions respectively; β = momentum correction factor for non-uniform vertical velocity profile; g = gravitational acceleration; $\bar{e}$ = depth mean eddy viscosity and C = Chezy bottom roughness value.

The depth integrated mass conservation equation governing the flow of water in a contact tank can be written as (Falconer, 1993):

$$\frac{\partial \eta}{\partial t} + \frac{\partial UH}{\partial x} + \frac{\partial VH}{\partial y} = 0 \tag{3}$$

The depth integrated transport or advection-diffusion equation governing the transport of chlorine in a contact tank can be written as (Ismail, 1996):

$$\begin{aligned} \frac{\partial}{\partial t}(HS) + \frac{\partial}{\partial x}(HUS) + \frac{\partial}{\partial y}(HVS) = \; & \frac{\partial}{\partial x}\left(D_{xx}H\frac{\partial S}{\partial x} + D_{xy}H\frac{\partial S}{\partial y}\right) \\ & + \frac{\partial}{\partial y}\left(D_{yx}H\frac{\partial S}{\partial x} + D_{yy}H\frac{\partial S}{\partial y}\right) - \frac{dS}{dt} \end{aligned} \tag{4}$$

where S = depth mean chlorine concentration; D_{xx}, D_{xy}, D_{yx} and D_{yy} are depth mean longitudinal dispersion and turbulent diffusion coefficients in the x, y directions defined by:

$$\begin{aligned} D_{xx} &= (K_lU^2 + K_tV^2)H\sqrt{g}/C\sqrt{U^2+V^2} \\ D_{xy} &= D_{yx} = (K_l - K_t)UVH\sqrt{g}/C\sqrt{U^2+V^2} \\ D_{yy} &= (K_lV^2 + K_tU^2)H\sqrt{g}/C\sqrt{U^2+V^2} \end{aligned} \tag{5}$$

where k_l = longitudinal dispersion coefficient and k_t = turbulent diffusion coefficient. $\frac{dS}{dt}$ is known as the decay term and if a first-order decay rate is assumed for chlorine then $\frac{dS}{dt} = \omega SH$ where $\omega = 0.038$ for free chlorine and $\omega = 0.00123$ for combined chlorine (Trussell and Chao, 1977).

3. Finite difference scheme for the transport equation

To obtain the depth integrated velocities required to solve the transport equation, i.e. equation (4), the hydrodynamic equations (1), (2) and (3) were expressed in a central implicit finite difference form. Full details of this unconditionally stable scheme and the solution procedure to obtain the water elevations and depth integrated velocities can be found in Falconer (1986).

QUICK (quadratic upstream interpolation for convective kinematics) schemes have been widely used to model the transport of a solute (for example Leonard, 1979; Falconer and Liu, 1988; Roelfzema and Gerritsen, 1989, Kaya, 1992). QUICK schemes have the advantage of computational simplicity and-at worst-weak grid scale oscillations. For the advective term in the transport equation the QUICK scheme uses quadratic interpolation to estimate the concentration value, on a space-staggered grid, at the same location as the velocity components. The diffusion term in the transport equation is expressed using conventional space centred expressions. A two-dimensional semi-time-centred implicit version of the QUICK scheme was used to represent the depth integrated transport equation in this study using the alternating direction implicit procedure.

For the first half time step, equation (4) can be expressed as :

$$SH_{i,j}^{n+1/2} - SH_{i,j}^{n} + \frac{\Delta t}{4\Delta x}\Big[\ UH_{i+1/2,j}^{n+1/2}(S_{i+1,j}^{n+1/2} + S_{i,j}^{n+1/2})$$

$$-UH_{i-1/2,j}^{n+1/2}(S_{i,j}^{n+1/2} + S_{i-1,j}^{n+1/2})\ \Big]$$

$$-\frac{\Delta t}{16\Delta x}\ (UH_{i+1/2,j}^{n+1/2}\nabla^2 S_{i+1/2,j}^{n} - UH_{i-1/2,j}^{n+1/2}\nabla^2 S_{i-1/2,j}^{n})$$

$$+\frac{\Delta t}{4\Delta y}\Big[\ VH_{i,j+1/2}^{n}(S_{i,j+1}^{n} + S_{i,j}^{n}) - VH_{i,j-1/2}^{n}(S_{i,j}^{n} + S_{i,j-1}^{n})\ \Big]$$

$$-\frac{\Delta t}{16\Delta y}\ (VH_{i,j+1/2}^{n}\nabla^2 S_{i,j+1/2}^{n} - VH_{i,j-1/2}^{n}\nabla^2 S_{i,j-1/2}^{n})$$

$$= \frac{\Delta t}{2\Delta x^2}\Big[\ HD^{n+1/2}_{xx_{i+1/2,j}}(S^{n+1/2}_{i+1,j} - S^{n+1/2}_{i,j}) - HD^{n+1/2}_{xx_{i-1/2,j}}(S^{n+1/2}_{i,j} + S^{n+1/2}_{i-1,j})$$

$$+\ HD^{n+1/2}_{xy_{i+1/2,j}}(S^n_{i+1/2,j+1/2} - S^n_{i+1/2,j-1/2})$$

$$-HD^n_{xy_{i-1/2,j}}(S^n_{i-1/2,j+1/2} - S^n_{i-1/2,j-1/2})\ \Big]$$

$$+\frac{\Delta t}{2\Delta y^2}\Big[\ HD^n_{yy_{i,j+1/2}}(S^n_{i,j+1} - S^n_{i,j}) - HD^n_{yy_{i,j-1/2}}(S^n_{i,j} - S^n_{i,j-1})$$

$$+\ HD^n_{yx_{i,j+1/2}}(S^n_{i+1/2,j+1/2} - S^n_{i-1/2,j+1/2})$$

$$-HD^n_{yx_{i,j-1/2}}(S^n_{i+1/2,j-1/2} - S^n_{i-1/2,j-1/2})\ \Big]$$

$$-\omega(SH)^n_{i,j} \tag{6}$$

where:

$$\nabla^2 S_{i+1/2,j} = S_{i+1,j} - 2S_{i,j} + S_{i-1,j} \quad \text{if } U \geq 0 \tag{7}$$
$$= S_{i,j} - 2S_{i+1,j} + S_{i+2,j} \quad \text{if } U < 0 \tag{8}$$

$$\nabla^2 S_{i-1/2,j} = S_{i,j} - 2S_{i-1,j} + S_{i-2,j} \quad \text{if } U \geq 0 \tag{9}$$
$$= S_{i-1,j} - 2S_{i,j} + S_{i+1,j} \quad \text{if } U < 0 \tag{10}$$

$$\nabla^2 S_{i,j+1/2} = S_{i,j+1} - 2S_{i,j} + S_{i,j-1} \quad \text{if } U \geq 0 \tag{11}$$
$$= S_{i,j} - 2S_{i,j+1} + S_{i,j+2} \quad \text{if } U < 0 \tag{12}$$

$$\nabla^2 S_{i,j-1/2} = S_{i,j-1} - 2S_{i,j} + S_{i,j+1} \quad \text{if } U \geq 0 \tag{13}$$
$$= S_{i,j-1} - 2S_{i,j} + S_{i,j+1} \quad \text{if } U < 0 \tag{14}$$

Likewise, for the second half time step, equation (4) can be expressed as:

$$SH^{n+1}_{i,j} - SH^{n+1/2}_{i,j} + \frac{\Delta t}{4\Delta x}\Big[\ UH^{n+1/2}_{i+1/2,j}(S^{n+1/2}_{i+1,j} + S^{n+1/2}_{i,j})$$

$$-UH^{n+1/2}_{i-1/2,j}(S^{n+1/2}_{i,j} + S^{n+1/2}_{i-1,j})\ \Big]$$

$$-\frac{\Delta t}{16\Delta x}\ (UH^{n+1/2}_{i+1/2,j}\nabla^2 S^{n+1/2}_{i+1/2,j} - UH^{n+1/2}_{i-1/2,j}\nabla^2 S^{n+1/2}_{i-1/2,j})$$

$$+\frac{\Delta t}{4\Delta y}\left[\ VH^{n+1}_{i,j+1/2}(S^{n+1}_{i,j+1} + S^{n+1}_{i,j}) - VH^{n+1}_{i,j-1/2}(S^{n+1}_{i,j} + S^{n+1}_{i,j-1})\ \right]$$

$$-\frac{\Delta t}{16\Delta y}\ (VH^{n+1}_{i,j+1/2}\nabla^2 S^{n+1/2}_{i,j+1/2} - VH^{n+1}_{i,j-1/2}\nabla^2 S^{n+1/2}_{i,j-1/2})$$

$$= \frac{\Delta t}{2\Delta x^2}\left[\ HD^{n+1/2}_{xx_{i+1/2,j}}(S^{n+1/2}_{i+1,j} - S^{n+1/2}_{i,j}) - HD^{n+1/2}_{xx_{i-1/2,j}}(S^{n+1/2}_{i,j} - S^{n+1/2}_{i-1,j})\right.$$

$$+\ \ HD^{n+1/2}_{xy_{i+1/2,j}}(S^{n+1/2}_{i+1/2,j+1/2} - S^{n+1/2}_{i+1/2,j-1/2})$$

$$\left. -HD^{n+1/2}_{xy_{i-1/2,j}}(S^{n+1/2}_{i-1/2,j+1/2} - S^{n+1/2}_{i-1/2,j-1/2})\ \right]$$

$$+\frac{\Delta t}{2\Delta y^2}\left[\ HD^{n+1}_{yy_{i,j+1/2}}(S^{n+1}_{i,j+1} - S^{n+1}_{i,j}) - HD^{n+1}_{yy_{i,j-1/2}}(S^{n+1}_{i,j} - S^{n+1}_{i,j-1})\right.$$

$$+\ \ HD^{n+1}_{yx_{i,j+1/2}}(S^{n+1/2}_{i+1/2,j+1/2} - S^{n+1/2}_{i-1/2,j+1/2})$$

$$\left. -HD^{n}_{yx_{i,j-1/2}}(S^{n+1/2}_{i+1/2,j-1/2} - S^{n+1/2}_{i-1/2,j-1/2})\ \right]$$

$$-\omega(SH)^{n+1/2}_{i,j} \tag{15}$$

where:

$$\nabla^2 S_{i+1/2,j} = S_{i+1,j} - 2S_{i,j} + S_{i-1,j} \quad \text{if } V \geq 0 \tag{16}$$
$$= S_{i,j} - 2S_{i+1,j} + S_{i+2,j} \quad \text{if } V < 0 \tag{17}$$

$$\nabla^2 S_{i-1/2,j} = S_{i,j} - 2S_{i-1,j} + S_{i-2,j} \quad \text{if } V \geq 0 \tag{18}$$
$$= S_{i-1,j} - 2S_{i,j} + S_{i+1,j} \quad \text{if } V < 0 \tag{19}$$

$$\nabla^2 S_{i,j+1/2} = S_{i,j+1} - 2S_{i,j} + S_{i,j-1} \quad \text{if } V \geq 0 \tag{20}$$
$$= S_{i,j} - 2S_{i,j+1} + S_{i,j+2} \quad \text{if } V < 0 \tag{21}$$

$$\nabla^2 S_{i,j-1/2} = S_{i,j-1} - 2S_{i,j} + S_{i,j+1} \quad \text{if } V \geq 0 \tag{22}$$
$$= S_{i,j-1} - 2S_{i,j} + S_{i,j+1} \quad \text{if } V < 0 \tag{23}$$

with i, j = grid locations in x, y coordinate directions; n and $n+1$= time step numbers; Δt = time step; Δx = grid size in x-direction and Δy = grid size in y-direction. Note that for both half time steps the decay term has been computed at the lower time step.

4. Physical experiments and model calibration

Teixeira and Shiono (1992) have conducted physical laboratory based experiments on the tank modelled numerically in this study (see Figures 1,2 and 3). However, they used a conservative tracer, instead of a non-conservative disinfectant. The instantaneous velocities were measured using a Laser Doppler Anemometer and a through flow rate of 1170 litres per second was used for all experimental runs. Concentration data were collected using two Model-10 Turner design fluorometers. A concentrated solution of 50 millilitres of a 100 parts per million Rhodamine-B dye was instantaneously injected at the inlet for each experimental run. This quantity of dye yielded a tank mean concentration, TRC_0, of 5.69 parts per billion. The tank mean concentration is defined as the mass of tracer injected into the tank divided by the volume of water in the tank. It was established that strong horizontal and vertical recirculation regions existed in the first two compartments (due to the configuration of the tank inlet and jet separation from the tank walls) and that the flow in compartments 3 to 8 was fairly uniform over the depth.

The tank planform area was overlain by 49 times 24 finite difference grid squares in the x, y directions respectively and with the length of each grid being 0.0425 metres. This proved to be a reasonably good fit, in that the domain and baffles were well represented. Although the finite difference scheme for the hydrodynamic equations was unconditionally stable, for accuracy requirements the time step, Δt, was chosen to satisfy the condition (Falconer, 1991) :

$$\frac{\Delta t}{\Delta x}\sqrt{gH_{\text{max}}} \leq 8 \tag{24}$$

where H_{max} = maximum depth. If the diffusion and decay terms are removed from equation (4), then Chen (1992) has shown that the corresponding QUICK scheme is stable provided that:

$$U_{\text{max}}\frac{\Delta t}{\Delta x} + V_{\text{max}}\frac{\Delta t}{\Delta y} \leq 2 \tag{25}$$

where U_{max} and V_{max} = maximum velocities in the x and y directions respectively. Pure advection presents a more severe stability situation than advection with diffusion. From computer simulations using the hydrodynamic equations (1), (2) and (3), it was found that $U_{\max}$ and $V_{\max}$ were approximately 18 and 110 millimetres per second respectively. The choice of $\Delta t = 0.14$ seconds and $\Delta x = 0.0425$ metres thereby satisfies both equations (24) and (25).

The numerical model was set up to run for steady through flow conditions. An inflow discharge of 1170 litres per second was specified and water elevations were specified at the outlet using the linear wave equation (Hebenstreit et al., 1980). A constant tracer source of 50 parts per million was included at the inlet boundary over a period of 10 seconds. This was in close agreement with the input conditions in the physical model. At the outlet the concentrations were calculated by backward differencing and assuming only advective transport to be significant. At the closed wall boundaries the normal velocity and dispersion coefficients were equated to zero, thereby allowing no tracer flux through the wall (Falconer, 1986). The numerical simulations were run for various values of β, k_s, k_l and k_t. Choosing $\beta = 1.017$, $k_s = 1$ mm, $k_l = 13$ and $k_t = 1$ gave good agreement with the experimental results and these values were then used for all subsequent runs. Figures 4(a)-4(c) compare the computed and measured tracer concentrations at 480 millimetres above the centre of the base of compartments 5 to 7 respectively. In these figures the tracer concentrations (TRC) have been non-dimensionalized by dividing by the tank mean concentration (TRC_0) and time has been non-dimensionalized by dividing by the theoretical contact time (T). The theoretical contact (or residence time) is defined as the volume of flow in the tank divided by the average flow through rate. It can be seen that the computed and measured concentrations are generally in close agreement at 480 millimetres above the centre of the base for compartments 5 to 7. Comparisons of the computed and measured concentrations at other locations in compartments 3 to 8 (not shown in this paper) yield similar conclusions. The agreement in compartments 1 and 2 is not as good, since the flow in compartments 1 and 2 was primarily three-dimensional in structure. Thus the QUICK scheme is shown to be capable of accurately modelling the transport of a conservative solute when the flow is uniform over the depth. Further details regarding the numerical modelling of a conservative tracer in the tank, using a 2-D semi-time-centred implicit QUICK scheme, can be found in Ismail and Falconer (1995).

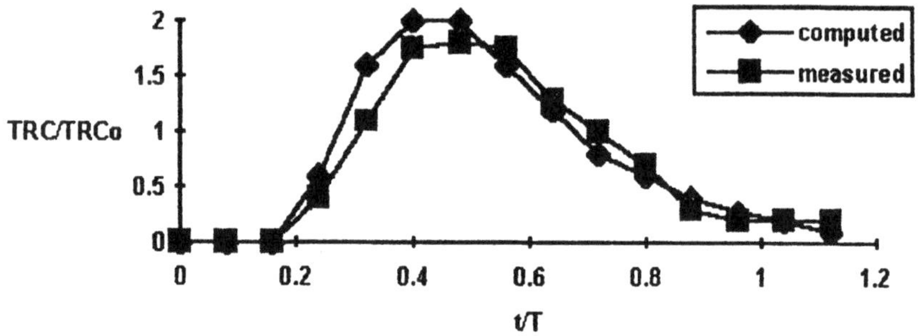

Figure 4a. Comparison of computed and measured tracer concentrations 480 mm above base of compartment 5

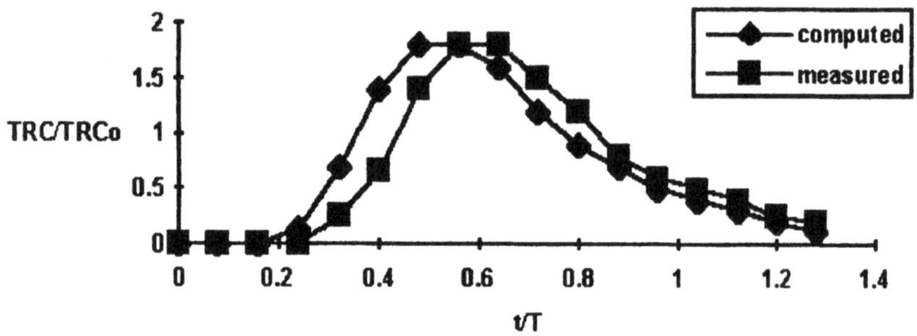

Figure 4b. Comparison of computed and measured tracer concentrations 480 mm above base of compartment 6

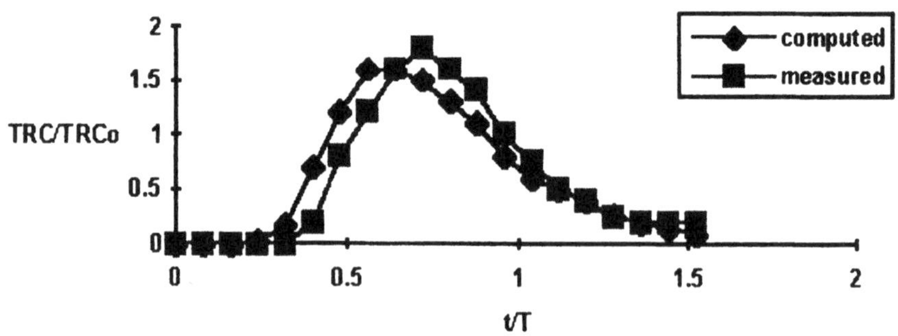

Figure 4c. Comparison of computed and measured tracer concentrations 480 mm above base of compartment 7

5. Model Implementation

The values of Δx and Δt were taken to be 0.0425 metres and 0.14 seconds respectively i.e. as in section 4. The numerical model was set up to run for steady through flow conditions. An inflow discharge of 1170 litres per second was specified and water elevations given at the outlet, using the linear wave equation. A continuos constant chlorine concentration of 6 milligrammes per litre was specified at the inlet. At the outlet the concentrations were calculated by backward differencing and assuming only advective transport to be significant. At the closed wall boundaries the normal velocity and dispersion coefficients were equated to zero, thereby allowing no chlorine flux through a wall. Values of $\beta = 1.017$, $k_s = 1$ mm, $k_l = 13$ and $k_t = 1$ were chosen. Figures 5(a) and 5(b) show the depth averaged concentrations assuming combined chlorine and free chlorine decay respectively. The model predicts that combined chlorine will decay to 0.5 milligrammes per litre in the last compartment of the tank and free chlorine will decay to 0.5 milligrammes per litre in the second compartment of the tank. Thus the model correctly predicts that combined chlorine is more persistent than free chlorine and that the chlorine concentration level is rapidly reduced by decay and kill of bugs. During the course of the simulations no numerical instabilities were detected.

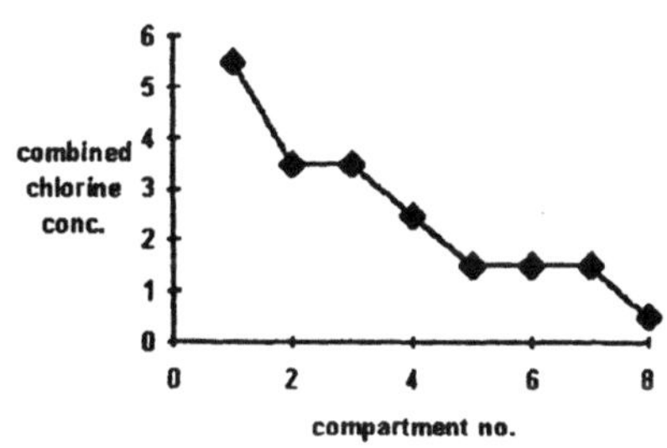

Figure 5a. Depth averaged combined chlorine concentrations

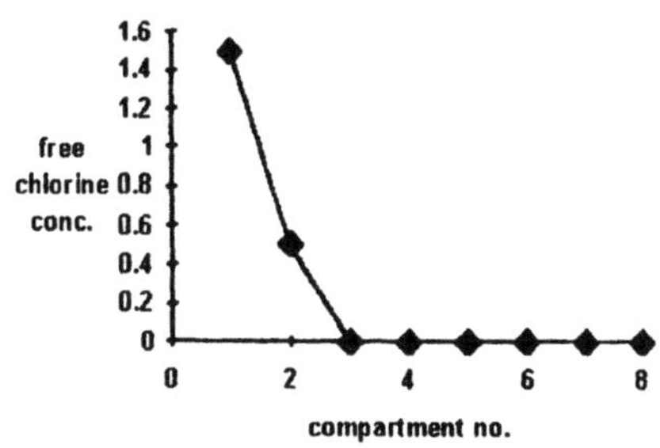

Figure 5b. Depth averaged free chlorine concentrations

6. Conclusions

In the paper some health problems associated with excess chlorine in a contact tank have been highlighted. A 2-D semi-time-centred implicit QUICK model has been developed to predict the decay of chlorine through a rectangular scaled serpentine contact tank with the numerical scheme leading to stable, accurate and computationally efficient results. No experimental data are yet available to verify the predictions for a non-conservative tracer but the scheme correctly predicts that combined chlorine is more persistent than free chlorine. Furthermore, the scheme gives good agreement with experimental data when the solute is a conservative solute and efforts are currently being made to obtain full scale field data.

References

1. Bull, R.J. (1986). *"Carcinogenic hazards associated with the chorination of drinking water"*, in Organic Carcinogens in Drinking Water, ed. Ram, N.M. et. al., John Wiley and Sons, London.
2. Chick, H. (1908). *"An investigation of the laws of disinfection"*, Journal of Hygiene Vol.8, No. 92.
3. Clark, R.M. et al. (1994). *"DBP control in drinking water: Cost and Performance"*, Journal of Enviromental Engineering, ASCE, Vol. 120 No. 4, pp. 759-782.
4. Chen, Y. (1992). *"Numerical modelling of solute transport processes using higher order accurate finite difference schemes: numerical treatment of flooding and drying"*, PhD Thesis, Department of Civil Engineering, University of Bradford, UK, pp.308.
5. Culp, G.L. (1974). *"New Concepts in Water Purification"*, Van Nostrand Reinhold.
6. Falconer, R.A. (1986). *"A water quality simulation study of a natural harbour"*, Journal of Waterway, Port, Coastal and Ocean Engineering, ASCE, Vol. 112, No. 2, pp 234-259.
7. Falconer, R.A. (1991). *"Review of modelling flow and pollutant transport in hydraulic basins"*, Proceedings of First International Conference on Water Pollution: Modelling, Measuring and Prediction, Southampton, UK, Computational Mechanics Publications, September 1991, pp. 519-531.
8. Falconer, R.A. (1993). *"Theory of nearly horizontal flows"* in Coastal, Estuarial and Harbour Engineers Reference Book(ed. Abbott, M.B. and Price, W.A.), Chapter 2, E and F.N.Spon, pp.27-36.
9. Falconer, R.A and Liu, S. (1988). *"Modelling solute transport using QUICK scheme"*, Journal of Enviromental Engineering, ASCE, Vol. 114, No 1, pp. 3-20.
10. Grizzle, J.M. (1984). *"Papillomas on fish exposed to chlorinated wastewater effluent"*, Journal of the National Cancer Institute, Vol. 73, pp. 1133-1142.
11. Hebenstreit, G.T. et al. (1980). *"Application of improved numerical techniques to the tsunami response of island systems"*, Journal of Physical Oceanography, Vol. 10, pp. 1134-1140.
12. Ismail, A.I.B.Md. (1996). *"Mathematical modelling of flow and disinfection processes in contact tanks"*, PhD thesis, Department of Civil and Enviromental Engineering, University of Bradford, pp. 221
13. Ismail, A.I.B.Md. and Falconer, R.A. (1996). *"Computational modelling of tracer transport in a contact tank"*, Proceedings of the Malaysian National Physics Conference, Universiti Pertanian Malaysia , November 1995(in press).
14. Leonard, B.P. (1979). *"A stable and accurate modelling procedure based on quadratic*

upstream modelling", Computer Methods in Applied Mechanical Engineering, Vol.19, pp. 59-98.

15. McGhee, T. J. (1991). *"Water Supply and Sewerage"*, McGraw Hill Inc, New York, Sixth edition.
16. Roelfzema, A and Gerritsen, H. (1989). *"Some applications of flow and transport modelling in three-dimensions"*, Proceedings of the Ninth Australian Conference on Coastal and Ocean Engineering, Adelaide, pp. 168-172.
17. Singer, P.C. (1994). *"Control of disinfection by-products in drinking water"*, Journal of Enviromental Engineering, ASCE, Vol. 120, No. 4, pp. 727-744.
18. Teixeira, E.C and Shiono, K. (1992). *"An investigation of the behaviour of a chlorine contact tank"*, Proceedings of the Sixth International Symposium on Application of Laser Technique to Fluid Mechanics, Lisbon, Portugal.
19. Trussell, R.R. and Chao, J.L. (1977). *"Rational design of chlorine contact facilities"*, Journal Water Pollution Control Federation", April.
20. White, G.C. (1979). *"Disinfection: the last line of defense for potable water"*, American Water Works Association., Vol 67, pp 410-413.

THE DYNAMICS OF A NEAR-GRAZING OSCILLATOR

J. MOLENAAR
Faculty of Mathematics and Computing Science
Eindhoven University of Technology
The Netherlands

AND

J. DE WEGER, D. BINKS, AND W. V.D. WATER
Faculty of Physics
Eindhoven University of Technology
The Netherlands

1. Introduction

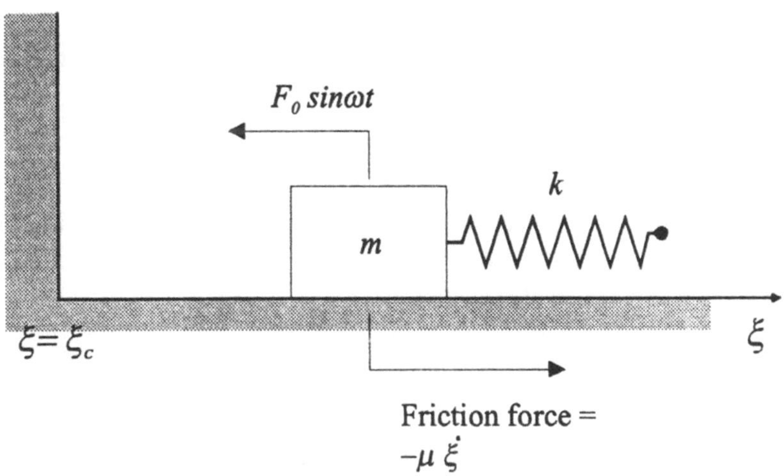

Figure 1. Sketch of the impact oscillator under study.

The driven harmonic oscillator in free space is a standard example of linear theory; its solution is explicitly known as a function of the initial values and the system parameters. This drastically changes if the oscillator may have impact with an obstacle. The impact introduces a nonlinearity into the system which gives rise to intricate bifurcations. Impact oscillators are found in mechanical systems such as fittings, railway bogies, motion of ships in a harbour and of suspension bridges. For a list of references, see Ref.(6). Recently, impacting systems are used in atomic force microscopy, with which the shape of surfaces can be measured very accurately from impacting on it on a microscopic level; see Ref.(8).

In this contribution we report some insights from current research in this field. Rather than on technical derivations the emphasis will be on the exposition of the physical background of the system. Because bifurcations occur when an impact becomes grazing, i.e. hits the obstacle at very low speed, we concentrate on the dynamics of grazing orbits. We show that grazing orbits give rise to a discontinuous distortion of phase space, which is similar to the one found in the famous horseshoe map. This indicates that bifurcations to chaotic behaviour are to be expected.
We also study the influence of the finite rigidity of the obstacle. This is in particular important for the comparison between numerical and experimental results. It is shown that a finite time of contact between the oscillator and the obstacle may influence the bifurcations considerably.

2. Mathematical model

We study the system sketched in Fig.1. Between impacts the harmonic oscillator is described by

$$m\ddot{\xi} + \mu\dot{\xi} + k\xi = F_0 \sin(\omega\tau), \tag{1}$$

where m is the mass of the oscillator, μ the friction coefficient, k the spring stiffness, F_0 the excitation amplitude, and ω the driving frequency. The nondimensional version of Eq.(1) reads as

$$\ddot{x} + \nu\dot{x} + \Omega^2(x+1) = f_0 \sin(2\pi t + \phi), \tag{2}$$

where $\nu := 2\pi\mu/m\omega$, $\Omega^2 := 4\pi^2 k/m\omega^2$ and $f_0 := 4\pi^2 F_0/m\omega^2\xi_c$. As dimensionless position we introduced $x := (\xi - \xi_c)/\xi_c$ with ξ_c the position of the obstacle. For time we used $t := (\omega\tau - \phi)/2\pi$. If we take for ϕ the solution of $\tan(\phi) = 2\pi\nu/(\Omega^2 - 4\pi^2)$, the particular solution has the simple form

$$p(t) = a \sin 2\pi t - 1, \tag{3}$$

with amplitude $a = f_0/\sqrt{(\Omega^2 - 4\pi^2)^2 + 4\pi^2\nu^2}$.

Experimentally the system is observed periodically and in the vicinity of the stop. For damped systems the particular solution is dominant. So it is convenient to take as moments of observation t_i the time that $p(t)$ is closest to the obstacle, i.e. $t_i = (i + 1/4)$, i = 1,2,..

3. Phase space analysis

Let us study the dependence of the solutions of (2) as a function of the amplitude f_0 for fixed value of the frequency ω. For f_0 such small that $a < 1$, the homogeneous solutions will damp out and the particular solution (3) is observed. For f_0 such that $a = 1$, $p(t)$ is grazing at times t_i. The corresponding elliptical orbit is drawn in Fig.2 in the phase plane $(x, v := \dot{x})$.

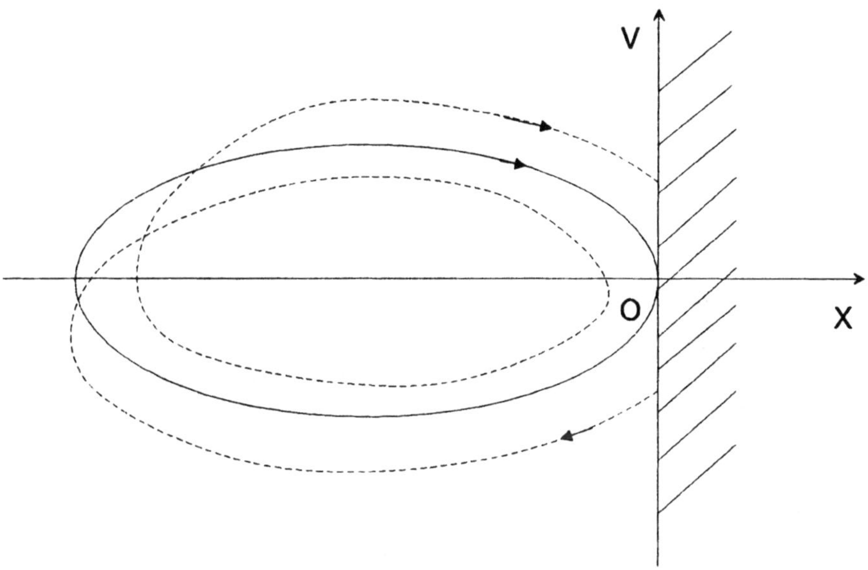

Figure 2. A grazing and a period-2 orbit in phase space.

Further increase of f_0 will lead to impacts. After some transients the orbit will be periodic or chaotic. As an example in Fig.2 a 2-periodic orbit is sketched with one impact per two cycles of the driving force. In Fig.3 the corresponding orbits are sketched in the (t, x) plane.

Using the notation $(x_i, v_i) := (x(t_i), v(t_i))$ we can describe the dynamics in phase space by the *Poincaré* map

$$PM : (x_i, v_i) \rightarrow (x_{i+1}, v_{i+1}). \tag{4}$$

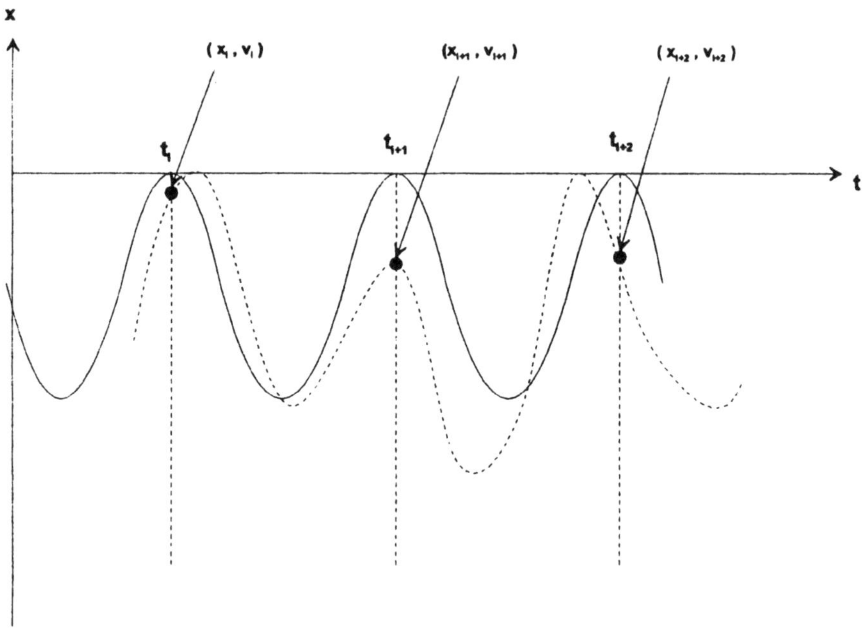

Figure 3. A grazing and a period-2 orbit in the space-time plane.

Before giving explicit expressions for this map, we present some introductory remarks.
Consider a point $P := (x_1, v_1)$, which lies on the grazing orbit and is close to the obstacle. See Fig.4.

The time of flight Δt from P to the origin O follows from Taylor expansion around O:

$$\Delta t = \sqrt{2x_1/A}. \tag{5}$$

Note that both x_1 and A are negative because the obstacle is assumed to be situated at $x \geq 0$. The acceleration A is taken constant as long as the oscillator is very close to the obstacle. In the dimensionless model its value is given by

$$A = \ddot{p}(t_i) = -4\pi^2. \tag{6}$$

An alternative expression for Δt can be found from expansion of the velocity around P:

$$\Delta t = v_1/|A|. \tag{7}$$

Combining (5) and (7) we find the parabola

$$v_1^2 - 2Ax_1 = 0. \tag{8}$$

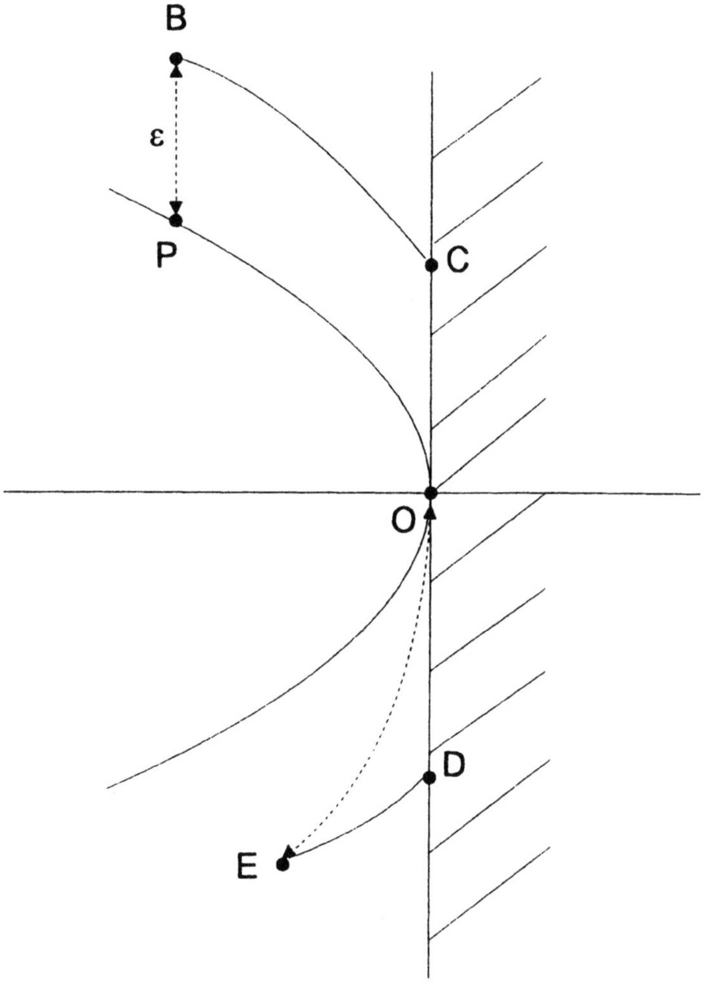

Figure 4. Detail of the phase space around a near-grazing impact.

Points inside this parabola do not impact, points outside it will do or have done. Next we consider a point $B := (x_1, v_1 + \epsilon)$ with $0 < \epsilon << 1$. So, B lies above and very close to the grazing orbit. The orbit starting at B will hit the obstacle at point C after a period $\Delta t(\epsilon)$, say. Taylor expansion around B yields

$$\Delta t(\epsilon) = \Delta t - \frac{\sqrt{2 v_1 \epsilon}}{|A|}, \tag{9}$$

Note that $\Delta t(\epsilon) < \Delta t$, because $v_B > v_P$. The velocity $v_{impact} = v_C$ at impact is at lowest order in ϵ given by

$$v_{impact} = \sqrt{2v_1\epsilon}. \tag{10}$$

We consider for the moment an elastic impact. Then the velocity is instantaneously reversed and the orbit jumps from C to D in the phase plane. In the remaining time $\sqrt{2v_1\epsilon}/|A|$ it travels from D to E where it arrives with velocity

$$v_E = v_D - \sqrt{2v_1\epsilon} = -2\sqrt{2v_1\epsilon} \tag{11}$$

In conclusion, the orbit starting at P has vanishing velocity after time of flight Δt, while after the same period the orbit starting at B has velocity v_E. The initial interval AP of length ϵ above the parabola thus evolves into an interval OE of length at least $|v_E| \approx \sqrt{\epsilon}$. So this interval is stretched considerably. The rate of stretching is measured by $|dv_E/d\epsilon| \approx 1/\sqrt{\epsilon}$, which becomes infinite in the limit of vanishing ϵ. This singular way of deformation is also typical for the famous horseshoe map (see, e.g., Ref.(7)).

This map is a transformation of the square onto itself. In Fig.5 the action of a generalized version of this map is depicted.

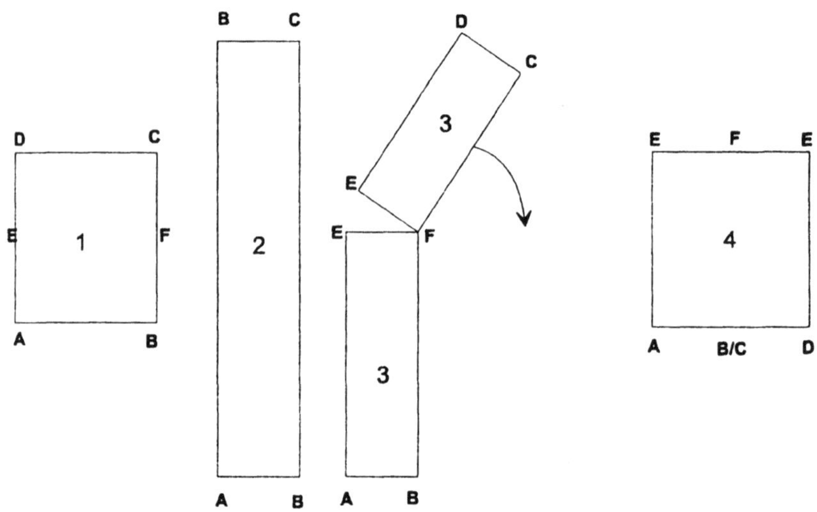

Figure 5. Sketch of the action of a generalized horseshoe map.

It is seen that a singular stretching takes place along the line EF. The horseshoe map is a generic example of a system with chaotic behaviour. It is therefore to be expected that the map PM introduced above will show similar behaviour.

4. Modelling of the impact

At impact the oscillator may loose energy. A possible approach is to consider the energy loss as taking place instantaneously at the moment of first contact. We model this by introducing a coefficient r with $0 < r \leq 1$:

$$\bar{v}_{impact} = r v_{impact}. \tag{12}$$

After that the obstacle will penetrate the obstacle and spend a finite time δt in it. Ignoring friction effects, which are very small around grazing, the equation of motion during penetration is

$$\ddot{x} + K^2 x = A, \tag{13}$$

with A given by (6), and K^2 the stiffness of the obstacle. In general we have $K >> \Omega$, with Ω defined in (2). The solution of (13) reads as

$$x(t - t_{impact}) = \frac{A}{K}(1 - cos(Kt)) + \frac{\bar{v}_{impact}}{\sqrt{K}} sin(Kt)). \tag{14}$$

From the condition $x(\delta t) = 0$ we obtain

$$\delta t = \frac{1}{K} \arccos \frac{1 - \lambda^2}{1 + \lambda^2}, \qquad \lambda := \frac{\bar{v}_{impact} K}{A}. \tag{15}$$

Plotting δt as a function of λ in Fig.6 we recognize two regimes:

$$\delta t = \begin{cases} \frac{\pi}{K} , & \lambda >> 1, \\ \frac{2\bar{v}_{impact}}{|A|} , & \lambda << 1. \end{cases} \tag{16}$$

So, for K fixed and $v_{impact} \downarrow 0$, δt is independent of K, while for v_{impact} fixed and $K \to \infty$, δt is proportional to $1/K$. The first limit corresponds to the grazing limit in case of an obstacle of finite hardness, as found in the experimental situation. The second limit refers to taking the wall infinitely hard; this can only be done in theory. This indicates that one should be careful in interchanging the limits $v_{impact} \to 0$ and $K \to \infty$. In measurements the influence of the softness of the wall may be considerable, see Ref.(4) .

5. The near-grazing map

The derivation of the map PM in (4) consists of two elements: the dynamics between impacts and the dynamics during impact. The latter aspect has

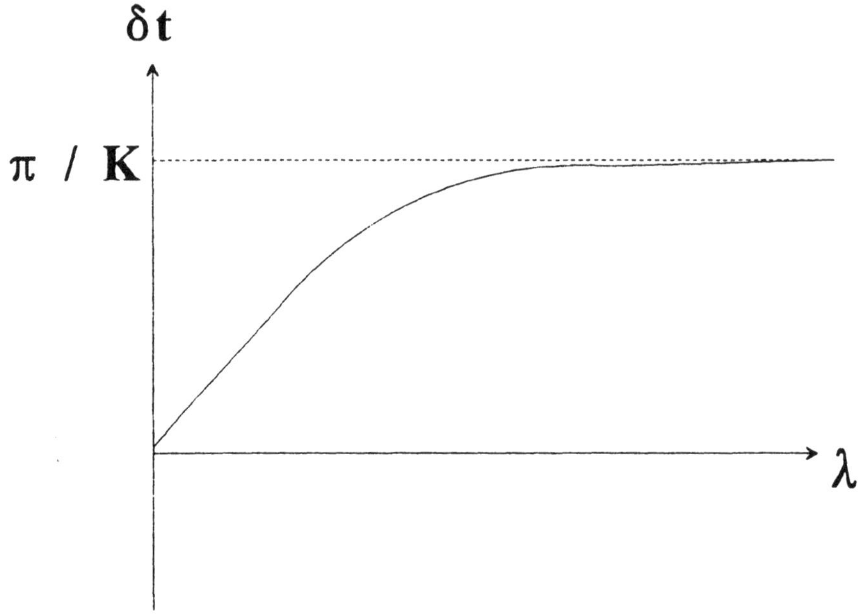

Figure 6. The flight time δt in the obstacle as a function of the parameter λ.

been dealt with in §4. Between impacts the system behaves linearly and has the solution

$$\mathbf{x}(t) = \mathbf{M}(t - t_0)\mathbf{x}(t_0) + (1 - \mathbf{M}(t - t_0))\mathbf{p}(t) \tag{17}$$

where $\mathbf{x}(t) := (x(t), \dot{x}(t))^T$ and $\mathbf{p}(t) := (p(t), \dot{p}(t))^T$. The evolution matrix $\mathbf{M}$ is explicitly known as a function of the system parameters:

$$\mathbf{M}(t) = (s_2 - s_1)^{-1} \begin{pmatrix} s_2 e^{s_1 t} - s_1 e^{s_2 t} & e^{s_2 t} - e^{s_1 t} \\ s_1 s_2 (e^{s_1 t} - e^{s_2 t}) & s_2 e^{s_2 t} - s_1 e^{s_1 t} \end{pmatrix} \tag{18}$$

with

$$\begin{aligned} s_1 &= (-\nu + \sqrt{\nu^2 - 4\Omega^2})/2 \\ s_2 &= (-\nu - \sqrt{\nu^2 - 4\Omega^2})/2. \end{aligned} \tag{19}$$

The derivation of the map PM in (4) is far from being straightforward. We refer for it to Ref.(5). It is convenient to write the map in terms of the variables

$$\begin{aligned} r_i &= l^{-1}(v_i^2 - 2Ax_i) \\ s_i &= -l^{-1}\left[M_{22}(((v_i^2 - 2Ax_i) + 2Ap_i) + 2AM_{12}(v_i - \dot{p}_i))\right] \end{aligned} \tag{20}$$

where the scaling factor l may be chosen arbitrarily, provided that $l \neq 0$. From (20) we see that r_i, s_i are defined via a nonlinear transformation of the phase space, such that r_i measures the distance to the grazing orbit and s_i is a linear combination of r_i and the velocity v_i.

The case $r_i > 0$ corresponds to the area outside the grazing orbit in phase space and thus gives rise to impact. If $r_i < 0$ the orbit passes the obstacle without impact. We assume that r_i, s_i, p_i and $\dot{p}_i$ are small and of the same order of magnitude. So, not only the general solution is assumed to be near grazing, but also the particular solution. Up to linear order in these quantities we find for $r_i < 0$ the map

$$\left\{ \begin{array}{lcl} r_{i+1} & = & \alpha r_i + s_i + \rho \\ s_{i+1} & = & -\gamma r_i \end{array} \right. \quad \text{for } r_i \leq 0 \tag{21}$$

where $\alpha := \mathrm{Tr}(\mathbf{M})$, $\gamma := \det(\mathbf{M})$ and $\rho := -2Al^{-1}(1 - \alpha + \gamma)$. For $r_i > 0$ we have to discern between the two regimes $\lambda \ll 1$ and $\lambda \gg 1$, according to the findings in §4. In the first regime we find the expression

$$\left\{ \begin{array}{lcl} r_{i+1} & = & \frac{-2A(r-1)M_{12}}{\sqrt{l}}\sqrt{r_i} + \beta r_i + s_i + \rho \\ s_{i+1} & = & -\zeta r_i \end{array} \right. \quad \lambda \ll 1 \tag{22}$$

where $\beta := \alpha + 4r(r-1)M_{11} + 2(r-1)(1-2r)M_{22} + (r-1)^2 M_{22}^2$ and $\zeta := \gamma + 2(r-1)\gamma + (r-1)^2 M_{22}^3$.

In the second regime we obtain

$$\left\{ \begin{array}{lcl} r_{i+1} & = & \frac{2A(r+1)M_{12}}{\sqrt{l}}\sqrt{r_i} + \sigma_r r_i + s_i + \delta \\ s_{i+1} & = & -\psi r_i \end{array} \right. \quad \text{for } \lambda \gg 1 \tag{23}$$

where $\sigma_r := \alpha - 2(r+1)M_{22} + (r+1)^2 M_{22}^2$, $\psi := \gamma - 2(r+1)\gamma + (r+1)^2 M_{22}^3$ and $\delta := \rho + 2AM_{12}\delta t / l$.

It is interesting to note that in (22) the square root singularity disappears in the limit $r \to 0$. This implies that, if the impact is elastic and the obstacle is not fully rigid, the near-grazing map is non-singular. It is just as if the oscillator does not notice the presence of the obstacle. This is in sharp contrast with the map derived for an infinitely hard obstacle. The latter case corresponds with the limit $\lambda \to \infty$ and thus with the map given by (23). Here, the square root singularity is present for all values of r, and thus also for elastic impacts.

For the analysis of these maps and the comparison with experimental data we refer to Ref.(5).

6. References

1. Nordmark, A.B. (1991), 'Non-periodic motion caused by grazing incidence in an impact oscillator', *J. Sound Vib.* **Vol. no. 145**, pp. 279–297
2. Chin, W., Ott, E., Nusse, H.E., and Grebogi, C. (1994), 'Grazing bifurcations in impact oscillators', *Phys. Rev. E*, **Vol. ,no 50**, pp.4427-4444
3. Weger, J. de, Binks, D., Molenaar, J., and Water, W.v.d. (1996), 'Generic behaviour of grazing impact oscillators', *Phys. Rev. Letters*, **Vol. no 76**, pp.3951–3954
4. Weger, J. de, Binks, D., Molenaar, J., and Water, W.v.d. (1996), 'Universal bifurcations in impact oscillators', in 'Proceedings of the IUTAM Symposium on Interaction between Dynamics and Control in Advanced mechanical Systems, Eindhoven, April 1996, The Netherlands', ed. Campen, D.H. van.
5. Weger, J. de, Binks, D., Molenaar, J., and Water, W.v.d. (1996), 'The dynamics of a near-grazing oscillator', *to be submitted to Phys. Rev. E.*
6. Budd, C.J. (1996), 'Non-smooth dynamical systems and the grazing bifurcations', in 'Nonlinear mathematics and its applications', ed. Aston, P.J., Cambridge University Press.
7. Wiggins, S. (1996), 'An introduction into the mechanisms and methods for the detection of chaos', in 'Nonlinear mathematics and its applications', ed. Aston, P.J., Cambridge University Press.
8. Q. Zhong, D. Inniss, K. Kjoller and V.B. Ellings (1993), 'Forces affecting the substrate in resonant tapping force microscopy', *Surf. Sci.* **Vol. no. 290**, L688; J.P. Spatz, S. Sheiko, M. Möller, R.G. Winkler, P.Reineker and O. Marti (1995), 'Fractured polymer/silica fiber surface studied by tapping mode atomic force microscopy', *Nanotechnology* **Vol. no. 6**, p. 40.

OSCILLATION CRITERIA FOR SECOND ORDER NONLINEAR DIFFERENTIAL EQUATIONS

S.M. NABABAN
Department of Mathematics, ITB
Jl. Ganeca 10, Bandung
Indonesia

Abstract. In this paper we consider the following second order nonlinear differential equation

$$u'' + f(t, u) = 0, \qquad t \geq 0.$$

Oscillation criteria for the above equation will be established by slightly modification of the method that has been used previously. The results obtained will improve and extend the previous results and some conditions imposed in the theorem will be weaker than used before.

1. Introduction

We consider the second order nonlinear differential equation

$$u'' + f(t, u) = 0, \qquad t \geq 0 \tag{1}$$

where the function f satisfies certain conditions to be given later.

Definition 1

a A solution $u(t)$ of (1) is called oscillatory in $[0, \infty)$ if for every $t_0 \in [0, \infty)$ there exists a $t_1 \geq t_0$ such that $u(t_1) = 0$.
b A solution $u(t)$ of (1) is called nonoscillatory in $[0, \infty)$ if there exists a $t_1 \geq 0$ such that $u(t) \neq 0$ for $t \geq t_1$.
c The differential equation (1) is called oscillatory in $[0, \infty)$ if every solution of (1) is oscillatory in $[0, \infty)$.

Oscillation criteria for the differential equation

$$u'' + a(t)f(u) = 0, \qquad t \geq 0 \tag{2}$$

including the Emden-Fowler equation, where $f(u) = |u|^\gamma \, sgn \, u$, has been developed by many authors, we mention, in particularly, the papers of [2,3,5,6,7,8]. Oscillation criteria for second order nonlinear inequalities which contained equation (1) has been developed in [4]. However, the conditions for function $f(t,u)$ in this paper are quite different with the conditions in [4] and also the oscillation criteria. In [1] the oscillation criteria for equation (1) has been discussed. Some condition in this paper is weaker than in [1] and results are another types of oscillation criteria for equation (1).
Our main results will contain and extend some previous results in [2]. Oscillation criteria for equation (1) will be established by using a slight modification of the method that has been used previously in [1].

2. Main Results

The following assumptions on function f will be retained in the sequel.

Assumption A. There exists a function $a \in C[0,\infty)$ and a function $\phi \in C^1(-\infty,\infty)$ such that

$$\frac{f(t,u)}{\phi(u)} \geq a(t), \qquad t \geq 0,$$

with $\phi' > 0$, $u\,\phi(u) > 0$, *for* $u \neq 0$ and

$$G(u) = \int_0^u \frac{d\tau}{\phi(\tau)} < \infty .$$

We now have the following theorems.

Theorem 2 *Assume that Assumption A holds and there exists a function $\rho \in C^1[0,\infty)$ with $\rho > 0$ and $\rho' \leq 0$ such that*

$$\limsup_{T\to\infty} \frac{1}{T}\int_0^T \rho(t) \int_0^t a(\tau)\, d\tau \, dt = \infty \tag{3}$$

Then equation (1) is oscillatory in $[0,\infty)$.

Proof. Suppose that equation (1) has a solution $u(t)$ which is nonoscillatory in $[0,\infty)$. Assume that $u(t) > 0$ on $[t_0,\infty)$ for some $t_0 \geq 0$. Let

$$v(t) = \frac{\rho(t)\, u'(t)}{\phi(u(t))}, \qquad t \geq t_0.$$

Then

$$\begin{aligned} v' &= \rho \frac{u''}{\phi(u)} + \frac{\rho' u'}{\phi(u)} - \frac{v^2}{\rho} \phi'(u) \\ &= \frac{\rho\, u''}{\phi(u)} + \frac{\rho'}{\rho} v - \frac{v^2}{\rho} \phi'(u) \\ &= - \frac{\rho\, f(t,u)}{\phi(u)} + \frac{\rho'}{\rho} v - \frac{v^2}{\rho} \phi'(u) \end{aligned}$$

By assumption A and $\rho > 0$, we obtain

$$v' \leq -\rho(t)\, a(t) + \frac{\rho'}{\rho}\, v\,, \qquad t \geq t_0$$

$$\frac{\rho v' - \rho' v}{\rho^2} \leq -a(t),$$

$$\frac{d}{dt}\left(\frac{v}{\rho}\right) \leq -a(t). \tag{4}$$

Integrating (4) on $[t_0, t]$, we get

$$\frac{v(t)}{\rho(t)} \leq \frac{v(t_0)}{\rho(t_0)} - \int_{t_0}^{t} a(\tau)\, d\tau.$$

Let $v(t_0) > 0$, in case $v(t_0) \leq 0$, we omit $\frac{v(t_0)}{\rho(t_0)}$ from this equation. Multiply by $\rho(t) > 0$ to obtain

$$v(t) \leq \frac{v(t_0)}{\rho(t_0)}\, \rho(t) - \rho(t) \int_{t_0}^{t} a(\tau)\, d\tau.$$

$$\frac{\rho(t)\, u'(t)}{\phi(u(t))} \leq \frac{v(t_0)}{\rho(t_0)}\, \rho(t) - \rho(t) \int_{t_0}^{t} a(\tau)\, d\tau.$$

Integrating on $[t_0, T]$, we obtain

$$\int_{t_0}^{T} \rho(t) \frac{u'(t)}{\phi(u(t))}\, dt \leq \frac{v(t_0)}{\rho(t_0)} \int_{t_0}^{T} \rho(t)\, dt - \int_{t_0}^{T} \rho(t) \int_{t_0}^{t} a(\tau)\, d\tau\, dt.$$

Since $\rho(t)$ is decreasing on $[0, \infty)$ and $\rho \geq 0$, we get

$$\int_{t_0}^{T} \rho(t)\, dG(u(t)) \leq \frac{v(t_0)}{\rho(t_0)}\, \rho(t_0)\, (T - t_0) - \int_{t_0}^{T} \rho(t) \int_{t_0}^{t} a(\tau)\, d\tau\, dt.$$

$$\rho(T)\, G(u(T)) - \rho(t_0)\, G(u(t_0)) - \int_{t_0}^{T} G(u(t)) \rho'(t)\, dt \leq v(t_0)\, (T - t_0) -$$

$$\int_{t_0}^{T} \rho(t) \int_{t_0}^{t} a(\tau)\, d\tau\, dt.$$

Since $\rho(T) > 0$, $G(u(T)) \geq 0$ and $\rho'(t) \leq 0$, we obtain

$$\int_{t_0}^{T} \rho(t) \int_{t_0}^{t} a(\tau)\, d\tau\, dt \leq v(t_0)\,(T - t_0) + C. \qquad \textit{where } C = \rho(t_0)\, G(u(t_0)).$$

Dividing by T, we get

$$\frac{1}{T} \int_{t_0}^{t} \rho(t) \int_{t_0}^{T} a(\tau)\, d\tau\, dt \leq v(t_0)\,(1 - \frac{t_0}{T}) + \frac{C}{T}.$$

Condition (3) gives a contradiction. Similar conclusion will be obtained when $u(t) < 0$ on $[t_0, \infty)$.

Remark : In case $\rho = 1$, we obtain the oscillation criteria in [2]. The condition for ρ is weaker than in [1] and condition (3) is another type of oscillation criterion for equation (1).

Theorem 3 *Assume that Assumption A holds and there exists a function* $\rho \in C^2\,[0, \infty)$ *with* $\rho > 0$, $\rho' \leq 0$ *and* $\rho'' \geq 0$ *such that*

$$\limsup_{T \to \infty} \frac{1}{T} \int_{0}^{T} \int_{0}^{t} \rho(\tau)\, a(\tau)\, d\tau\, dt = \infty. \tag{5}$$

Then equation (1) is oscillatory in $[0, \infty)$.

Proof: Suppose that $u(t)$ is a nonoscillatory solution of equation (1) in $[0, \infty)$. Let $u(t) > 0$ on $[t_0, \infty)$ for some $t_0 \geq 0$. Let

$$v(t) = \frac{\rho(t)\, u'(t)}{\phi(u(t))}, \qquad t \geq t_0.$$

Arguing as in the proof of Theorem 2, we obtain the equation (4), i.e.

$$\frac{d}{dt}(\frac{v}{\rho}) \leq -a(t), \qquad t \geq t_0$$

$$\rho\, \frac{d}{dt}(\frac{v}{\rho}) \leq -\rho(t)\, a(t) \tag{6}$$

Integrating (6) on $[t_0, t]$, we get

$$\int_{t_0}^{t} \rho(\tau)\, d(\frac{v}{\rho}) \leq -\int_{t_0}^{t} \rho(\tau)\, a(\tau)\, d\tau.$$

$$v(t) - v(t_0) - \int_{t_0}^{t} \frac{v(\tau)\,\rho'(\tau)}{\rho(\tau)}\,d\tau \le -\int_{t_0}^{t} \rho(\tau)\,a(\tau)\,d\tau.$$

$$v(t) \le v(t_0) - \int_{t_0}^{t} \rho(\tau)\,a(\tau)\,d\tau + \int_{t_0}^{t} \frac{u'(\tau)}{\phi(u(\tau))}\,\rho'(\tau)\,d\tau.$$

$$v(t) \le v(t_0) - \int_{t_0}^{t} \rho(\tau)\,a(\tau)\,d\tau + \int_{t_0}^{t} \rho'(\tau)\,d(G(u(\tau)))$$

$$v(t) \le v(t_0) - \int_{t_0}^{t} \rho(\tau)\,a(\tau)\,d\tau + \rho'(t)\,G(u(t)) - \rho'(t_0)\,G(u(t_0)) -$$

$$\int_{t_0}^{t} \rho''(\tau)\,G(u(\tau))\,d\tau.$$

Since $\rho' \le 0$, $\rho'' \ge 0$, and $G(u(t)) \ge 0$, we have

$$v(t) \le C_1 - \int_{t_0}^{t} \rho(\tau)\,a(\tau)\,d\tau,$$

where $C_1 = v(t_0) - \rho'(t_0)\,G(u(t_0))$.

$$\frac{\rho(t)\,u'(t)}{\phi(u(t))} \le C_1 - \int_{t_0}^{t} \rho(\tau)\,a(\tau)\,d\tau. \tag{7}$$

Integrating on $[t_0, T]$, we obtain

$$\int_{t_0}^{T} \rho(t)\,dG(u(t)) \le C_1\,(T - t_0) - \int_{t_0}^{T}\int_{t_0}^{t} \rho(\tau)\,a(\tau)\,d\tau\,dt.$$

$$\rho(T)\,G(u(T)) - \rho(t_0)\,G(u(t_0)) - \int_{t_0}^{T} G(u(t))\,\rho'(t)\,dt \le C_1\,(T - t_0) -$$

$$\int_{t_0}^{T}\int_{t_0}^{t} \rho(t)\,a(\tau)\,d\tau\,dt.$$

Since $\rho > 0$, $\rho' \le 0$ and $G(u(t)) > 0$, we obtain

$$\int_{t_0}^{T}\int_{t_0}^{t} \rho(\tau)\,a(\tau)\,d\tau\,dt \le C_2 + C_1\,(T - t_0),$$

where $C_2 = \rho(t_0)\,G(u(t_0))$.
Dividing by T we have

$$\frac{1}{T}\int_{t_0}^{T}\int_{t_0}^{t} \rho(\tau)\,a(\tau)\,d\tau\,dt \le \frac{C_2}{T} + C_1\,(1 - \frac{t_0}{T}),$$

From condition (5) we get a contradiction.

Remark. Condition (5) improve the result in [2] and also another type of oscillation criterion for equation (1).
The following theorem weaken the oscillatory criterion in [1], we need only one condition in Theorem 1 of [1].

Theorem 4 *Assume that Assumption A holds and there exists a function* $\rho \in C^2\,[0,\infty)$ *with* $\rho > 0$, $\rho' \leq 0$ *and* $\rho'' \geq 0$ *such that*

$$\int_{\delta}^{\infty} \rho(t)\; a(t)\; dt \;=\; \infty, \tag{8}$$

for some $\delta \geq 0$. *Then equation (1) is oscillatory in* $[0,\infty)$.

Proof: Suppose that $u(t)$ is a nonoscillatory solution of equation (1) in $[0,\infty)$. Let $u(t) > 0$, on $[t_0,\infty)$ for some $t_0 \geq 0$. Let

$$v(t) \;=\; \frac{\rho(t)\; u'(t)}{\phi(u(t))}, \qquad t \geq t_0.$$

Arguing similar to proof of theorem 3, we obtain

$$\frac{\rho(t)\; u'(t)}{\phi(u(t))} \;\leq\; C_1 \;-\; \int_{t_0}^{t} \rho(\tau)\; a(\tau)\; d\tau, \tag{9}$$

where $C_1 \;=\; v(t_0) \;-\; \rho'(t_0)\; G(u(t_0))$.
From condition (8), it follows that there exists a $t_1 \geq t_0$ such that

$$\int_{t_0}^{t} \rho(\tau)\; a(\tau)\; d\tau \;\geq\; 2\,|C_1|, \qquad t \geq t_1.$$

Hence, (9) implies the inequality

$$\frac{\rho(t)\; u'(t)}{\phi(u(t))} \;\leq\; -\,\frac{1}{2}\int_{t_0}^{t} \rho(\tau)\; a(\tau)\; d\tau, \qquad t \geq t_1.$$

Integrating on $[t_1, T]$, we have

$$\rho(T)\; G(u(T)) \;-\; \rho(t_1)\; G(u(t_1)) \;-\; \int_{t_1}^{T} G(u(t))\; \rho'(t)\; dt \;\leq$$

$$-\,\frac{1}{2}\int_{t_1}^{T}\int_{t_0}^{t} \rho(\tau) a(\tau)\; d\tau\; dt.$$

Since $\rho(T) > 0$, $G(u(T)) > 0$ and $\rho'(t) \leq 0$, we obtain

$$\int_{t_1}^{T} \int_{t_0}^{t} \rho(\tau)\, a(\tau)\, d\tau\, dt \leq 2\, \rho(t_1)\, G(u(t_1)) = K.$$

From condition (8), we can choose a $t_2 \geq t_1$ such that

$$\int_{t_0}^{t} \rho(\tau)\, a(\tau)\, d\tau \geq K., \qquad t \geq t_2.$$

It follows that

$$\int_{t_1}^{T} \int_{t_0}^{t} \rho(\tau)\, a(\tau)\, d\tau\, dt \geq \int_{t_2}^{T} \int_{t_0}^{t} \rho(\tau)\, a(\tau)\, d\tau\, dt \geq K\,(T - t_2).$$

Letting $T \longrightarrow \infty$, we arrive at a contradiction . Similar conclusion will be obtained when $u(t) < 0$, on $[t_0, \infty)$. Hence, equation (1) is oscillatory in $[0, \infty)$.

Using stronger condition for ϕ, we prove the following theorem which is another type of oscillation criteria for equation (1) and different from one obtained in [2, 3, 4, 5, 6, 7, 8].

Theorem 5 *Assume that Assumption A holds. Then equation (1) is oscillatory in* $[0, \infty)$ *if*

$$(i) \quad \limsup_{T\to\infty} \frac{1}{T} \int_0^T \int_0^t s a(s) ds\, dt = \infty. \tag{10}$$

$$(ii) \int_{\alpha}^{\infty} \frac{dy}{\phi(y)} < \infty \text{ and } \int_{-\infty}^{-\alpha} \frac{dy}{\phi(y)} < \infty \text{ for some } \alpha > 0 \tag{11}$$

Proof. Suppose that $u(t)$ is a nonoscillatory solution of equation (1) in $[0, \infty)$. Assume that $u(t) > 0$ on $[t_0, \infty)$ for some $t_0 \geq 0$. Cloose t_0 such that $u(t_0) \geq \alpha$. Define the function

$$v(t) = \frac{u'(t)}{\phi(u(t))}, t \geq t_0$$

Then

$$v' = \frac{u''}{\phi(u)} - \frac{(u')^2 \phi'(u)}{\phi^2(u)} = -\frac{f(t,u)}{\phi(u)} - v^2 \phi'(u)$$

By Assumption A we obtain

$$v'(t) \leq -a(t)$$

$$tv'(t) \le -ta(t), t \ge t_0.$$

Integrating on $[t_0.t]$,we have

$$tv(t) - \int_{t_0}^{t} v(s)ds \le t_0 v(t_0) - \int_{t_0}^{t} sa(s)ds$$

$$t\frac{u'(t)}{\phi(u(t))} - \int_{t_0}^{t} \frac{u'(s)}{\phi(u(s))}ds \le t_0 v(t_0) - \int_{t_0}^{t} sa(s)ds.$$

Let $G(u) = \int_\alpha^u \frac{dy}{\phi(y)}$. Then by the first part of condition (11) we obtain

$$t\frac{d}{dt}G(u(t)) \le t_0 v(t_0) + \int_{u(t_0)}^{\infty} \frac{dy}{\phi(y)} - \int_{t_0}^{t} sa(s)ds \le C_1 - \int_{t_0}^{t} sa(s)ds. \quad (12)$$

where $C_1 = t_0 v(t_0) + \int_{u(t_0)}^{\infty} \frac{dy}{\phi(y)}$.
Integrating in $[t_0, T]$, we obtain

$$\int_{t_0}^{T} t\frac{d}{dt}G(u(t))dt \le C_1(T - t_0) - \int_{t_0}^{T}\int_{t_0}^{t} sa(s)ds\ dt$$

$$TG(u(T)) - t_0 G(u(t_0)) - \int_{t_0}^{T} G(u(t))dt \le C_1(T - t_0) - \int_{t_0}^{T}\int_{t_0}^{T} sa(s)ds\ dt.$$

Observe that $\int_{t_0}^{T} G(u(t))dt = \int_{t_0}^{T}\left(\int_\alpha^{u(t)} \frac{dy}{\phi(y)}\right) dt \le \int_{t_0}^{T}\left(\int_\alpha^{\infty} \frac{dy}{\phi(y)}\right) dt = C_2(T - t_0)$
where $C_2 = \int_\alpha^\infty \frac{dy}{\phi(y)}$.
Hence $TG(u(T)) \le t_0 G(u(t_0)) + (C_1 + C_2)(T - t_0) - \int_{t_0}^{T}\int_{t_0}^{t} sa(s)ds\ dt$.

Dividing by T we obtain

$$G(u(T)) \le \frac{C_0}{T} + (C_1 + C_2)\left(1 - \frac{t_0}{T}\right) - \frac{1}{T}\int_{t_0}^{T}\int_{t_0}^{t} sa(s)ds\ dt,$$

where $C_0 = t_0 G(u(t_0))$.
Since $G(u(T)) > 0, \forall T \ge t_0$, we have

$$\frac{1}{T}\int_{t_0}^{T}\int_{t_0}^{t} sa(s)ds\ dt \le \frac{C_0}{T} + (C_1 + C_2)\left(1 - \frac{t_0}{T}\right).$$

By taking lim sup and $T \to \infty$, from (10) we arrive at a contradiction. Similar result will be obtained when $u(t) < 0$ on $[t_0, \infty)$, by using (10) and the second condition in (11).

References

1. COLES, W.J.(1992) *A Nonliniear Oscillation Theorem*, International Conference on Differential Equations; (H.A. Antosiewicz, Ed) , pp 193-202, Academic Press, New York.
2. KAMENEV, I.V.(1971) *Some Specially Nonlinear Oscillation Theorems*, Mat Zametki 10, pp 129 - 134 .
3. KWONG, M.K. and WONG, J.S.W. (1983) *An Oscillation Criteria for Second Order Sublinear Differential Equations*, SIAM J.Math. Anal, **Vol 14, No 3**, pp 474-476.
4. NABABAN, S.M.(1982) *Oscillation Criteria for Second Order Nonlinear Inequalities*, Sains Malaysiana , 11, (1), 39-43.
5. PHILOS, CH.G. (1983)*Oscillation of Sublinear Differential Equations of Second Order* , Nonlinear Analysis, Theory, Methods & Application, **Vol 7, No 10** , pp 1071 -1080.
6. WONG, J.S.W.(1975) *Oscillation Theorems for Second Order Nonlinear Differential Equations*, Bulletion of Institute of Math, Acad . Sinica, **Vol 3, No 2**, pp 283 - 309.
7. WONG, J.S.W.(1989) *A Sublinear Oscillation Theorem*, Journal of Math. Anal & Appl, 139, 408 - 412.
8. WONG, J.S.W.(1992) *An Oscillation Criterion for Second Order Sublinear Differential Equations*, Journal of Math. Anal & Appl, 171, 346-351.

RADIATION AND DIFFRACTION OF SOUND FROM BODIES MOUNTED ON AN INFINITE BAFFLE USING BOUNDARY ELEMENT METHOD

BENJAMIN SOENARKO
Department of Engineering Physics
Bandung Institute of Technology
Jl. Ganesha 10, Bandung 40132
Indonesia.

Abstract. This paper presents a Boundary Element formulation for solving the radiation and diffraction of sound from bodies mounted on an infinite baffle. The method is based on the Helmholtz integral equation wherein the sound pressure in the field can be expressed as an integral of the pressure and particle velocity over the surface of the body. By using an appropriate Green's function the integration over the infinite plane is avoided. Numerical implementation is done by discretizing the surface into elements and integrations are evaluated using simple Gaussian formula. A special coefficient for the pressure in the field is derived due to the contact phenomenon between the body and the infinite plane. It turns out that the information of the acoustic variables on the contact plane are not required. The elements on the contact plane are thus dummy elements that are needed only for the calculation of the coefficient of the pressure in the acoustic field. The formulation are applicable for any arbitrary shape body mounted on an infinite baffle.

1. Background

The radiation and diffraction effects by an object in an infinite baffle may be studied to get a general idea about the phenomena on different kinds of diaphragma shape used for direct-radiator loudspeaker. The radiation from a concave sound source in an infinite baffle was studied by Ohie et al [1], where the space inside the truncated cone was approximated by the stairwise combination of thin cylinders, and cylindrical wave functions were

used in each cylinder. Shindo el al [2] used finite element method to solve the same kind of problem.
The difficulty of solving the radiation problem involving arbitrary shape diphragma in an infinite baffle arises from the fact that the method of separation of variables cannot be applied to get the solution. Examples of these problems involve the radiation from a finite cylinder [3-5] and the radiation from a piston set in a sphere [6]. Fenlon [7] discussed a weighted residual method for the solution. Suzuki and Tichy [8,9] used the least square error method to solve the radiation and diffraction of sound from a convex or a concave dome in an infinite baffle.
The solution for problems involving arbitrary shape bodies in infinite baffle presented in this paper is based on the Helmholtz integral equation in which the sound pressure in the field can be expressed as an integral of the pressure and particle velocity over the surface of the body. The numerical implementation of the formulation involves discretization of the surface of the body. Since only the surface of the body is dealt with, the method is oftenly called the Boundary Element Method or the Boundary Integral Equation (BIE) method. The reduction in the computational dimension of the problem by one (e.g., three-dimensional problems are solved on the two-dimensional boundary of the domain with the BIE method), is often beneficial. The BIE method has been used extensively [10-17] for the numerical solution of radiation and scattering problems in which the radiating body or scattering obstacle has an arbitary shape. In the present paper, a BEM formulation involving arbitrary shape body in an infinite baffle is presented using a half space Green's function [18] with a modified coefficient of the field pressure. Examples are shown for a spherical shape dome wherein good agreement was obtained between the BEM solution and the analytical solution proposed by Suzuki and Tichy [8,9]. The computing time with the BEM solution turns out to be relatively much faster than the analytical solution due to the number of terms required in the series expansion of the analytical solution.

2. The Bem Formulation

Consider a body B with surface S_0 mounted on an infinite baffle S_H surrounded by a half infinite acoustic medium B'. Calculation of the radiated or scattered field from S_0 includes three possible locations of the field point P, i.e. P outside of B, P on B and P at the point of contact between B and S_H . Using the half space Green's function [18]: $\Psi_H = \exp(iKR)/R + \exp(-iKR')/R'$, where R' corresponds to the position of point P', i.e. the image of P, in the Helmholtz Integral Equation:

$$\int_{S_0} [\phi(Q)\frac{\partial}{\partial \nu}\Psi(P,Q) - \Psi(P,Q)\frac{\partial}{\partial \nu}\phi(Q)]dS(Q) = C(P)\phi(P) \qquad (1)$$

and removing the singularity when $P = Q$, the following result is obtained:

$$\int_{S_0} [\phi\frac{\partial \Psi_H}{\partial \nu} - \Psi_H\frac{\partial \phi}{\partial \nu}]dS = 4\pi\phi(P) \qquad (2)$$

which means that $C(P) = 4\pi$ for this case. When P is on the surface S_0 of B, a proper limiting process [15] yields :

$$C(P) = 4\pi + \int_{S_0+S_B} \frac{\partial}{\partial \nu}(\frac{1}{R})dS \qquad (3)$$

where S_B is the portion of B that is in contact with the infinite plane S_H. Equation (3) shows that the calculation of $C(P)$ must include the integral over the closed region bounded by S_0 and S_B, whereas the left hand side of Eq.(1) involves only the boundary of the body B, i.e. S_0. The most crucial situation is the case wherein field point P is at the point of contact between S_0 and S_H as shown in Fig.1. Using a proper limiting process [15,19] the following results are obtained:

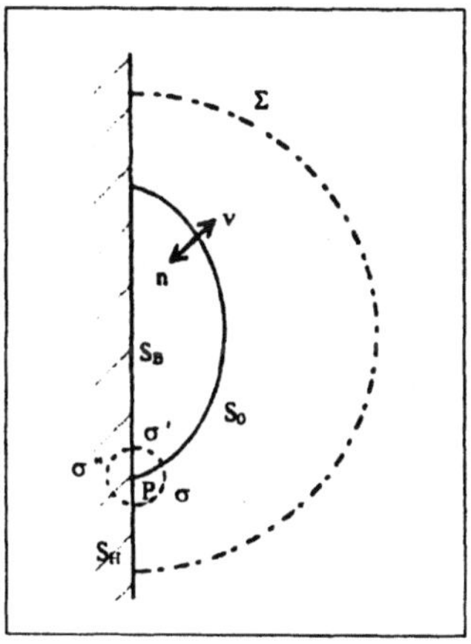

Figure 1. A field point P located at the point contact between S_0 and S_H

$$\int_{S_0+\sum+S_H+\sigma} (\phi\frac{\partial \Psi_H}{\partial v} - \Psi_H\frac{\partial \phi}{\partial v})dS = 0 \qquad (4)$$

$$\int_{\sum} (\phi\frac{\partial \Psi_H}{\partial \nu} - \Psi_H\frac{\partial \phi}{\partial \nu})dS = 0 \qquad (5)$$

$$\int_{S_H}(\phi\frac{\partial\Psi_H}{\partial\nu}-\Psi_H\frac{\partial\phi}{\partial\nu})dS=0 \tag{6}$$

$$\lim_{R\to 0}\int_{\sigma}(\phi\frac{\partial\Psi_H}{\partial\nu}-\Psi_H\frac{\partial\phi}{\partial\nu})dS=-2\phi(P)\lim_{R\to 0}\int_{\sigma}\frac{dS}{R^2}$$
$$=-2\hat{C}(P)\phi(P) \tag{7}$$

where $\hat{C}(P)=\lim_{R\to 0}\int_{\sigma}\frac{dS}{R^2}$ may be interpreted as the solid angle made by the surface S_0 and the plane S_H at the point P. Equations (4-7) may be combined to give:

$$\int_{S_0}(\phi\frac{\partial\Psi_H}{\partial\nu}-\Psi_H\frac{\partial\phi}{\partial\nu})dS=2\hat{C}(P)\phi(P) \tag{8}$$

Now consider the coefficient $C(P)=C_f(P)$ for a full space problem involving the body bounded by S_0 and S_B. In this case $C_f(P)$ may be interpreted as the exterior solid angle made by S_0 and S_B which can be expressed by:

$$C_f(P)=\lim_{R\to 0}\int_{\sigma+\sigma''}\frac{dS}{R^2} \tag{9}$$

However $\lim_{R\to 0}\int_{\sigma''}\frac{dS}{R^2}=2\pi$ and noting that $\hat{C}(P)=\lim_{R\to 0}\int_{\sigma}\frac{dS}{R^2}$, Eq.(9) can be expressed as $C_f(P)=\hat{C}(P)+2\pi$ or $\hat{C}(P)=C_f(P)-2\pi$. But from the full space formulation [15] it can be shown that

$$C_f(P)=4\pi+\int_{S_0+S_B}\frac{\partial}{\partial\nu}(\frac{1}{R})dS \tag{10}$$

Therefore

$$\hat{C}(P)=2\pi+\int_{S_0+S_B}\frac{\partial}{\partial\nu}(\frac{1}{R})dS \tag{11}$$

Using Eq. (11) in Eq. (8) yields

$$\int_{S_0}(\phi\frac{\partial\Psi_H}{\partial\nu}-\Psi_H\frac{\partial\phi}{\partial\nu})ds=$$
$$2[2\pi+\int_{S_0+S_B}\frac{\partial}{\partial\nu}(\frac{1}{R})dS]\phi(P) \tag{12}$$

or

$$\int_{S_0}(\phi\frac{\partial\Psi_H}{\partial\nu}-\Psi_H\frac{\partial\phi}{\partial\nu})dS=C_H(P)\phi(P) \tag{13}$$

where

$$C_H(P)=2[2\pi+\int_{S_0+S_B}\frac{\partial}{\partial\nu}(\frac{1}{R})dS] \tag{14}$$

is the geometrical coefficient for contact point P.

Equation (14) shows that the calculation of $C_H(P)$ involves the boundary surface S_B which is a part of the infinite plane S_H, in addition to the boundary surface of the body S_0.

Another expression for Eq. (14) is:

$$C_H(P) = 2[C_f(P) - 2\pi] \tag{15}$$

which clearly shows the interpretation of the geometrical coefficient at point P. Again $C_f(P)$ indicates the solid angle bounded by S_H and S_0 in the space (exterior), 2π is the solid angle bounded by the infinite plane S_H. Factor 2 appearing in Eq.(15) is due to the contribution of point P', i.e. the image of point P in the integral formula, as a result of the half space phenomenon.

In general, for any surface S_H with a reflection coefficient R_p, Eq. (14) takes the form:

$$C(P) = (1 + R_P)[2\pi + \int_{S_0+S_B} \frac{\partial}{\partial \nu}(\frac{1}{R})ds] \tag{16}$$

Equation (14) or (16) is useful for calculating the acoustic field of a radiating or scattering body mounted on an infinite baffle using the BEM.

3. Test Cases

To verify the BEM formulation a number of test cases were run. The first test case is the radiation from a convex hemispherical dome of radius A mounted in an infinite baffle. The dome vibrates with a uniform velocity over its surface. The baffle is assumed to be rigid. Figure 2 shows the pressure response at a distance of 40A from the center of the dome perpendicular to the infinite baffle. It is shown that a good agreement was obtained between the BEM and the analytical solution proposed by Ref.[8]. A polar plot of the radiation pattern is depicted in Fig.3. It can be seen that the BEM result agrees very well with the analytical solution.

The second test case is the diffraction problem in which the radiated waves from a ring source are diffracted by a hemispherical convex dome. The ring source is of radius $1.5A$ surrounding the dome concentrically on the infinite plane. Figure 4 shows the result of the BEM and the analytical solution in which the pressure response is normalized to the pressure of the ring source without the convex dome. A good agreement between the two results can be observed in the depicted plots. To show the directivity pattern a polar plot is presented at a distance of $40A$ from the center of the dome. The polar plot is depicted in Fig.5 in which both the BEM and the analytical results show a good agreement.

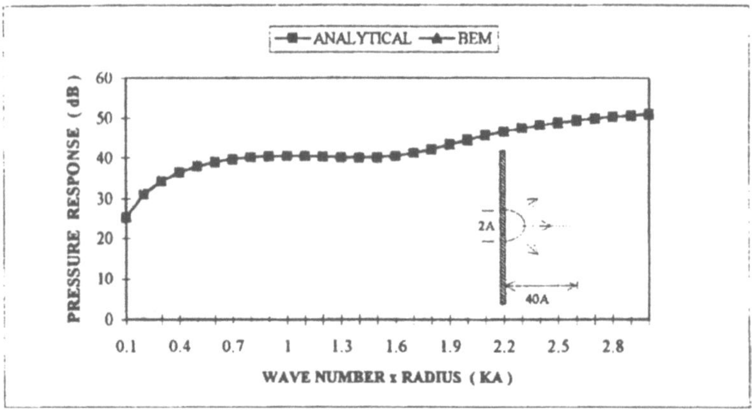

Figure 2. Pressure response of a radiating convex dome mounted on an infinite baffle at 40A distance

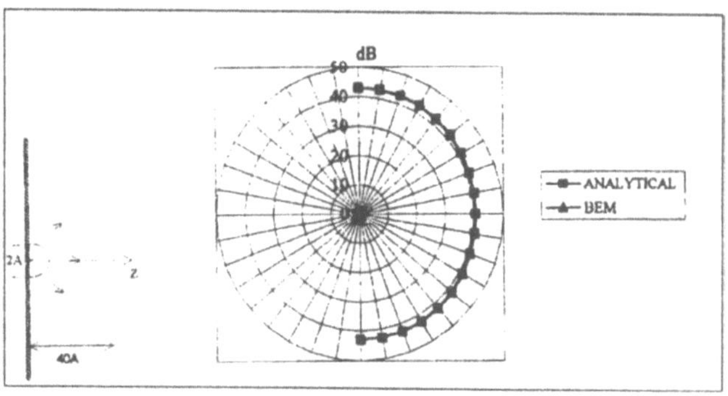

Figure 3. Polar plot of the radiated sound field from a convex dome in an infinite baffle ($KA = 1$)

The problem of a radiating cylinder of radius A in an infinite baffle is solved using the present formulation. The lateral (curved) surface of the cylinder is vibrating with a uniform radial velocity. The ends of the cylinder are motionless. The results are shown in Fig.6 in which the pressure response at a distance of $40A$ is plotted versus KA where K is the wave number. Analytical solution may be found for this problem although it is not in the closed form [20], whereas the BEM yields a solution with considerable accuracy as tested by other examples. The polar plot of the radiated field is shown in Fig.7 for $KA = 1$. It can be seen that the acoustic field is almost uniform around this location as expected for low frequency radiation.

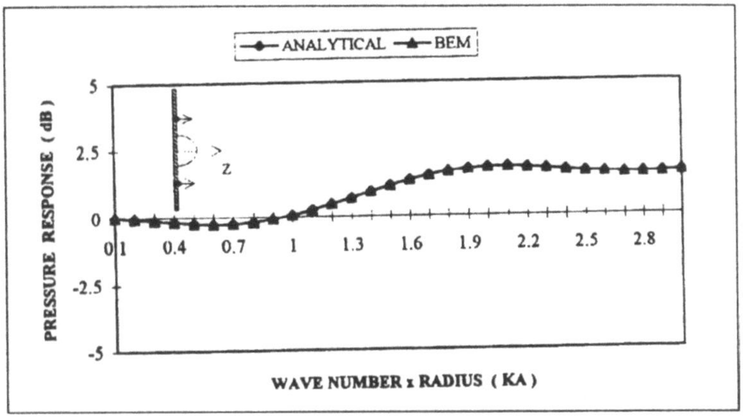

Figure 4. Diffraction from a convex dome in an infinite baffle with a ring source at $40A$ distance

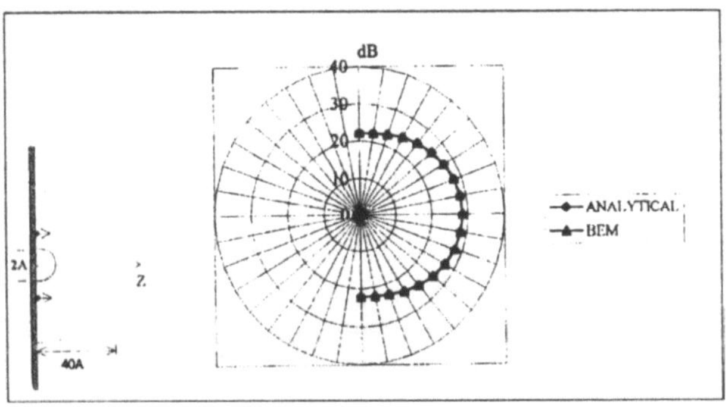

Figure 5. Polar plot of the diffracted field from a convex dome in an infinite baffle with a ring source ($KA = 1$)

The problem of a diffraction from a cylinder in an infinite baffle is shown in Fig.8. A ring source of radius $1.5A$ is located on the infinite plane surrounding the cylinder. This problem does not have analytical solution available. However, using the BEM the solution can be easily found with a great deal of confidence as demonstrated in several test cases shown. Figure 8 shows the polar plot for $KA = 1$.

4. Summary

A BEM formulation for radiating or scattering body mounted in an infinite baffle has been presented. It is shown that for the contact points

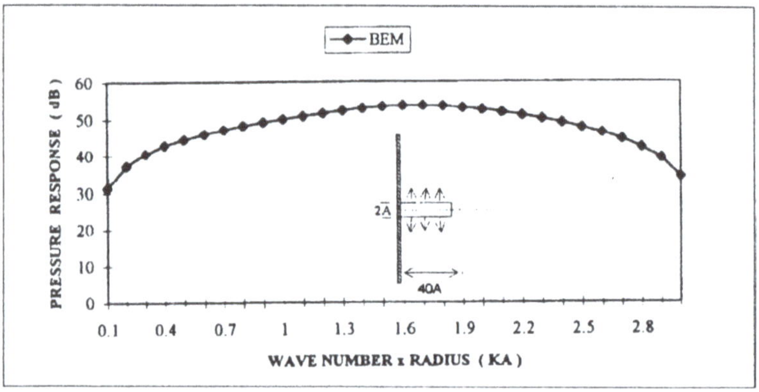

Figure 6. Pressure response of a radiating cylinder mounted an infinite baffle a at $40A$ distance

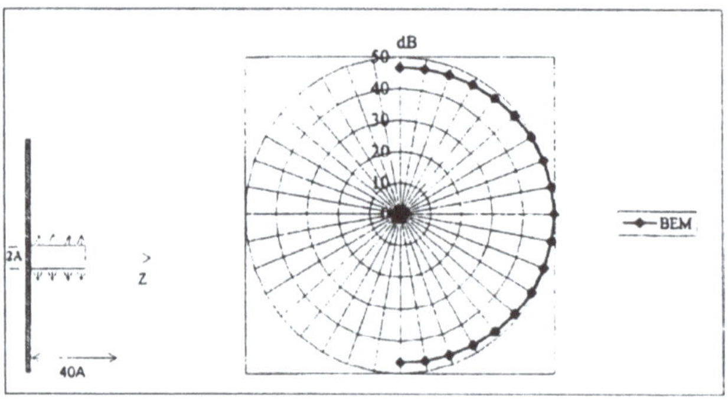

Figure 7. Polar plot of the radiation from a vibrating cylinder mounted on an infinite baffle ($KA = 1$)

between the body and the infinite plane, a special geometrical coefficient in the Helmholtz formula is needed. The derived formulation shows that no acoustic variable is involved in the contact plane for evaluating the integral equation. However discretization of the contact plane is needed for the calculation of the geometrical coefficient. Test cases were shown involving hemispherical dome and cylinder mounted in an infinite baffle. Comparison with available analytical solution has shown a good agreement which confirms the validity of the proposed formulation.

The present formulation can be extended to include axisymmetric bodies. For this special case, the surface integrals are reduced to line integrals and an integral over the angle of revolution. The integration over the angle

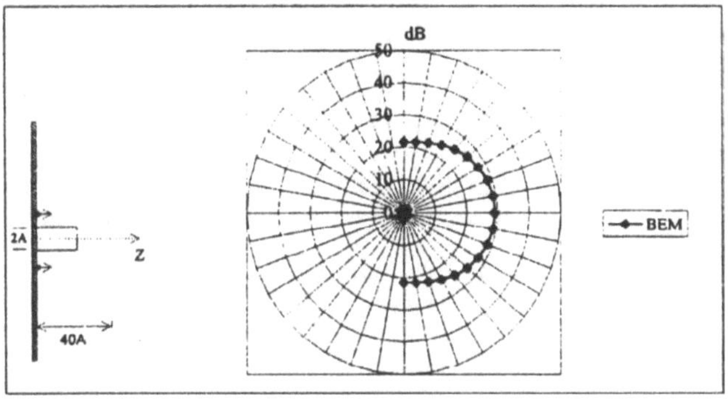

Figure 8. Polar plot of the diffracted field from a cylinder in an infinite baffle with a ring source ($KA = 1$)

can be computed partly analytically in terms of elliptic integrals and partly numerically using simple Gaussian quadrature formula. The complete formulation of this feature is a part of our ongoing research;the results for publication are forthcoming.

Acknowledgement

The author would like to thank the Department of Mathematics of ITB for allowing the use of their RS 6000 workstation to run his BEM program. A special thanks is also acknowledged to Mr.Heksaputra for providing the plots of the test cases.

References

1. Ohie, S., Suzuki, H., and Shindo, T., " Sound radiation from a concave radiator in an infinite baffle", J. Acoust. Soc. Jpn. 34, pp. 429-435, 1978.
2. Shindo, T., Kyono, N., Yashima, O., Yamabuchi, T., and Kagawa, Y., "The role of the dust cap in the cone-type loudspeaker", J. Audio Eng. Soc. 28, pp. 482-489, 1980.
3. Brigham, A. and Borg, M. F., "An approximation solution to the acoustic radiation of a finite cylinder", J. Acoust. Soc. Am. 32, pp. 971-981, 1960.
4. Wiliams, W., Parke, N. G., Moran, D. A., and Sherman, C. H., "Acoustic radiation from a finite cylinder", J. Acoust. Soc. Am. 36, pp. 2316-2322, 1964.
5. McCormik, J. M., and Baron, M. L., "Sound radiation from submerged cylindrical shells of finite length", Trans. ASME Ser. B87, pp. 393-405, 1965.
6. Skudrzyk, E., The Foundations of Acoustics, Springer- Verlag, New York, 1971.
7. Fenlon, F. H., "Calculation of the acoustic radiation field at the surface of finite cylinder by the method of weighted residuals", Proc. IEEE. 57. pp. 291-306, 1969.
8. Suzuki, H., and Tichy, J., "Sound radiation from convex and concave domes in an infinite baffle", J. Acoust. Soc. Am. 69(1), pp. 41-49. 1981.
9. Suzuki, H., and Tichy, J., "Diffraction of sound by a convex or a concave dome in an infinite baffle", J. Acoust. Soc. Am. 70(5), pp. 1480-1487, 1981.
10. Chertock, G., "Sound radiation from vibrating surfaces", J. Acoust. Soc. Am, vol.

36, pp. 1305-1313, 1964.
11. Schenck, H. A., "Improved integral formulation for acoustic radiation problems", J. Acoust. Soc. Am. vol. 44, pp. 41-58, 1968.
12. Engblom, J. J., and Nelson, R. B., "Consistent formulation of sound radiation from arbitrary structure", ASME Journal of Applied Mechanics, vol. 97, pp. 295-300, 1975.
13. Meyer, W.L., Bell, W.A., Zinn,B.T., and Stallybrass, M.P., "Boundary integral solution of three dimensional acoustic radiation problems", J. Sound. vib. vol 59, pp. 245-262, 1978.
14. Koopman, G.H., and Benner,H., "Method for computing the sound power of machines based on the Helmholtz integral", J. Soc. Am. vol. 71, pp. 78-89, 1982.
15. Seybert, A.F., Soenarko, B., Rizzo, F.J., and Shippy,D.J., "An advanced computational method for radiation and scattering of acoustic waves in three dimensions", J. Acoust. Soc. Am., vol 77, pp. 362-368, 1985.
16. Seybert, A.F., Soenarko, B., Rizzo, F.J., and Shippy, D.J., "A Special integral equation formulation for acoustic radiation and scattering for axisymmetric bodies and boundary conditions", J. Acoust. Soc. Am., vol. 80, pp. 1241- 1247, 1986.
17. Latcha, M.A., and Akay, A., "Application of the Helmholtz integral in acoustics ", ASME Journal of Vibration, Acoustics, Stress, and Reliability in Design, vol. 108, pp. 447-453, 1986.
18. Seybert, A.F. and Soenarko, B., "Radiation and scattering of acoustic waves from bodies of arbitrary shape in a three-dimensional half space", ASME Journal of Vibration, Acoustics, Stress, and Reliability in Design, vol. 110, pp. 112-117, 1988.
19. Soenarko, B., " An advanced boundary element formulation for acoustics radiation and scattering in three dimensions", Ph.D. dissertation, University of Kentucky, 1983.
20. Morse, P.M., "Vibration and Sound", Acoustical Society of America,1976

A DIFFERENTIAL EQUATION FOR FILTERING OF A STOCHASTIC DYNAMICAL SYSTEM

TRAN HUNG THAO
Institute of Mathematics, NCSR
P.O. BOX 631 BOHO
HANOI, VIETNAM

1. Introduction

In this paper we consider a problem of filtering for a stochastic dynamical system $(X_t, t \geq 0)$ from observations $(\theta_t, t \geq 0)$ of the form

$$\theta_t = \int_0^t h(s, X_s)ds + Y_t$$

where $h(t, x)$ is some non-anticipating function and Y_t is an annealing noice that is a Lévy process. This problem arises from Simulated Annealing Theory applied to Filtering.
The paper is organized as follows. In section 2, we give the dynamics for the system in emphasizing on Lévy process. Section 3 stands for annealing process that is the solution of a diffusion equation and in Section 4, after introducing the concept of Lévy annealing noice we established a differential equation for the filter $\pi_t(f)$ of the considered dynamical system.

2. Dynamics

Let Ω be the space of all functions ω from the interval $[0, T]$ $(0 < T < \infty)$ into a locally compact Hausdorff space S:

$$\Omega = a = \{\omega : [0, T] \to S\}$$

and we denote $\omega(t) = X_t(\omega)$.

Let W be the space of all real-valued functions defined on $[0, 1]$

$$W = \{w : [0, T] \to \mathbb{R}^1\}.$$

Also we denote the value of w at $t \in [0, T]$ by

$$w(t) = \theta_t(w).$$

Denote by $\mathcal{F}_t$ the σ-field generated by $(X_s; 0 \leq s \leq t)$ and by $\mathcal{W}_t$ the σ-field generated by $(\theta_s; 0 \leq s \leq t)$.

Suppose that P is a probability measure defined on the measurable space $(\Omega \times W, \mathcal{F}_T \times \mathcal{W}_T)$. The stochastic dynamical system considered here is the following process:

$$\{\Omega \times W, \mathcal{F}_T \times \mathcal{W}_T, P(X_t, \theta_t)\} \tag{2.1}$$

Consider an element Y_t of W such that $(W, \mathcal{W}, Q; Y_t)$ is a Lévy process, where $W = \bigvee_{t=0}^{T} w_t$ and Q is a probability on (W, w). According the Levy decomposition we have:

$$Y_t = M_t + \int_0^t b(s)ds + \int_0^t \int_{|u|\leq 1} uJ_c(ds, du) + \int_0^t \int_{|u|>1} J(ds, du) \tag{2.2}$$

where M_t is a continuous martingale such that

$$d < M_t, M_t >= a(t)dt,$$

$J(dt, du)$ a Poisson random measure with density λ

$$E[J(dt, du)] = \lambda(t, du)dt$$

and $J_c(dt, du)$ the compensator of $J(dt, du)$:

$$J_c(dt, du) = J(dt, du) - \lambda(t, du)dt.$$

3. Diffusion Annealing Process

We assume that we are given a Gibbs distribution g on $\mathbb{R}$ with a density with respect to the Lebesgue measure dx, given by

$$g(y) = \frac{1}{Z_T} \exp\{-\frac{1}{T}U(y)\} \tag{3.1}$$

$$Z_T = \int_{-\infty}^{\infty} \exp\{-\frac{1}{T}U(y)\}dx \tag{3.2}$$

where $U(y)$ is a real-valued function on $\mathbb{R}$ so that $Z_T < \infty; T$ is a positive constant and considered as the temperature in an annealing process.

We seek a diffusion process Y_t on $\mathbb{R}$ having g as its unique invariant measure. This process will be specified via a stochastic differential diffusion equation of the form

$$dY(t) = b(Y(t))dt + \sigma(Y(t))dB(t).$$

where $b(y)$ and $\sigma(y)$ are the drift and the diffusion coefficients, respectively, and $B(t)$ is the standard Brownian motion on $\mathbb{R}$.

According to Langevin and Smoluchowski, the corresponding diffusion equation is

$$dY(t) = U'(Y(t))dt + \sqrt{2\tau}dB(t). \tag{3.3}$$

And one can prove that under some assumptions opposed on U, this equation has a solution $Y(t)$ whose density $p(t, y)$ weakly tends to $g(y)$ when $t \to \infty$. The solution $Y(t)$ of (3.3) is of course a diffusion process expressing the state of the annealing material in the temperature τ and at the time t, it is called a diffusion annealing process. On the other hand, one observe that as the temperature τ tends to 0 the Gibbs distribution (3.1) concentrates on the global minima of $U(y)$ expressing the minimum energy of the system.

4. State Estimate of a System with Lévy Annealing Noises

In the Simulated Annealing one are interested in problems of state estimation of a dynamical system perturbed by some noises related to an annealing process. We consider a dynamical system (X_t, θ_t) defined as in Section 2, where X_t is the system process and θ_t is the observation defined by.

a. $\theta_t = \int_0^t h(s, X_s)ds + Y_t,$ (4.1)

b. $h(t, x)$ is a positive and non-anticipating function such that

$$E\int_0^t h^2(s, X_s)ds < \infty \quad \text{for each } t, \tag{4.2}$$

c. Y_t is a Levy process. (4.3)

Lemma 1 *The observation process θ_t itself also a Lévy process.*

Proof. Because Y_t is a Lévy process we have the following decomposition:

$$Y_t = M_t + \int_0^t b(s)ds + \int_0^t \int_{|u|\leq 1} uJ_c(ds, du) + \int_0^t \int_{|u|\geq 1} uJ(ds, du)$$

Then θ_t has also a Lévy decomposition

$$\theta_t = M_t + \int_0^t \bar{b}(s)ds + \int_0^t \int_{|u|\leq 1} J_c(ds, du) + \int_0^t \int_{|u|\geq 1} uJ(ds, du)$$

where $\bar{b} = b + h$. Thus θ_t is a Lévy process.

Lemma 2 *The first term M_t in the Lévy decomposition of the observation process θ_t is a martingale. Furthermore it is a Brownian motion.*

Proof. In virtue of the Lévy-Khintchine formula, the Fourier transform of M_t is:

$$\psi_M(u) = \frac{\sigma^2}{2}u^2 - i\alpha u \tag{4.4}$$

Then M_t is a Brownian motion which is a martingale.

Definition 4.1.
The noise Y_t is called a Lévy annealing noise if it satisfies the following diffusion annealing equation:

$$dY_t = -U'(Y_t)dt + \varepsilon_t dM_t \tag{4.5}$$

where $U \in C_b^2(\mathbb{R}), \varepsilon_t$ is some parameter tending to 0 when $t \to \infty$ and M_t is the first composant in the Lévy decomposition of θ_t which is a Brownian motion as shown in Lemma 2.

Theorem 1 *If $U \in C_b^2$ and*

$$Z_t = \int_{-\infty}^{\infty} e^{\frac{1}{\varepsilon_t}U(y)}dy < \infty \tag{4.6}$$

then the Lévy annealing noise Y_t is lightly different from a Gibbs distribution when $t \to \infty$, i.e.

1. *If ϵ_t is fixed, $\epsilon_t = \epsilon_0(const)$, then the distribution of Y_t converges weakly when $t \to \infty$ to a Gibbs distribution defined by the density: $g(y) = \frac{1}{Z_0}exp - \frac{1}{\epsilon_0}U(y)$,*
2. *If ϵ_t is a nonnegative function of t decreasing to 0 when $t \to \infty$ then $\|p_t(y) - g_t(y)\| \to 0$ when $t \to \infty$, where $p_t(y)$ is the distribution of Y_t and $g_t(y)$ is the density of a Gibbs distribution defined by $g_t(y) = \frac{1}{Z_t}e^{-\frac{1}{\epsilon_t}U(y)}$, and $\|.\|$ denotes the total variation norm.*

Proof. Y_t is the solution of the diffusion equation

$$dY_t = -U'(Y_t)dt + \sqrt{2\tau}dB_t, \tag{A}$$

where B_t is a Brownian motion ($M_t = B_t$) and $\tau = \frac{1}{2}\epsilon_t^2$ is the temperature parameter.

1. If $\epsilon_t = \epsilon_0(const)$ then (A) is a Langevin equation of form

$$dY_t = -U'(Y_t)dt + \sqrt{2\tau_0}dB_t,$$

and it is well-known that its solution Y_t converges in law to a random variable of Gibbs distribution $g(y)$.

2. Suppose $\epsilon_t = \epsilon(t)$, a function decreasing to 0 of t. The solution Y_t of (A) is a Markov process whose density p_t satisfies the Fokker-Plank equation

$$\frac{\partial p}{\partial t} = \tau(t)\Delta p + div(\nabla U.p), \tag{B}$$

where $\tau(t) = \frac{\epsilon^2(t)}{2}$, and we know also that under same additional assupmtions on the regularity of U, the total variation norm $\|p_t - g_t\|$ tends to 0 when $t \to \infty$, where $g_t = \frac{1}{Z_t}e^{-\frac{1}{2\tau}U}$.

Definition 4.2.
Let $f(x)$ be a bounded measurable function on $S\ :\ f\ :\ S \to \mathbb{R}$, we say $\pi_t(f)$ the best state estimation of the system X via f the following conditional expectation

$$\pi_t(f) = E_P(f(X_t)|\mathcal{W}_t) \tag{4.7}$$

where $\mathcal{W}_t$ is the σ-field $\mathcal{W}_t = \sigma(\theta_s; s \leq t)$ which records all information about X up to the t.

Theorem 2 *Suppose that $f : S \to \mathbb{R}$ is a bounded measurable function on S such that there is a bounded measurable and non-anticipating function $g : [0,T] \times W \times S \to \mathbb{R}$ for which*

$$f(X_t) - f(X_0) - \int_0^t g(s,\theta_s,X_s)ds$$

is a martingale with respect to $(\mathcal{W}_t, P_{\mathcal{W}})$ then the best estimation is given by the following equation

$$d\pi_t(f) = \pi_t(g)dt\ +\ \pi_t(f - \pi(h))d\bar{M}_t\ +\ \int \pi(f(\frac{\varrho}{\pi(\varrho)} - 1)\bar{J}_c(dt,du) \tag{4.8}$$

where ϱ is some non-anticipating function on $[0,T] \times W \times S$, $d\pi_t = dM_t - a(t)\pi_t(h)dt$, $\bar{J}_c(dt,du) = J(dt,du) - \pi_t(\varrho)\lambda(t,du)dt$.

Proof. The observation θ is in fact a Levy process (Lemma 4.1), so under assumptions of Theorem 4.2, we can apply results by Grigelionis [1] to obtain the equation (4.8) for best estimation $\pi_t(f)$.

References

1. B. Grigelionis, *On Non-linear Filtering theory and Absolute Continuity of Measure corresponding to Stochastic Processes.* Lecture Notes in Math. 330, Proc. USSR Japan Symp. on Prob, Springer Berlin, (1973), 80-94.
2. Tran Hung Thao, *Optimal State Estimation for A Stochastic Dynamical System from Point Process Observation.* Sonderdruck aus Methods of Operations Research. Ulm, 62(1989), 421-430.

GENERAL THEORY OF RELATIVITY WITHOUT CHRISTOFFEL SYMBOLS

TRIYANTA
Department of Physics-ITB
Jalan Ganesha 10 Bandung 40132
Indonesia

Abstract. One peculiarity of the Fock-Schwinger gauge in gauge theory is that one may derive the so-called inversion formula. Such a formula may also be obtained in general theory of relativity if one restricts a Fock-Schwinger gauge-like condition to Christoffel symbols. The formula then may be applied to reformulate expressions in general theory of relativity without Christoffel symbols. An analysis of the Christoffel symbol-free geodesic differential equation, at the lowest approximation, will be shown.

1. Introduction

Reformulation of a dynamical equation in different but simpler expressions are of importance since these then enable one to solve the equation easier. Due to such a reason, a quantity which is then called a potential, for example, is introduced [1,2]. This quantity is defined, generally speaking, such that its derivative, e.g. gradient in mechanics and electrostatics and curl in magnetostatic, leads to a well known physical quantity such as electric and magnetic fields. Because of such a definition the potential contains a number of degrees of freedom. This, however, can be reduced by imposing an arbitrary condition, the so-called gauge condition or simply the gauge, to the potential. Among gauge conditions already introduced by physicists the Fock-Schwinger (FS) gauge[1] ,

$$x^{\mu}A_{\mu}(x)=0 \tag{1}$$

has one peculiar property that cannot be found in other gauge conditions[2]. The property that we mean here is the possibility to derive inversion formulae. As is already well known, (non-Abelian) gauge potentials $A^{\alpha\mu}(x)$ are defined through

$$F^{a\mu\nu}=\partial^{\mu}A^{a\nu}-\partial^{\nu}A^{a\mu}-igf^{abc}A^{b\mu}A^{c\nu} \tag{2}$$

where $F^{a\mu\nu}, g$, and $f^{abc}, a, b, c = 1, 2, \cdots N^2-1$, are field strengths, coupling constant and antisymmetric structure constants of the SU(N) gauge group respectively. This equation expresses field strengths in terms of potentials. The inversion formula is nothing but the reverse expression of the above equation, i.e. a formulation of potentials in terms of field strengths. This formula cannot be derived in all known gauge conditions but in the FS gauge[3] .

In general theory of relativity there are two quantities that are related one another just like the expression (2). These are the Riemann curvature tensors $R^{\alpha}{}_{\beta\gamma\delta}(x)$ and the Christoffel symbols (of second kind) $\Gamma^{\alpha}_{\mu\nu}(x)$ with their relationship is given by

$$R^{\alpha}{}_{\mu\nu\rho}=\partial_{\nu}\Gamma^{\alpha}_{\mu\rho}-\partial_{\rho}\Gamma^{\alpha}_{\mu\nu}+\Gamma^{\alpha}_{\nu\sigma}\Gamma^{\sigma}_{\mu\rho}-\Gamma^{\alpha}_{\rho\sigma}\Gamma^{\sigma}_{\mu\nu}. \tag{3}$$

Because of similarity between both expressions one may conclude that an inversion formula in general theory of relativity may also be obtained if the Christoffel symbol is restricted by the FS gauge condition in general theory of relativity:

$$x^{\mu}\Gamma^{\alpha}_{\mu\nu}(x)=0. \tag{4}$$

---

[1] The gauge condition (1) has various names: the Fock-Schwinger gauge, the Schwinger gauge, the complete Lorentz covariant gauge, the coordinate gauge, the fixed-point gauge, the Constrom-Dubovikov-Smilga (CDS) gauge, the Poincare gauge, the homogeneous gauge, and the multipolar gauge [3].

[2] x^{μ} is the space-time coordinate and A^{μ} is the potential or the gauge field. Throughout the paper summation over repeated Lorentz (Greek) or space (Latin) indices is understood. Unlees indicated explisity, the Greek indices take the values 0,1,2,3 while the space indices take the values 1,2,3.

[3] Actually there is another gauge, the so-called fixed axial gauge [4], that also has an inversion formula. This formula, however, cannot be written in a compact form like the one in the FS gauge.

By utilitizing the inversion formula all expressions in the general theory of relativity may be rewritten in such a form where the Christoffel symbols disappear. The organization of the paper is as follows. In the next section inversion formulae in the general theory of relativity will be derived, both in integral and differential expressions. Section III is devoted for Christoffel symbol-free formulations of the theory. In this section the explicit expressions of the Christoffel symbol-free covariant derivatives as well as the geodesic differential equation will be shown. Analysis up to the lowest approximation to the geodesic differential equation will also be given. Section IV is given for Conclusion.

2. Inversion Formulae

To obtain the inversion formulae one must restrict the Christoffel symbols $\Gamma^{\mu}_{\alpha\beta}$ with the FS condition (4). Accordingly the Riemann curvature tensors $R^{\mu}{}_{\alpha\beta\gamma}$ have to obey

$$x^{\mu} R^{\alpha}{}_{\beta\mu\nu}(x) = x^{\mu}\partial_{\mu}\Gamma^{\alpha}_{\beta\nu}(x) + \Gamma^{\alpha}_{\beta\nu}(x). \tag{5}$$

By introducing a parameter $\lambda \in [0,1]$ via a scale transformation $x^{\mu} \to \lambda x^{\mu}$ the above expression becomes

$$\lambda x^{\mu} R^{\alpha}{}_{\beta\mu\nu}(\lambda x) = \lambda \frac{d}{d\lambda}\Gamma^{\alpha}_{\beta\nu}(\lambda x) + \Gamma^{\alpha}_{\beta\nu}(\lambda x) = \frac{d}{d\lambda}\lambda\Gamma^{\alpha}_{\beta\nu}(\lambda x). \tag{6}$$

Integrating both sides of this expression with respect to λ from 0 to 1 one has

$$\Gamma^{\alpha}_{\beta\nu}(x) = \int_0^1 d\lambda\ \lambda x^{\mu} R^{\alpha}{}_{\beta\mu\nu}(\lambda x). \tag{7}$$

Notice that the above equation is an expression of the Christoffel symbols $\Gamma^{\mu}_{\alpha\beta}$ in terms of Riemann curvature tensors $R^{\mu}{}_{\alpha\beta\gamma}$, i.e. the converse of the equation (3). Thus the equation (7) is the inversion formula that we look for. To convince oneself that $\Gamma^{\mu}_{\alpha\beta}$ and $R^{\mu}{}_{\alpha\beta\gamma}$ in the formula really describe the Christoffel symbol and the Riemann curvature tensor respectively one must be able to derive the equation (3) from (7). This is a straightforward derivation and one needs to impose the Bianchi identities

$$\nabla_{\nu} R^{\alpha}{}_{\beta\mu\rho} + \nabla_{\mu} R^{\alpha}{}_{\beta\rho\nu} + \nabla_{\rho} R^{\alpha}{}_{\beta\nu\mu} = 0 \tag{8}$$

to obtain the inversion formula [5]. Hence the Bianchi identities (8) are the necessary and sufficient conditions for the inversion formula (7) to be the relations between Christoffel symbols and their Riemann curvature tensors. Notice that the form of the formula (7) has exactly the same with that in non-Abelian gauge theory[2]. This can be understood since the relationship between the gauge potential and the field strength (2) and between the Christoffel symbol and the Riemann curvature tensor (3) are similar. Let us now consider an arbitrary, well behaved, function $f(x)$. The following identity applies for it:

$$\int_0^1 f(\lambda x)d\lambda = \lambda f(\lambda x)|_{\lambda=0}^1 - \int_0^1 \lambda \tfrac{d}{d\lambda} f(\lambda x)d\lambda = f(x) - x^\mu \partial_\mu \int_0^1 f(\lambda x)d\lambda$$

It is obvious that this identity is equivalent to

$$f(x) = (1 + x \cdot \partial) \int_0^1 f(\lambda x)d\lambda. \tag{9}$$

Using the above identity the inversion formula (7) may be written as

$$\Gamma^\alpha_{\beta\nu}(x) = (1 + x \cdot \partial)^{-1} x^\mu R^\alpha{}_{\beta\mu\nu}(x) \tag{10}$$

where $(1+x\cdot\partial)^{-1}$ is the inverse of the differential operator $(1+x\cdot\partial) = (1+x^\mu\partial_\mu)$. The idea to write the non-integral expression of the inversion formula is inspired by our research experience in gauge field theory, specifically on the FS gauge [3]. When deriving the FS gauge propagator, i.e. the Green function for the gauge field in the FS gauge, we obtained that the expression of the FS gauge propagator in coordinate space, not in momentum space, contains operators of the form $(n+x\partial)^{-m}$ where n and m are integers. These operators are nothing but the inverse of $(n + x\partial)^m$. Because operations of integration and differentiation can be thought of as reciprocal to one another one can conclude that one of the operators $(n + x\partial)^{-m}$ has to be equivalent to an operation of integration. It turns out that, see equation (9), the operator that is equivalent to the integration is $(1 + x\partial)^{-1}$.

Let us now Taylor-expand the Riemann curvature tensor $R^\alpha{}_{\beta\mu\nu}(\lambda x)$ around $x = 0$,

$$R^\alpha{}_{\beta\mu\nu}(\lambda x) = R^\alpha{}_{\beta\mu\nu}(0) + \sum_{n=1}^{\infty} \frac{\lambda^n}{n!} x^{\alpha_1} \cdots x^{\alpha_n} \left(\partial_{\alpha_1} \cdots \partial_{\alpha_n} R^\alpha{}_{\beta\mu\nu}(x)|_{x=0}\right). \tag{11}$$

Because of the FS gauge condition (4) the above equation equals

$$R^\alpha{}_{\beta\mu\nu}(\lambda x) = R^\alpha{}_{\beta\mu\nu}(0) + \sum_{n=1}^{\infty} \frac{\lambda^n}{n!} x^{\alpha_1} \cdots x^{\alpha_n} \left(\nabla_{\alpha_1} \cdots \nabla_{\alpha_n} R^\alpha{}_{\beta\mu\nu}(x)|_{x=0}\right), \tag{12}$$

namely, the ordinary derivatives ∂_μ may be replaced by the covariant derivatives Δ_μ . The proof can be seen in the Appendix. By inserting (12) into (7) one has the inversion formula in the form of infinite series

$$\Gamma^\alpha_{\beta\nu}(x) = \frac{1}{2} x^\mu R^\alpha{}_{\beta\mu\nu}(0) + \sum_{n=1}^{\infty} \frac{1}{n!(n+2)} x^\mu x^{\alpha_1} \cdots x^{\alpha_n} \nabla_{\alpha_1} \cdots \nabla_{\alpha_n} R^\alpha{}_{\beta\mu\nu}(0). \tag{13}$$

Notice that the above formula may also be derived from the equation (10). This is done by inserting (12), with $\lambda = 1$, into (10) and by utilizing the identities

$$(1 + x.\partial)^{-1} x^{\alpha_1} x^{\alpha_2} \cdots x^{\alpha_n} = \frac{1}{n+1} x^{\alpha_1} x^{\alpha_2} \cdots x^{\alpha_n}. \tag{14}$$

3. Christoffel Symbol-free Formulations

Let us now utilize the inversion formula (10) to reformulate the general theory of relativity such that the theory becomes free from the Christoffel symbols. The idea to reformulate into such a form comes from the fact that the field equation in general theory of relativity, namely the Einstein field equation, is an equation for Riemann curvature tensor, it is not an equation for Christoffel symbols. Thus it is, perhaps, worth advantage if one can formulate all expressions in the theory in terms of Riemann curvature tensor, and free from the Christoffel symbol. And these can be done if one works in the FS gauge since one may benefit from the existence of the inversion formula. In the theory of relativity expressions or equations which contain the Christoffel symbols are the covariant derivatives and the geodesic differential equation. By recalling the inversion formula (10) the covariant derivative ∇_μ in the FS gauge (for vectors V^μ and V_μ, tensors of order two $V^{\mu\nu}, V_{\mu\nu}$, and $V^{\mu\nu}$, and tensors of higher order $V^{\mu\cdots}_{\nu\cdots}$) can be reexpressed not in terms of Christoffel symbols but in terms of Riemann curvature tensors:

$$\begin{aligned}
\nabla_\alpha V^\mu &= \partial_\alpha V^\mu - (1 + x\partial)^{-1} x^\sigma R^\mu{}_{\alpha\nu\sigma} V^\nu \\
\nabla_\alpha V_\mu &= \partial_\alpha V_\mu + (1 + x\partial)^{-1} x^\sigma R^\nu{}_{\alpha\mu\sigma} V_\nu \\
\nabla_\alpha V^{\mu\nu} &= \partial_\alpha V^{\mu\nu} - (1 + x\partial)^{-1} x^\sigma R^\mu{}_{\alpha\lambda\sigma} V^{\lambda\nu} - (1 + x\partial)^{-1} x^\sigma R^\nu{}_{\alpha\lambda\sigma} V^{\mu\lambda} \\
\nabla_\alpha V^\mu_\nu &= \partial_\alpha V^\mu_\nu - (1 + x\partial)^{-1} x^\sigma R^\mu{}_{\alpha\lambda\sigma} V^\lambda_\nu + (1 + x\partial)^{-1} x^\sigma R^\lambda{}_{\alpha\nu\sigma} V^\mu_\lambda \\
\nabla_\alpha V_{\mu\nu} &= \partial_\alpha V_{\mu\nu} + (1 + x\partial)^{-1} x^\sigma R^\lambda{}_{\alpha\mu\sigma} V_{\lambda\nu} + (1 + x\partial)^{-1} x^\sigma R^\lambda{}_{\alpha\nu\sigma} V_{\mu\lambda}
\end{aligned} \tag{15}$$

$$\nabla_\alpha V^{\mu\cdots}_{\nu\cdots} = \partial_\alpha V^{\mu\cdots}_{\nu\cdots} - (1+x\partial)^{-1} x^\sigma R^\mu{}_{\alpha\lambda\sigma} V^{\lambda\cdots}_{\nu\cdots} - \cdots$$
$$+ (1+x\partial)^{-1} x^\sigma R^\lambda{}_{\alpha\nu\sigma} V^{\mu\cdots}_{\lambda\cdots} + \cdots$$

Up to first order the above derivatives have the forms

$$\nabla_\alpha V^\mu = \partial_\alpha V^\mu - \frac{1}{2} x^\sigma R^\mu{}_{\alpha\nu\sigma}(0) V^\nu$$

$$\nabla_\alpha V_\mu = \partial_\alpha V_\mu + \frac{1}{2} x^\sigma R^\nu{}_{\alpha\mu\sigma}(0) V_\nu \tag{16}$$

$$\nabla_\alpha V^{\mu\nu} = \partial_\alpha V^{\mu\nu} - \tfrac{1}{2} x^\sigma R^\mu{}_{\alpha\lambda\sigma}(0) V^{\lambda\nu} - \tfrac{1}{2} x^\sigma R^\nu{}_{\alpha\lambda\sigma}(0) V^{\mu\lambda}$$

$$\nabla_\alpha V^\mu_\nu = \partial_\alpha V^\mu_\nu - \tfrac{1}{2} x^\sigma R^\mu{}_{\alpha\lambda\sigma}(0) V^\lambda_\nu + \tfrac{1}{2} x^\sigma R^\lambda{}_{\alpha\nu\sigma} V^\mu_\lambda$$

$$\nabla_\alpha V_{\mu\nu} = \partial_\alpha V_{\mu\nu} + \tfrac{1}{2} x^\sigma R^\lambda{}_{\alpha\mu\sigma}(0) V_{\lambda\nu} + \tfrac{1}{2} x^\sigma R^\lambda{}_{\alpha\nu\sigma}(0) V_{\mu\lambda}$$

$$\nabla_\alpha V^{\mu\cdots}_{\nu\cdots} = \partial_\alpha V^{\mu\cdots}_{\nu\cdots} - \tfrac{1}{2} x^\sigma R^\mu{}_{\alpha\lambda\sigma}(0) V^{\lambda\cdots}_{\nu\cdots} - \cdots + \tfrac{1}{2} x^\sigma R^\lambda{}_{\alpha\nu\sigma} V^{\mu\cdots}_{\lambda\cdots} + \cdots$$

One property of the metric tensor $g^{\mu v}$ is that its covariant derivative vanishes, $\nabla_\alpha g^{\mu\nu} = 0$. Since covariant derivatives in the FS gauge have to obey the equation (15) the metric tensor in this gauge can be written in terms of Riemann tensor as follows

$$g^{\mu\nu} = \Box^{-1}(2+x\partial)^{-1} \partial^\alpha x^\sigma (R^\mu{}_{\alpha^\nu\sigma} + R^\nu{}_{\alpha^\mu\sigma}) \tag{17}$$

where $\Box^{-1}$ is the invers of $\partial^\mu \partial_\mu$.

In the geometry of curved spacetime a differential equation for a curve having an extremal length is the geodesic equation,

$$\frac{d^2x^\mu}{ds^2} + \Gamma^\mu_{\alpha\beta} \frac{dx^\alpha}{ds} \frac{dx^\beta}{ds} = 0, \tag{18}$$

where ds is the infinitesimal distance along the geodesic line. Physically the above equation describes the motion of infinitesimally small test particle moving in the gravitational field. It is not difficult to rewrite the geodesic equation into

$$\ddot{x}^k + (\Gamma^k_{\alpha\beta} - \Gamma^0_{\alpha\beta} \dot{x}^k) \dot{x}^\alpha \dot{x}^\beta = 0. \tag{19}$$

Here a dot indicates a differentiation with respect the coordinate $x^0 = ct$. Since $c \approx 3 \,.\, 10^8\ m/s$, at the lowest approximation, the above equation will reduce to the Newton law, $\ddot{x}^k \simeq -\Gamma^k_{oo}$, with Γ^k_{oo} acts just like a Newtonian force per mass unit and, accordingly, the corresponding Newtonian potential is given by (g_{oo} is the 00-component of the metric tensor)

$$\phi(x) = -\frac{c^2}{2}(1 - g_{00}(x)). \tag{20}$$

In the FS gauge the Christoffel symbols may disappear from the geodesic equation. In this gauge the differential equation (19) becomes

$$\ddot{x}^k - \left[(1 + x\partial)^{-1} x^\sigma R^k{}_{\alpha\beta\sigma} - \dot{x}^k (1 + x\partial)^{-1} x^\sigma R^0{}_{\alpha\beta\sigma}\right] \dot{x}^\alpha \dot{x}^\beta = 0 \tag{21}$$

At the lowest approximation, the above equation has the form

$$\ddot{x}^k \approx (1 + x\partial)^{-1} x^\sigma R^k{}_{oo\sigma} \tag{22}$$

which means that $(1 + x\partial)^{-1} x^\sigma R^k{}_{oo\sigma}$ acts as a Newtonian force per unit of mass. Further approximation will give

$$\ddot{x}^k \approx \frac{1}{2}\partial^k (x\partial - 1) g_{00}(x). \tag{22}$$

Thus the corresponding Newtonian potential in the FS gauge reads

$$\phi_{FS}(x) = \frac{c^2}{2}(x\partial)^{-1}(x\partial - 1) g_{00}(x). \tag{23}$$

At the Newtonian limit (lowest order) the relationship between the potential in the FS gauge ϕ_{FS} and the more familiar potential ϕ can be obtained by combining (24) and (20). One has

$$\phi_{FS} = \phi - (x\partial)^{-1}\phi + \frac{c^2}{2}. \tag{24}$$

Since ϕ is a Newtonian gravitational potential, namely $\phi = K/r, K$ a negative constant, it turns out that the corresponding ϕ_{FS} equals $2K/r + c^2/2$. This is an example of the application of the inversion formula: the formula enables us to obtain a new (that is the FS) potential, other than the familiar one. Such an application in electrostatics and magnetostatic as well as in cosmology has also been carried out [5,6]. The result (25) says that, in contrast to the original Newtonian potential ϕ, the Newtonian FS potential ϕ_{FS} is approximately finite around the centre of gravitation and becomes very large as we move away from it.

4. Conclusion

In this paper the Christoffel symbols have been restricted by a gauge condition, an analogous treatment to the gauge fields in the gauge theory. The condition we choose is the FS gauge condition (4). As a consequence of imposing the FS gauge is that there is the inversion formula, that is the expression of the Christoffel symbols in terms of the Riemann curvature tensors. The formula has been shown in the forms of integral, inverse differential operators, and infinite series. The existence of the formula enables us to reformulate the general theory of relativity without Christoffel symbols. Here, the expression of the metric tensor in terms of Riemann tensors has been obtained and discussion of the geodesic equation, at the lowest approximation, results in the so-called Newtonian gravitation FS potential with its value: approximately finite around the centre of gravitation and very large at points far away from it.

5. Appendices

The proof of the equation (12) is as follows. Consider the following expression

$$x^{\alpha_1}\cdots x^{\alpha_n}[\partial_{\alpha_1}\partial_{\alpha_2}\cdots\partial_{\alpha_n}R^{\alpha}{}_{\beta\mu\nu}(x)-\nabla_{\alpha_1}\nabla_{\alpha_2}\cdots\nabla_{\alpha_n}R^{\alpha}{}_{\beta\mu\nu}(x)] \quad (A-1)$$

The equation (12) holds if the expression (A-1) vanishes for every n. For $n=1$

$$x^{\alpha_1}[\partial_{\alpha_1}-\partial_{\alpha_1}-\Gamma^{\alpha}_{\alpha_1\lambda}R^{\lambda}{}_{\beta\mu\nu}(x)+$$
$$\Gamma^{\lambda}_{\alpha_1\beta}R^{\alpha}{}_{\lambda\mu\nu}(x)+\Gamma^{\lambda}_{\alpha_1\mu}R^{\alpha}{}_{\beta\lambda\nu}(x)+\Gamma^{\lambda}_{\alpha_1\nu}R^{\alpha}{}_{\mu\nu\lambda}(x)]=0 \quad (A-2)$$

because the condition (4). For n=2 the expression (A-1) reads

$$x^{\alpha_1}x^{\alpha_2}\{\partial_{\alpha_1}\partial_{\alpha_2}-[\partial_{\alpha_1}(\nabla_{\alpha_2}R^{\alpha}{}_{\beta\mu\nu}(x))+\Gamma^{\alpha}_{\alpha_1\lambda}R^{\lambda}{}_{\beta\mu\nu}(x)]$$
$$-\Gamma^{\lambda}_{\alpha_1\beta}R^{\alpha}{}_{\lambda\mu\nu}(x)-\Gamma^{\lambda}_{\alpha_1\mu}R^{\alpha}{}_{\beta\lambda\nu}(x)-\Gamma^{\lambda}_{\alpha_1\nu}R^{\alpha}{}_{\beta\mu\lambda}(x)]\} \quad (A-3)$$

The last four terms vanish because of the condition (4). The remaining terms also vanish

$$x^{\alpha_1}x^{\alpha_2}\partial_{\alpha_1}\{\Gamma^{\alpha}_{\alpha_2\lambda}R^{\lambda}{}_{\beta\mu\nu}(x)-\Gamma^{\lambda}_{\alpha_2\beta}R^{\alpha}{}_{\lambda\mu\nu}(x)-\Gamma^{\lambda}_{\alpha_2\nu}R^{\alpha}{}_{\beta\lambda\nu}(x)-\Gamma^{\lambda}_{\alpha_2\nu}R^{\alpha}{}_{\beta\mu\lambda}(x)\}$$
$$=x^{\alpha_1}\partial_{\alpha_1}\{x^{\alpha_2}[\Gamma^{\alpha}_{\alpha_2\lambda}R^{\lambda}{}_{\beta\mu\nu}(x)-\Gamma^{\lambda}_{\alpha_2\beta}R^{\alpha}{}_{\lambda\mu\nu}(x)-\Gamma^{\lambda}_{\alpha_2\mu}R^{\alpha}{}_{\beta\lambda\nu}(x)-\Gamma^{\lambda}_{\alpha_2\nu}R^{\alpha}{}_{\beta\mu\lambda}(x)]\}$$
$$-x^{\alpha_2}[\Gamma^{\alpha}_{\alpha_2\lambda}R^{\lambda}{}_{\beta\mu\nu}(x)-\Gamma^{\lambda}_{\alpha_2\beta}R^{\alpha}{}_{\lambda\mu\nu}(x)-\Gamma^{\lambda}_{\alpha_2\mu}R^{\alpha}{}_{\beta\lambda\nu}(x)-\Gamma^{\lambda}_{\alpha_2\nu}R^{\alpha}{}_{\beta\mu\lambda}(x)]=0 \quad (A-4)$$

Thus (A-1) is also zero for $n = 2$. By recalling (A-4) it can be easily shown that the remaining term for $n = 3$ has the form

$$x^{\alpha_1}x^{\alpha_2}x^{\alpha_3}\partial_{\alpha_1}\partial_{\alpha_2}[\Gamma^{\alpha}_{\alpha_3\lambda}R^{\lambda}{}_{\beta\mu\nu}(x) - \Gamma^{\lambda}_{\alpha_3\beta}R^{\alpha}{}_{\lambda\mu\nu}(x) - \Gamma^{\lambda}_{\alpha_3\mu}R^{\alpha}{}_{\beta\lambda\nu}(x) - \Gamma^{\lambda}_{\alpha_3\nu}R^{\alpha}{}_{\beta\mu\lambda}(x)] = 0. \qquad (A-5)$$

It is zero because of (4) and (A-4). Note that the left-hand side of (A-5) can be written as

$$[x^{\alpha_1}x^{\alpha_2}\partial_{\alpha_1}\partial_{\alpha_2}x^{\alpha_3} - x^{\alpha_1}x^{\alpha_3}\partial_{\alpha_1} - x^{\alpha_2}x^{\alpha_3}\partial_{\alpha_2}]. [\Gamma^{\alpha}_{\alpha_3\lambda}R^{\lambda}{}_{\beta\mu\nu} - \Gamma^{\lambda}_{\alpha_3\beta}R^{\alpha}{}_{\lambda\mu\nu}(x) - \Gamma^{\lambda}_{\alpha_3\mu}R^{\alpha}{}_{\beta\lambda\nu}(x) - \Gamma^{\alpha}_{\alpha_3 v}R^{\alpha}{}_{\beta\mu\lambda}(x)]. \qquad (A-6)$$

The similarity between (A-4) and (A-5) enables us to prove for higher n by mathematical induction. Suppose that this expression holds for n

$$x^{\alpha_1}x^{\alpha_2}\cdots x^{\alpha_n}x^{\sigma}\partial_{\alpha_1}\partial_{\alpha_2}\cdots\partial_n \left[\Gamma^{\alpha}_{\sigma\lambda}R^{\lambda}{}_{\beta\mu\nu}(x) - \Gamma^{\lambda}_{\sigma\beta}R^{\alpha}{}_{\lambda\mu\nu}(x) - \Gamma^{\lambda}_{\sigma\mu}R^{\alpha}{}_{\beta\mu\nu}(x) - \Gamma^{\lambda}_{\sigma\nu}R^{\alpha}{}_{\beta\mu\lambda}(x)\right]. \qquad (A-7)$$

Accordingly, for $n+1$

$$x^{\alpha_1}x^{\alpha_2}\cdots x^{\alpha_n}x^{\alpha_{n+1}}x^{\sigma}\partial_{\alpha_1}\partial_{\alpha_2}\cdots\partial_n\partial_{n+1}$$
$$\left[\Gamma^{\alpha}_{\sigma\lambda}R^{\lambda}{}_{\beta\mu\nu}(x) - \Gamma^{\lambda}_{\sigma\beta}R^{\alpha}{}_{\lambda\mu\nu}(x) - \Gamma^{\lambda}_{\sigma\mu}R^{\alpha}{}_{\beta\lambda\nu}(x) - \Gamma^{\lambda}_{\sigma\nu}R^{\alpha}{}_{\beta\mu\lambda}(x)\right].$$
$$x^{\alpha_1}x^{\alpha_2}\cdots x^{\alpha_n}x^{\alpha_{n+1}}\partial_{\alpha_1}\partial_{\alpha_2}\cdots\partial_n\partial_{n+1}$$

$$x^{\sigma}\left[\Gamma^{\alpha}_{\sigma\lambda}R^{\lambda}{}_{\beta\mu\nu}(x) - \Gamma^{\lambda}_{\sigma\beta}R^{\alpha}{}_{\lambda\mu\nu}(x) - \Gamma^{\lambda}_{\sigma\mu}R^{\alpha}{}_{\beta\lambda\nu}(x) - \Gamma^{\lambda}_{\sigma\nu}R^{\alpha}{}_{\beta\mu\lambda}(x)\right].$$
$$-x^{\alpha_1}x^{\alpha_2}\cdots x^{\alpha_n}x^{\alpha_{n+1}}\partial_{\alpha_1}\partial_{\alpha_2}\cdots\partial_n$$

$$\left[\Gamma^{\alpha}_{\alpha_{n+1}\lambda}R^{\lambda}{}_{\beta\mu\nu}(x) - \Gamma^{\lambda}_{\alpha_{n+1}\beta}R^{\alpha}{}_{\lambda\mu\nu}(x) - \Gamma^{\lambda}_{\alpha_{n+1}\mu}R^{\alpha}{}_{\beta\lambda\nu}(x) - \Gamma^{\lambda}_{\alpha_{n+1}\nu}R^{\alpha}{}_{\beta\mu\lambda}(x)\right].$$
$$-x^{\alpha_1}x^{\alpha_2}\cdots x^{\alpha_n}x^{\alpha_{n+1}}\partial_{\alpha_1}\partial_{\alpha_2}\cdots\partial_{n-1}\partial_{n+1}$$

$$\left[\Gamma^{\alpha}_{\alpha_n\lambda}R^{\lambda}{}_{\beta\mu\nu}(x) - \Gamma^{\lambda}_{\alpha_n\beta}R^{\alpha}{}_{\lambda\mu\nu}(x) - \Gamma^{\lambda}_{\alpha_n\mu}R^{\alpha}{}_{\beta\lambda\nu}(x) - \Gamma^{\lambda}_{\alpha_n\nu}R^{\alpha}{}_{\beta\mu\lambda}(x)\right]$$
$$-\cdots$$
$$-x^{\alpha_1}x^{\alpha_2}\cdots x^{\alpha_n}x^{\alpha_{n+1}}\partial_{\alpha_1}\partial_{\alpha_3}\cdots\partial_{n+1}$$
$$\left[\Gamma^{\alpha}_{\alpha_1\lambda}R^{\lambda}{}_{\beta\mu\nu}(x) - \Gamma^{\lambda}_{\alpha_1\beta}R^{\alpha}{}_{\lambda\mu\nu}(x) - \Gamma^{\lambda}_{\alpha_1\mu}R^{\alpha}{}_{\beta\lambda\nu}(x) - \Gamma^{\lambda}_{\alpha_1\nu}R^{\alpha}{}_{\beta\mu\lambda}(x)\right] = 0 \qquad (A-8)$$

Thus expansion (12) is valid.

References

1. Jackson J.D.(1983), *Classical Electrodynamics*, 2nd ed., Wiley Eastern limited.
2. Rohrlich, F. (1990), *Classical Charged Particles*, Addison-Wesley Publishing Company, Inc.
3. Delbourgo, R. and Triyanta (1992), *Equivalence between the Fock-Schwinger Gauge and the Feynman Gauge*,**Int.J.Mod.Phys.A Vol.27 No.23**, 5833-54, and references there in.
4. Halpern, M.B. (1979), *Field Strength and Dual Variable Formulations of Gauge Theory*, **Phys.Rev.D 19(2)**, 517-530
5. Wibowo, B.M.W (1995), *Simbol Christoffel Gauge Fock-Schwinger*,**Sarjana Degree Final Project**, Physics Department ITB.
6. Triyanta (1992), *The Fock-Schwinger gauge Potentials for some Simple Systems*, **KFI Vol.3 No.2**, 46-55.

A BACK-TURNING JET FORMED BY A UNIFORM SHALLOW STREAM HITTING A VERTICAL WALL

L. H. WIRYANTO
Department of Mathematics
Institut Teknologi Bandung
Indonesia.

AND

E. O. TUCK
The University of Adelaide
Australia

Abstract. A 2-D free uniform flow hitting a vertical wall is considered. The fluid is assumed to be inviscid and incompressible, and the flow is irrotational. Therefore, the problem reduces to a nonlinear boundary-value problem for Laplace's equation. This problem is solved numerically by using a boundary integral equation instead of the series method used by previous researchers. Our result confirms previous calculations, but with a slightly different value for the minimum Froude number.

1. Introduction

Jet flows under gravity can be obtained in many engineering problems such as weir flows and jets emerging from a nozzle. The first type of problem can be seen in papers such as by Vanden-Broeck & Keller [1]. Meanwhile, the second type of problem has been computed by Vanden-Broeck & Keller [2], Goh & Tuck [3], and Tuck [4].

In the case of a free uniform stream hitting a vertical wall, a jet is formed after the oncoming flow forced to go up along the wall reaches its maximum height at a stagnation point. The jet then falls back upon the oncoming flow under the effect of gravity. However, we assume that the falling jet does not contaminate the oncoming flow, i.e. the jet lies on another Riemann sheet. This problem was solved by Dias & Christodoulides [5]. Solutions are found only in the supercritical case, i.e. when the stream speed U exceeds $\sqrt{gh}$ (g is gravity and h is the stream depth) with the Froude number $F = U/\sqrt{gh}$ greater than a critical value F_c. Dias and Christodoulides gave the estimate $F_c = 2.961$.

The jet seems to move farther upstream from the wall in the limit as F tends to F_c. Dias & Tuck [6] computed this limiting flow directly and gave the extrapolated critical value $F_c = 2.994$.

The authors in both papers [5] and [6] used truncated series methods constructed from the singular points with the coefficients determined by collocation. In this paper we recompute Dias and Christodoulides' problem by using a boundary integral equation. This method has been applied to many problems such as withdrawing fluid through a sink by Hocking & Forbes [7] and ploughing flows by Tuck [8]. Here, the jet can be considered similar to the sink in [7], but since the jet is bounded by two streams which are free surfaces, one of the singularities is in the integration domain. To avoid this, the jet is approached from the stagnation point for the upper free surface and from upstream for the lower free surface.

The character of the splashlike flow corresponding to the Froude number confirms previous calculations, but with a slightly different value $F_c = 2.9831$ for the minimum Froude number. This result is discussed in Sec. 4 after we formulate the integral equation in Sec. 2 and the numerical procedure in Sec. 3.

2. Formulation of the problem

We consider the 2-D steady irrotational flow of an incompressible inviscid fluid. A fast uniform stream of finite depth water is generated far to the right, and flows leftward. A vertical wall bounds the flow on the left side and a corner is formed with a horizontal bottom at C. Hence a falling jet is formed which is bounded by two free surfaces after the flow hits the wall and goes up reaching a stagnation point (see Fig. 1(a)). The system of (x, y) coordinates is defined with the y axis along the vertical wall and the x axis chosen as the level of the free surface upstream.

The pressure is assumed to be constant on the free surfaces so that Bernoulli's equation yields

$$\frac{1}{2}(\psi_x^2 + \psi_y^2) + gy = \text{constant} \tag{1}$$

on free surfaces. Here, $\psi = \psi(x, y)$ is the streamfunction. The problem is nondimensionalized by taking h as the unit length and U as the unit velocity so that the net discharge becomes unity. In terms of these dimensionless variables, the condition (1) becomes

$$\frac{1}{2}F^2(\psi_x^2+\psi_y^2)+y=\frac{1}{2}F^2. \tag{2}$$

Our task is to solve the Laplace partial differential equation for $\psi(x,y)$, subject to the nonlinear boundary condition (2) on the (unknown) free surfaces, and $\psi = 1$ on the bottom surface $y = -1$ and the vertical wall $x = 0$.

Introducing the complex variables $z = x + iy$ and $f = \phi + i\psi$, where ϕ is the velocity potential, the flow domain in the f plane is an infinite strip of height 1, bounded by the streamlines $\psi = 1$ and 0 (see Fig. 1(b)). The first streamline is along the horizontal and vertical walls and the upper free surface (ICSJ in Fig. 1(a)) and $\psi = 0$ presents the lower free surface from I to J. The uniform stream and the free-falling jet are the beginning and the ending of the flow. So, they lie at $\phi \to -\infty$ and $\phi \to +\infty$ respectively.

To derive the integral equation for this problem we follow Tuck [8]. The transformation

$$s = e^{-\pi f} \tag{3}$$

maps the strip $0 \leq \psi \leq 1$ to the lower half of the s-plane (Fig. 1(c)). We choose that $f = i$ corresponds to the stagnation point S, so that the free surfaces lie along the real s-axis where the upper free surface is along $-1 \leq s \leq 0$ and the lower free surface is along $s \geq 0$. The remainder of the real s axis corresponds to the vertical and bottom walls which are separated by the corner point C (in the z-plane) corresponding to $-s_c$.

The complex potential f is determined by solving for $\gamma = \tau - i\alpha$, which is defined in relation to the velocity field by

$$\frac{df}{dz} = -e^{\gamma}. \tag{4}$$

This says that τ corresponds to the magnitude of the velocity ($|df/dz| = e^{\tau}$) at any point and α corresponds to the angle of streamlines where along the walls we have

$$\alpha(s) = \begin{cases} 0, & \text{if } s < -s_c \\ -\frac{\pi}{2}, & \text{if } -s_c < s < -1. \end{cases} \tag{5}$$

The relation between τ and α on the free surfaces $-1 < s^* < 0$ and $0 < s^* < \infty$ can be derived by using the Cauchy integral formula since γ tends to zero as $|s| \to \infty$. The real part of the integral gives

$$\tau(s^*) = -\frac{1}{2}\log\left|\frac{1+s^*}{s_c+s^*}\right| + \frac{1}{\pi}\mathrm{PV}\int_{-1}^{\infty}\frac{\alpha(s)}{s-s^*}ds \tag{6}$$

after substituting the value of α in (5). PV denotes the Cauchy principal value for the integration.

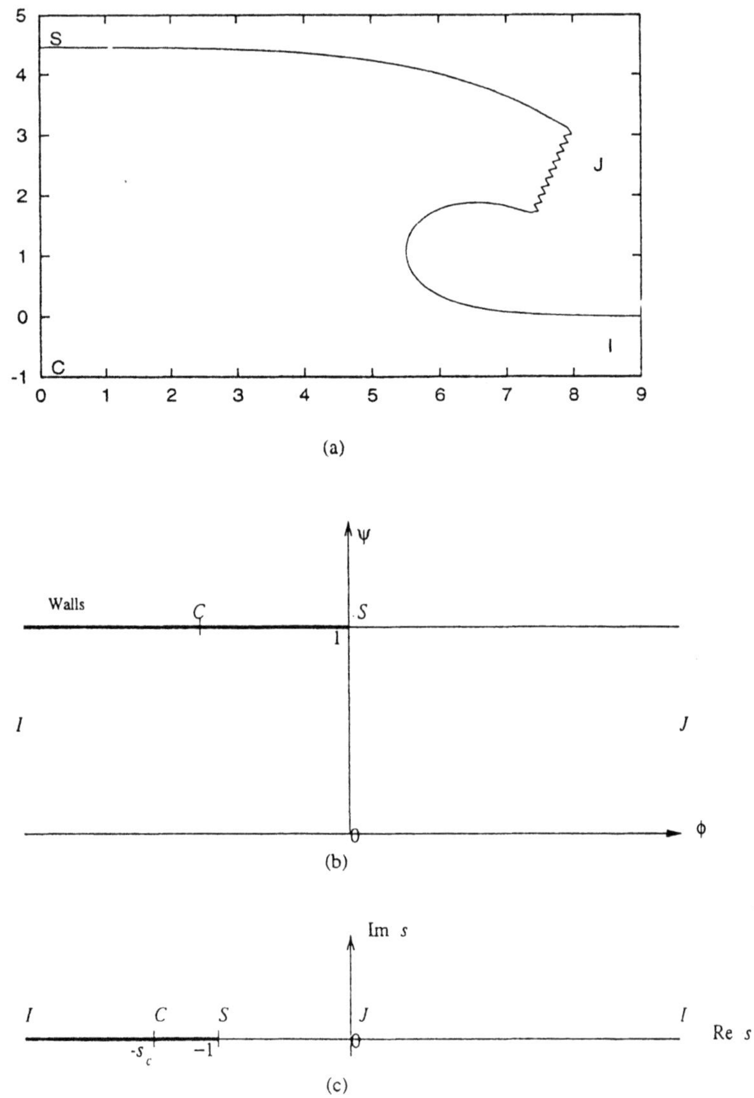

Figure 1. (a) Sketch of the flow and of the coordinates. The profile of the free surfaces was computed for $F = 2.9848$. (b) The complex potential plane $f = \phi + i\psi$. (c) The complex s-plane. The bold line corresponds to the bottom and vertical walls.

On the other hand, the values of x and y on the free surfaces are obtained by using (3) and (4) in the derivative with respect to s

$$\frac{dz}{ds} = \frac{e^{-\gamma}}{\pi s}. \tag{7}$$

This gives

$$\frac{dx}{ds} = \frac{1}{\pi s} e^{-\tau} \cos\alpha \tag{8}$$

$$\frac{dy}{ds} = \frac{1}{\pi s} e^{-\tau} \sin\alpha. \tag{9}$$

Hence, x and y can be determined by the condition at the stagnation point $x(-1) = 0$ and $y(-1) = \frac{1}{2}F^2$ after knowing α, which can be done by substituting both (6) and the integration from (9) into Bernoulli's equation, which becomes

$$\frac{1}{2}F^2 e^{2\tau} + y = \frac{1}{2}F^2. \tag{10}$$

Equation (10) combined with (6) and (9) is in effect on integral equation for $\alpha(s)$.

Numerically, (10) can be solved by choosing collocation points on the free surfaces so that we obtain a system of nonlinear equations that can be solved by Newton's method. The major difficulty occurs with integration from (9), in which the factor πs in the dominator makes it essentially impossible to integrate through $s = 0$. Instead, we use

$$y(s^*) = \frac{1}{2}F^2 + \frac{1}{\pi}\int_{-1}^{s^*} \frac{e^{-\tau}\sin\alpha}{s} ds, \quad \text{for } -1 < s^* < 0 \tag{11}$$

and

$$y(s^*) = -\frac{1}{\pi}\int_{s^*}^{s_{max}} \frac{e^{-\tau}\sin\alpha}{s} ds, \quad \text{for } s^* > 0 \tag{12}$$

where we truncate the domain of the integration in (6) at a finite value s_{max} representing the upstream zero level.

This completes the formulation of the problem. We seek the function $\alpha(s)$ satisfying the integral equation (10) after substituting (6) and (11) for the upper free surface or (12) for the lower free surface.

3. Numerical procedure

The numerical method for the solution of the governing equation derived in section 2 is described briefly in this section. The free surfaces are presented as a set of discrete points at the M surface grid points $s_0, s_1, \cdots, s_{NN}, s_{NN+1}, \cdots, s_M$. Here, $s_0 = -1$ is the surface stagnation point and $s_{NN} = -\epsilon$ ($\epsilon > 0$ and small) is the final jet point on the upper free surface. These NN grid spaces are reflected to the lower free surface so that we have the same grid of points in both intervals [-1,$-\epsilon$] and [ϵ,1]. Then the remaining interval (1,s_{max}] of the lower free surface is discreted into N grid spaces with $s_M = s_{max}$ chosen sufficiently far upstream and $M = 2NN + N + 1$.

To get better accuracy, the space around singular points, i.e. the jet at $s = 0$ and the stagnation point at $s = -1$, is made dense. The grid points on [-1,-ϵ] are defined by

$$s_k = \bar{s} - \frac{1}{2}\delta \cos(\frac{k}{NN}\pi), \quad \text{for } k = 0, 1, \cdots, NN, \tag{13}$$

where $\bar{s} = \frac{1}{2}(-\epsilon - 1)$ and $\delta = -\epsilon + 1$. For the range $(1, s_{max})$, we make the intervals equally spaced in the potential function ϕ, i.e.

$$s_{2NN+1+l} = e^{\pi\phi_l}, \quad \text{for } l = 1, 2, \cdots, N, \tag{14}$$

where $\phi_l = \frac{l}{N}t$ with $s_{max} = e^{\pi t}$. Then, the M collocation points are chosen as the midpoints

$$s_j^* = \frac{s_j + s_{j-1}}{2}, \quad \text{for } j = 1, 2, \cdots, NN, NN+2, \cdots, M \tag{15}$$

and $s_{NN+1}^* = -\frac{1}{2}\epsilon$.

The Cauchy principal value integral (6) is approximated by piece-wise continuous behaviour of the numerator for each segment, giving

$$\text{PV}\int_{-1}^{\infty} \frac{\alpha(s)}{s - s_j^*} ds \approx \sum_{l=1}^{M} (\alpha_l - \alpha_{l-1}) + \left\{\alpha_l + (\alpha_l - \alpha_{l-1})\frac{s_j^* - s_{l-1}}{s_l - s_{l-1}}\right\} \log\left|\frac{s_l - s_j^*}{s_{l-1} - s_j^*}\right| \tag{16}$$

where $\alpha_l = \alpha(s_l)$ with $\alpha_0 = -\pi$. As angles α_j^* corresponding to the collocation points, we also use the average of α_j and α_{j-1} except for $j = NN+1$, where a linear extrapolation is applied from α_{NN}^* and α_{NN}. Therefore, the piece-wise continuous approximation can be applied to (11) and (12) to evaluate $y_j = y(s_j^*)$.

The M equations are obtained by substituting the values of τ from (6) with Cauchy principal-value integral (16), and the values of y at each collocation point s_j^* $(j = 1, 2, \cdots, M)$. However, we must determine M+1 unknowns $\alpha_1, \alpha_2, \cdots, \alpha_M$ and F for given $s_c > 1$. Hence, an extra equation is required to get a closed system. This equation is constructed from the distance of the stagnation point to the bottom wall giving

$$1 + \frac{1}{2}F^2 = -\frac{1}{\pi}\int_{-s_c}^{-1} \frac{e^{-\tau}}{s} ds. \tag{17}$$

This integration is evaluated by linearizing the numerator except for the segments bounded by -s_c or -1 where τ cannot be evaluated from (6). Instead, step functions are applied with the grid of points in the integration domain made dense around $-s_c$ and -1 similar to (13) to reduce the error. Then, this system of M+1 equations is solved by Newton's method.

The calculation is started with a small value M and an initial guess α distributed following s in $[-\frac{3}{2}\pi, -\pi]$ for the upper free surface and in $[-\frac{3}{2}\pi, 0]$ for the lower free surface. The angle of the jet is given by $-\frac{3}{2}\pi$, and the stagnation point and the far upstream are given by $-\pi$ and 0 respectively. Once a solution is obtained, it is used as the initial guess for a bigger value M. Here, we include the collocation points in (-1,-ϵ) and (ϵ, s_{max}) as the new grid points followed by the angles α.

The convergence of the procedure is achieved for $M \to \infty$ and $s_{max} \to \infty$. This is done by computing the Froude number for 3 different values M where each time we double the previous values of NN and N, for fixed s_{max}. From these three Froude numbers, we determine three unknowns, F_∞, A and $\beta(\approx 2)$, satisfying

$$F = F_\infty + A\left(\frac{M_{max}}{M}\right)^\beta, \tag{18}$$

where M_{max} is the biggest value of the three Ms. When this is solved, the approximated Froude number at $M = \infty$ is F_∞. Then, this is repeated for two other values of s_{max}, each chosen also double from the previous s_{max}. The accuracy of the Froude number is obtained up to the fourth decimal for $M = 153$ ($NN = 28$ and $N = 96$) and $s_{max} = 250000$. These values are used to compute solutions with $\epsilon = 10^{-4}$, where this value is chosen so that ϵ is in the same order of the minimum grid between s_j and s_j^*. The convergence of the procedure with respect to ϵ is tested the same way as the other parameters M and s_{max}. This gives the same accuracy.

4. Numerical results

We used the numerical scheme described in section 3 to compute solutions for various values of s_c. The Froude number and the profile of free surfaces are part of the solution which will be presented here.

In our scheme, we determine the lower free surface by integrating from upstream. The position of the final point of the jet must be consistent with the final point from the upper free surface. This is checked by plotting streamlines between both free surfaces as shown in Fig. 2 with streamlines corresponding to $\psi = 0.25$, 0.5, and 0.75 for $F = 2.9844$.

In relating the Froude number to s_c, we found that the Froude number decreases by giving smaller s_c. This approaches the minimum value $F_c = 2.9831$ as $s_c \to 1$ which is extrapolated from the Froude numbers computed up to $s_c = 1.01$. This value is smaller than the extrapolated result 2.994 that Dias and Tuck [6] estimated, but it is bigger than their computed value 2.978 using up to 730 terms of their truncated series.

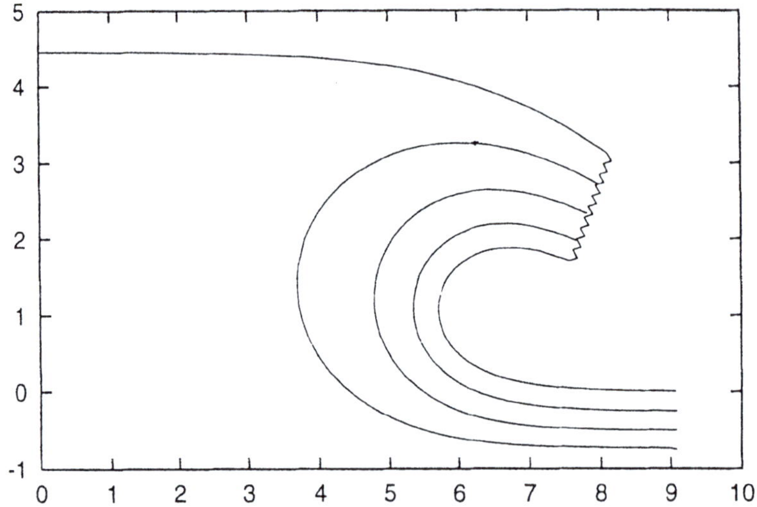

Figure 2. The free surfaces and the three streamlines corresponding to $\psi = 0.25$, 0.5, and 0.75 for $F = 2.9844$.

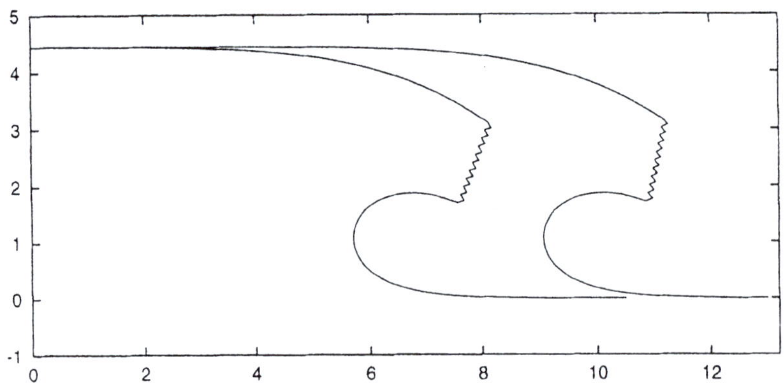

Figure 3. The free surfaces for $F = 2.9844$ (left jet) and $F = 2.9832$ (right jet).

Figure 3 shows the free surfaces for two different values of the Froude number. The smaller Froude numbers produce a splash which seems to move farther upstream from the vertical wall. This is obtained from the minimum distance d of the lower free surface from the vertical wall. A plot of d (scaled with the upstream water depth h) vs F is shown in Fig. 4. As F approaches 2.9831, the distance d increases very rapidly and approaches infinity. This character agrees with the result in [5], but with bigger F. Dias and Christodoulides gave $F = 2.962$ and 2.965 corresponding to $d = 9.9$ and 5.7 respectively, but we found 2.9832 and 2.9844.

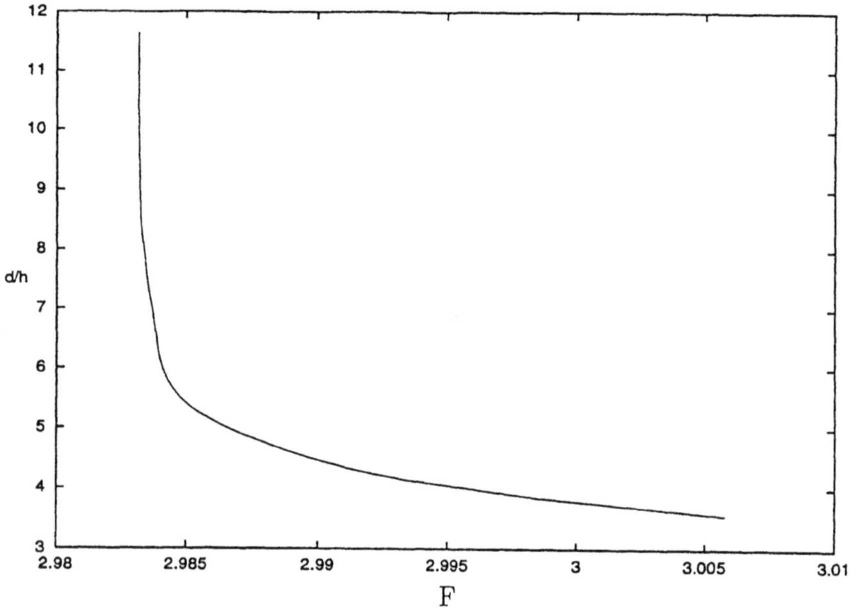

Figure 4. Plot of d/h vs F.

References

1. J.-M. Vanden-Broeck and J.B. Keller, Weir flows, *J. Fluid Mech.* 176 (1987) 283-293.
2. J.-M. Vanden-Broeck and J.B. Keller, Jets rising and falling under gravity, *J. Fluid Mech.* 124 (1982) 335-345.
3. K.H.M. Goh and E.O. Tuck, Thick waterfalls from horizontal slots, *J. Eng. Math.* 19 (1985) 341-349.
4. E.O. Tuck, Efflux from a slit in a vertical wall, *J. Fluid Mech.* 176 (1987) 253-264.
5. F. Dias and P. Christodoulides, Ideal jets falling under gravity, *Phys. Fluids A* 3 (1991) 1711-1717.
6. F. Dias and E.O. Tuck, Steady breaking waves, *Phys. Fluids A* 5 (1993) 277-279.
7. G.C. Hocking and L.K. Forbes, Subcritical free-surface flow caused by a line source in a fluid of finite depth, *J. Eng. Math.* 26 (1992) 455-466.
8. E.O. Tuck, Ploughing flows (1996) (Submitted to Euro. J. Applied Math.).

A FIFTH -ORDER FIVE-STAGE RK-METHOD BASED ON HARMONIC MEAN

NAZEERUDDIN YAACOB AND BAHROM SANUGI
Department of Mathematics
Universiti Teknologi Malaysia
Sekudai

Abstract. In this paper a fifth-order five-stage Runge-Kutta formula based on Harmonic mean is developed for solving initial value problem of the type $y' = f(y)$, $y(x_0) = y_0$. The stability analysis reveals that the region of absolute stability for this method is larger than that of Taylor series's method of order five. Numerical examples indicate that this method is superior when compared to some existing methods including the 5-stage GM-RK method, Taylor Series's method of order 5, and RK5(6) method of Nystrom version.

1. Introduction

Butcher [1987] provides a general outline on the attainable orders for explicit Runge-Kutte methods. It was proved that there does not exist a five-stage methods of order five. This fact is actually true for linear methods but is was thought it might not be true for nonlinear methods. In a number of publications, e.g. Sanugi [1986], Sanugi & Evans [1994], Evans & Sanugi [1986], Wazwaz [1994] and Evans & Yaacob [1995], the existence of nonlinear Runge-Kutta methods based on various means formulation such as Geometric and Heronian means in addition to the Arithmetic mean which is more usually employed was revealed. In particular, Sanugi [1986] has established that R-stage methods of order $p = R$ were possible in the nonlinear formulation strategy for R=1(1)4.

Later, Sanugi and Yaacob [1995] have developed a new fifth-order five-stage RK method that was based on the Geometric mean. However, the presence of four square root signs in the formula could cause some difficulties during computation unless the function is monotonically increasing or decreasing. To overcome this problem we now try to use the Harmonic mean to replace the Geometric mean. The derivation of the method is given in detail along with error analysis as well as the absolute stability consideration. Numerical examples are provided which shows that this new method is well placed among the fifth order methods and it is superior when compared with fifth order Taylor series method, fourth-order five-stage Merson's method including a fifth-order sixth-stage Nystrom's method.

2. Derevation of a New Fifth-Order Five-Stage RK Method Based On Harmonic Mean

Sanugi & Evans [1994] have developed the fourth-order, four-stage RK formula which is given by

$$y_{n+1} = y_n + \frac{h}{3}\sum_{i=1}^{3}\left(\frac{2k_ik_{i+1}}{k_i + k_{i+1}}\right) + O(h^5),\ k_j + k_{j+1} \neq 0 \tag{1}$$

for some particular values of the parameters.
Now, let us introduce our new fifth order scheme which is defined as

$$y_{n+1} = y_n + h\sum_{i=1}^{4} w_i\left(\frac{2k_ik_{i+1}}{k_i + k_{i+1}}\right) + O(h^6),\ k_j + k_{j+1} \neq 0 \tag{2}$$

where

$$\begin{aligned}
k_1 &= f(x_n, y_n) &(3)\\
k_2 &= f(x_n + a_1h, y_n + a_1hk_1) &(4)\\
k_3 &= f(x_n + (a_2 + a_3)h, y_n + a_2hk_1 + a_3hk_2) &(5)\\
k_4 &= f(x_n + (a_4 + a_5 + a^*)h, y_n + a_4hk + a_5hk + a^*hk) &(6)\\
k_5 &= f(x_n + (a_6 + a_7 + a_8 + a^{**})h, y_n + a_6hk_1 + a_7hk_2 + \\
&\quad a_8hk_3 + a^{**}hk_4) &(7)
\end{aligned}$$

where $w_1,\ w_2,\ w_3,\ w_4$ are the weights chosen in such a way so that $\sum_{i=1}^{4} w_i = 1$, and $a^*,\ a^{**},\ a_j,\ j = 1,\ 2,\ \cdots,\ 8$ are parameters to be determined and $\frac{2k_ik_{i+1}}{k_i+k_{i+1}}$ is defined as the Harmonic Mean for k_i and k_{i+1}

Following the argument given by Sanugi [1986], there is no loss of generality in restricting ourselves to a system of the form

$$y'(x) = f(y(x)) \tag{8}$$

in which x does not appear as an argument since, if necessary, an additional differential equation can be added for which the solution is x. An example is $y' = 1$ with initial condition $y(0) = 0$. Thus, equation (4)-(7) may be written more concisely as follows.

$$k_1 = f(y_n) \tag{9}$$
$$k_2 = f(y_n + a_1 hk_1) \tag{10}$$
$$k_3 = f(y_n + a_2 hk_1 + a_3 hk_2) \tag{11}$$
$$k_4 = f(y_n + a_4 hk_1 + a_5 hk_2 + a^* hk_3) \tag{12}$$
$$k_5 = f(y_n + a_6 hk_1 + a_7 hk_2 + a_8 hk_3 + a^{**} hk_4) \tag{13}$$

From these equations (2) and (10) through (13) it is clear that we have 14 parameters to be determined before the method is explicitly defined.
In order to obtain a fifth order method, we equate all the coefficients of $h,\ h^2,\ h^3,\ h^4,\ h^5$ in (2) to the 12 coefficients of Taylor expansion of $y(x_n + h)$ at $x = x_n$. This is conveniently done using MATHEMATICA. This underdetermined system of non-linear equations gives us two degrees of freedom. If we decide to compute k_4 at the mid-point $x = x_n + \frac{h}{2}$ and k_5 at the end point $x = x_n + h$, then we would have two additional conditions, i.e.

$$a^* = \frac{1}{2} - a_4 - a_5 \quad \text{and} \quad a^{**} = 1 - a_6 - a_7 - a_8.$$

This arrangement leaves us with 12 equations to be solved for the 12 unknowns $a_1, ..., a_8,\ w_1, ..., w_4$. The lengthy algebraic manipulations are omitted in this discussion. For comparison purpose the Taylor expansion for $y(x_n + h)$ up to the sixth order term is given by

$$\begin{aligned} y(x_n + h) &= y_n + hf + \frac{1}{2}h^2 ff_y + \frac{1}{6}h^3(ff_y^2 + f^2 f_{yy}) \\ &+ \frac{1}{24}h^4(f^3 f_{yyy} + 4f^2 f_y f_{yy} + ff_y^3) \\ &+ \frac{1}{120}h^5(ff_y^4 + 11f^2 f_y^2 f_{yy} + 4f^3 f_{yy}^2 + 7f^3 f_y f_{yyy} + f^4 f_{yyyy}) \\ &+ \frac{1}{720}h^6(f^5 f_{yyyyy} + 11f^4 f_y f_{yyyy} + 15f^4 f_{yy} f_{yyy} \\ &+ 32f^3 f_y^2 f_{yyy} + 34f^3 f_y f_{yy}^2 \\ &+ 26f^2 f_y^3 f_{yy} + ff_y^5) + O(h^7) \end{aligned} \tag{14}$$

The system consiting of 12 equations obtained is then solved using the NAG subroutine on HP9000 computer machine. With various settings, we managed to obtain three sets of solutions for the parameters, viz.

Set 1

w1 =2.9979246086 ; w2 = -3.5709059558; w3= 1.27087332673; w4 = 0.3021480799; a1=-0.1597159217; a2 = 0.1222087100; a3 = -0.0438837951; a4 = -9.6560859331; a5 = 2.6892083053; a* = 7.4668776278; a6 = 50.2840419661; a7 = -12.4357169343; a8 = -404968987566 and a** = 3.6485737248.

Set 2

w1 = 0.1990193382; w2=-0.2535846961; w3 =0.5740453004; w4=0.4805200575; a1=1.2808930996; a2=-0.1216330996; a3 = 0.3153843797; a4=0.1914004985; a5=0.1771161874; a*=0.1314833141; a6=0.8116923889; a7=-0.0668709652; a8=-0.1223829683 and a**=0.3775615446.

Set 3

w1=-8.1882682025; w2=7.6727352875; w3=1.2093464966; w4=0.3061864185; a1=-0.2048829749; a2=-0.0746541126; a3=0.05556210274; a4=9.5751406930; a5 =-0.6322425347; a* = -8.4428981580; a6 = -52.5437298011; a7 =-0.3379090366; a8 = 51.0542147380 and a** =2.8274240998.

It was tested that among these three sets of parameters, Set 2 gives the most accurate solution to the problems chosen. Hence by chosing the parameters in Set 2 formula (2) becomes

$$\begin{aligned} y_{n+1} &= y_n + h(0.1990193382(\frac{2k_1k_2}{k_1+k_2}) - 0.2535846961(\frac{2k_2k_3}{k_2+k_3}) \\ &\quad +0.5740453004(\frac{2k_3k_4}{k_3+k_4}) + 0.4805200575(\frac{2k_4k_5}{k_4+k_5})). \qquad (15) \end{aligned}$$

where

$$\begin{aligned} k_1 &= f(y_n) && (16) \\ k_2 &= f(y_n + 1.2808930996hk_1) && (17) \\ k_3 &= f(y_n - 0.1216330944hk_1 + 0.3153843796hkk_2) && (18) \\ k_4 &= f(y_n + 0.1914004985hk_1 + 0.1771161874hk_2 \\ &\quad +0.1314833141hk_3) && (19) \\ k_5 &= f(y_n + 0.8116923889hk_1 - 0.0668709652hk_2 \\ &\quad -0.1223829683hk_3 + 0.3775615446hk_4) && (20) \end{aligned}$$

or more concisely,

$$y_{n+1} = y_n + h(\frac{4343}{21822}(\frac{2k_1k_2}{k_1+k_2}) - \frac{7304}{28803}(\frac{2k_2k_3}{k_2+k_3})$$

$$+ \frac{26231}{45695}(\frac{2k_3k_4}{k_3+k_4}) + \frac{32450}{67531}(\frac{2k_4k_5}{k_4+k_5})) \tag{21}$$

where

$$\begin{aligned}
k_1 &= f(y_n) && (22)\\
k_2 &= f(y_n + \frac{256609}{200336}hk_1) && (23)\\
k_3 &= f(y_n - \frac{19889}{103913}hk_1 + \frac{28797}{236753}hk_2) && (24)\\
k_4 &= f(y_n + \frac{10285}{32611}hk_1 + \frac{9715}{54851}hk_2 + \frac{15811}{120251}hk_3) && (25)\\
k_5 &= f(y_n + \frac{178397}{219784}hk_1 - \frac{7572}{113233}hk_2 - \frac{18010}{147161}hk_3 + \frac{61501}{162890}hk_4) && (26)
\end{aligned}$$

For short we shall call this new formula as HaM-RK5(5).

3. Error Analysis

Subtracting (15) from (14) we obtained the local truncation error (LTE) of the method, viz.

$$\begin{aligned}
\text{LTE} &= (-0.0000959103ff_y^5 + 0.00874182f^2f_y^3f_{yy}\\
&- 0.0399141f^3f_yf_{yy}^2\\
&+ 0.0163115f^3f_y^2f_{yyy} - 0.00601939f^4f_{yy}f_{yyy}\\
&- 0.00157171f^4f_yf_{yyyy} + 0.0000329614f^5f_{yyyyy})h^6 + O(h^7) && (27)
\end{aligned}$$

If, in addition, the function f is linear in y then the derivatives of f with respect to y of order two and above all vanish, i.e. $f_{yy} = f_{yyy} = f_{yyyy} = f_{yyyyy} = 0$, so that the local truncation error assumes a much simpler form, i.e.

$$\text{LTE} = -0.0000959103ff_y^5h^6 + O(h^7). \tag{28}$$

In terms of strict error bound (with respect to y only) due to Lotkin (see Lambert [1973]) we have the local truncation error which can be expressed as

$$\begin{aligned}
&|\text{LTE}| < 0.0726874h^6P^5Q\\
\text{or } &|\text{LTE}| < \left(\frac{928}{12767}\right)h^6P^5Q && (29)
\end{aligned}$$

where

$$|f(x,y)| < Q \text{ and } \left|\frac{\partial^{i+j}f(x,y)}{\partial x^i\partial x^j}\right| < \frac{P^{i+j}}{Q^{j-1}}, \qquad i+j \le p(= order)$$

P, Q are constants.

4. Stability Aanlysis

Applying the test equation $y' = \lambda y$, λ a complex constant to (21) yields the stability polynomial Q(z) given by

$$Q(z) = \frac{y_{n+1}}{y_n} = \sum_{i=0}^{5} \frac{z^i}{i!} + 0.0014848z^6 + O(z^7). \tag{30}$$

This confirms the order of the method.
The absolute stability region of HaM-RK5(5) method is shown in Figure 4.1. For comparison, the absolute stability regions of HaM-RK5(5) and Taylor(5) methods are shown in Figure 4.2.

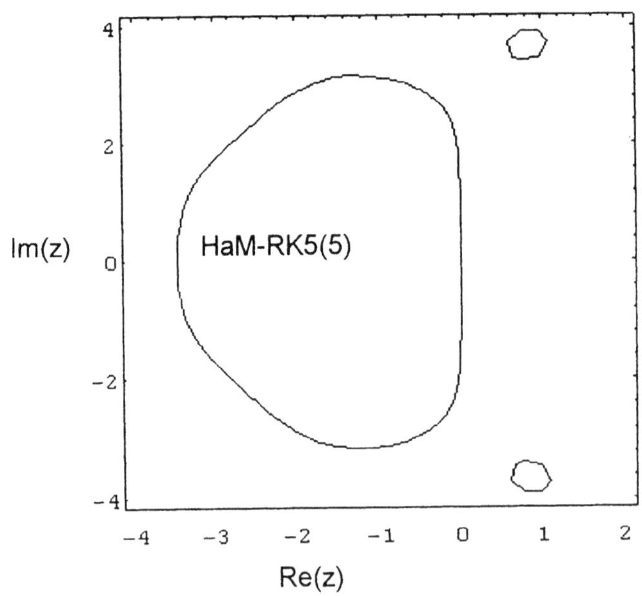

Figure 4.1: Absolute Stability Region for HaM-RK5(5)

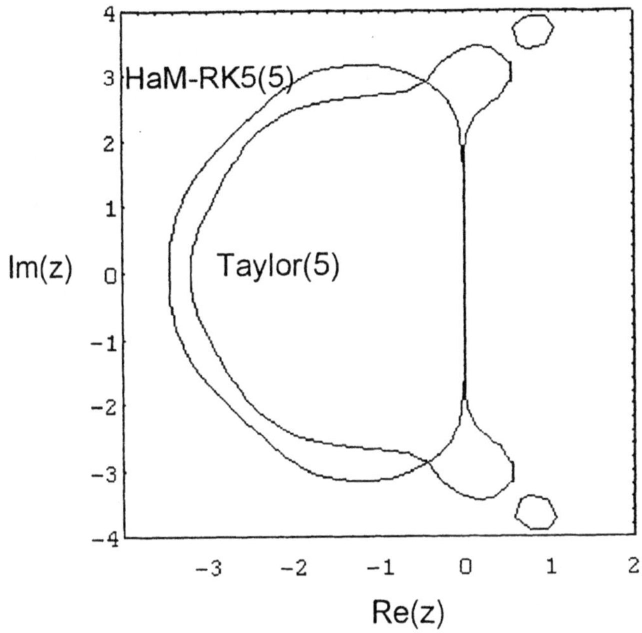

Figure 4.2: Comparison of absolute stability regions Taylor(5) and HaM-RK5(5)

5. Numerical Examples

We consider two initial value problems.
Problem 1

$$y' = y, \qquad y(0) = 1, \qquad 0 \leq x \leq 1$$

which has an exact solution $y = \exp(x)$.

Problem 2

$$y' = -y, \qquad y(0) = 1, \qquad 0 \leq x \leq 1$$

with exact solution $y = \exp(-x)$.
We will employ the fifth order HaM-RK method given by (21) using the step size $h = 0.1$. We also solve the problems using four other methods, i.e linear fourth-order five-stage RK-Merson method, 5th order Taylor series method, linear fifth-order six-stage GM-RK Nystrom method (see Lambert,[1973]) and nonlinear fifth-order five-stage GM-RK method (with new set of parameters).

The corresponding absolute errors in the methods for the respective problems are given in Tables (5.1) and (5.2)

TABLE 5.1 (h=0.1)

X	YEXACT	ABSERROR RK4(5)-MERSON	ABSERROR TAYLOR(5)	ABSERROR RK5(6)-NYSTROM	ABSERROR GM5(5)-SANUGI	ABSERROR HaM5(5)-YAACOB
0.0	0.1000D+01	0.0000D+00	0.0000D+00	0.0000D+00	0.0000D+00	0.0000D+00
0.1	0.1105D+01	0.1530D-07	0.1409D-08	0.1409D-08	0.5647D-09	0.2674D-08
0.2	0.1221D+01	0.3381D-07	0.1990D-08	0.3114D-08	0.2780D-08	0.3960D-08
0.3	0.1349D+01	0.5605D-07	0.9163D-09	0.5163D-08	0.1098D-07	0.8631D-09
0.4	0.1492D+01	0.8260D-07	0.5902D-08	0.7608D-08	0.1409D-07	0.1159D-08
0.5	0.1649D+01	0.1141D-06	0.2904D-08	0.1051D-07	0.2279D-07	0.1609D-08
0.6	0.1822D+01	0.1513D-06	0.1076D-07	0.1394D-07	0.3229D-07	0.4904D-08
0.7	0.2014D+01	0.1951D-06	0.1766D-07	0.1797D-07	0.4135D-07	0.8776D-08
0.8	0.2226D+01	0.2464D-06	0.2140D-07	0.2270D-07	0.4316D-07	0.5188D-08
0.9	0.2460D+01	0.3064D-06	0.3002D-07	0.2822D-07	0.6081D-07	0.4896D-08
1.0	0.2781D+01	0.3763D-06	0.3768D-07	0.3466D-07	0.7361D-07	0.3071D-08

TABLE 5.2 (h=0.1)

X	YEXACT	ABSERROR RK4(5)-MERSON	ABSERROR TAYLOR(5)	ABSERROR RK5(6)-NYSTROM	ABSERROR GM5(5)-SANUGI	ABSERROR HaM5(5)-YAACOB
0.0	0.1000D+01	0.0000D+00	0.0000D+00	0.0000D+00	0.0000D+00	0.0000D+00
0.1	0.9048D+00	0.1252D-07	0.1369D-08	0.1369D-08	0.1924D-08	0.9450D-09
0.2	0.8187D+00	0.2266D-07	0.3701D-08	0.2478D-08	0.5378D-08	0.3221D-08
0.3	0.7408D+00	0.3076D-07	0.6580D-08	0.3363D-08	0.9245D-08	0.3829D-08
0.4	0.6703D+00	0.3710D-07	0.5863D-08	0.4058D-08	0.9028D-08	0.4150D-08
0.5	0.6065D+00	0.4196D-07	0.5638D-08	0.4589D-08	0.1124D-07	0.2730D-08
0.6	0.5488D+00	0.4556D-07	0.7101D-08	0.4983D-08	0.1413D-07	0.4296D-08
0.7	0.4966D+00	0.4810D-07	0.5799D-08	0.5260D-08	0.1515D-07	0.5833D-08
0.8	0.4493D+00	0.4974D-07	0.7022D-08	0.5440D-08	0.1481D-07	0.5483D-08
0.9	0.4066D+00	0.5063D-07	0.7149D-08	0.5537D-08	0.1493D-07	0.6381D-08
1.0	0.3679D+00	0.5091D-07	0.7703D-08	0.5567D-08	0.1544D-07	0.5264D-08

From the results we find that this new HaM-RK5(5) formula is far more accurate than that of GM-RK5(5) especially for the above two problems. We also find that HaM-RK5(5) gives more accurate solution than Taylor series, method of order 5 as well as fifth-order six-stage Nystrom's method.

6. Conclusion

In this paper we have developed a new fifth-order Runge-Kutta formula based on the Harmonic mean of several combination of the functional values. For the particular problems considered in the example, this new formula is better than its counterpart GM-RK5(5), Taylor(5) and RK-Nystrom5(6) in terms of accuracy. The fact that a fifth order accuracy was achieved in this formulation as against the classical fact of its non-exitence in the linear formulation opens further investigations for other methods in the family. Beside these two new formulae (HaM and GM) of fifth-order and five-stage we believe that other nonlinear form of RK formulae are posible enabling us to define some new other fifth order methods which are more competitive in terms of accuracy and stability while preserving the advantage of saving one function over their linear counterparts. We also found that the formulae are not suitable for functions f involving independent variable x and dependent variable y simultaneously, making it more suitable for linear systems of the form $y' = Ay + b$ only. Further investigation is recommended.

References

1. Butcher, J.C. (1987) The Numerical Analysis of Ordinary Differential Equations : Runge Kutta and General Linear Methods, John Wiley & Sons, U.K..
2. Sanugi, B.B. (1986) *New Numerical Strategies For Initial Value Type Ordinary Differential Equations*, Ph.D. Thesis, Loughborough University of Tecnology, U.K..
3. Sanugi, B.B. (1994) & D.J. Evans, *A new fourth order Runge-Kutta formula based on harmonic mean*, Int. Jour. Comp. Math., Vol. 50, 113-118.
4. Evans, D.J. & Sanugi, B.B. (1986) *A new fourth order Runge-Kutta method for intial value problems*, in S.O Fatunla (ed.), Computational Mathematics II, Boole Press.
5. Wazwaz, A.M. (1994) *A Comparsion of Modified Runge-Kutta Formula Based On A Variety of Means,* Int. Jour. Comp. Math.,Vol.50, 105-112.
6. Evans, D.J. and Yaacob, N. (1995) *A Fourth Order Runge-Kutta Based On The Heronian Mean Formula*,Int. Jour. Comp. Math., Vol.58, 103-115.
7. Sanugi, B.B. and Yaacob, N. (1995) *A New Fifth Order Five-stage Runge-Kutta Method For Intial Value Type problems in ODEs*, Int. Jour. Comp. Math., (In Press).
8. Lambert, J.D. (1973) Computational Methods in Ordinary Differential Equations, John Wiley & Sons Ltd..

MIX
Papier aus verantwortungsvollen Quellen
Paper from responsible sources
FSC® C105338

If you have any concerns about our products,
you can contact us on
ProductSafety@springernature.com

In case Publisher is established outside the EU,
the EU authorized representative is:
Springer Nature Customer Service Center GmbH
Europaplatz 3, 69115 Heidelberg, Germany

Printed by Libri Plureos GmbH
in Hamburg, Germany